D0848643

Primate Paternalism

Van Nostrand Reinhold Primate Behavior and Development Series
Terry L. Maple, **Series Editor**

AGING IN NONHUMAN PRIMATES, edited by Douglas M. Bowden

CAPTIVITY AND BEHAVIOR: Primates in Breeding Colonies, Laboratories, and Zoos, edited by J. Erwin, Terry L. Maple and G. Mitchell

ORANG-UTAN BEHAVIOR, by Terry L. Maple

GORILLA BEHAVIOR, by Terry L. Maple and Michael P. Hoff

BEHAVIORAL SEX DIFFERENCES IN NONHUMAN PRIMATES, by G. Mitchell

THE MACAQUES: Studies in Ecology, Behavior and Evolution, edited by Donald G. Lindburg

PRIMATE PATERNALISM, edited by David M. Taub

Primate Paternalism

DAVID MILTON TAUB, Editor

Yemassee Primate Center
Yemassee, South Carolina
and
Department of Psychiatry and Behavioral Sciences
Medical University of South Carolina
Charleston, South Carolina

Van Nostrand Reinhold
Primate Behavior and Development Series

Van Nostrand Reinhold Company
Scientific and Academic Editions

New York Cincinnati Stroudsburg
Toronto London Melbourne

To the memory of my beloved mother, Majorie Theo Taub,
for her parental investment;
To Pam and Alison for their love and inspiration;
To Dr. Mac, who, for better or worse, was responsible for my seeking to become an anthropologist; and
To "Mo," "RD," and the "B" Brothers for being themselves.

Copyright © 1984 by **Van Nostrand Reinhold Company Inc.**
Library of Congress Catalog Card Number: 83-16480
ISBN: 0-442-27217-0

All rights reserved. No part of this work covered by the copyright hereon may be reproduced or used in any form or by any means—graphic, electronic, or mechanical, including photocopying, recording, taping, or information storage and retrieval systems—without permission of the publisher.

Manufactured in the United States of America.

Published by Van Nostrand Reinhold Company Inc.
135 West 50th Street
New York, New York 10020

Van Nostrand Reinhold Company Limited
Molly Millars Lane
Wokingham, Berkshire RG11 2PY, England

Van Nostrand Reinhold
480 Latrobe Street
Melbourne, Victoria 3000, Australia

Macmillan of Canada
Division of Gage Publishing Limited
164 Commander Boulevard
Agincourt, Ontario MIS 3C7, Canada

15 14 13 12 11 10 9 8 7 6 5 4 3 2 1

Library of Congress Cataloging in Publication Data
Main entry under title:
Primate paternalism.
Includes bibliographical references and index.
1. Primates—Behavior. 2. Parental behavior in animals.
3. Mammals—Behavior. I. Taub, David Milton.
QL737.P9P6728 1984 599.8′0451 83-16480
ISBN 0-442-27217-0

Contents

Preface / vii

Contributors / xi

Empirical Studies

1: Father-Offspring Relationships in Laboratory Families of Saddle-Back Tamarins *(Saguinus fuscicollis)*, *Mary Scott Cebul and Gisela Epple* / 1

2: Male Caretaking Behavior Among Wild Barbary Macaques *(Macaca sylvanus)*, *David Milton Taub* / 20

3: Male-Infant Interactions Among Free-Ranging Stumptail Macaques, *Alejandro Estrada* / 56

4: Adult Male-Immature Interactions in Captive Stumptail Macaques *(Macaca arctoides)*, *Euclid O. Smith and Patricia G. Peffer-Smith* / 88

5: Free-Living Rhesus Monkeys: Adult Male Interactions with Infants and Juveniles, *Stephen H. Vessey and Douglas B. Meikle* / 113

6: Social Relations of Males and Infants in a Troop of Japanese Monkeys: A Consideration of Causal Mechanisms, *Harold Gouzoules* / 127

7: Why Males Use Infants, *Shirley C. Strum* / 146

8: Triadic Interactions Among Male and Infant Chacma Baboons, *Curt Busse* / 186

9: Ontogeny of Infant-Adult Male Relationships during the First Year of Life for Yellow Baboons *(Papio cynocephalus), David M. Stein* / 213

10: Adult Male-Infant Interactions in the Chimpanzee *(Pan troglodytes), Diana Davis* / 244

Theoretical Considerations

11: The Evolution of Male Parental Investment: Effects of Genetic Relatedness and Feeding Ecology on the Allocation of Reproductive Effort, *Jeffrey A. Kurland and Steven J. C. Gaulin* / 259

12: Significance of Paternal Investment by Primates to the Evolution of Male-Female Associations, *William J. Hamilton III* / 309

13: The Evolutionary Role of Socio-ecological Factors in the Development of Paternal Care in the New World Family Callitrichidae, *A. G. Pook* / 336

Reviews

14: Interactions Between Adult Males and Infants in Prosimians and New World Monkeys, *Jerry L. Vogt* / 346

15: Adult Male-Infant Interactions in Old World Monkeys and Apes, *David Milton Taub and William K. Redican* / 377

16: Observational Studies of Father-Child Relationships in Humans, *Michael E. Lamb* / 407

Index / 431

About the Editor / 441

Preface

In the modern discipline of "primatology," Itani is generally credited with providing the first detailed report of adult male interactions with infants among nonhuman primates, Japanese macaques, about 25 years ago. From then until now, there has been a persistent and growing interest in and reports of "paternal" behavior among nonhuman primates. These reports have generated intense interest because among mammals in general males tend to be minor actors in the care and rearing of offspring. However, when one surveys carefully this ever growing literature, as a number of investigators have done in recent years (c.f., Hrdy, Redican, Redican and Taub, and Chapters 14 and 15), one is struck by a number of detrimental common factors that characterize the reports of this phenomenon. First, with some few exceptions, the reports are qualitative and descriptive: They provide scant information on the frequency or rate at which the different types of interactions occur, the differential distribution of interactions between male actors and infant recipients, or the nature of the relationship (if any) between the males and the mothers of the infants with which they interact. Second, these reports generally do not provide basic sociometric information about the participants in the interactions, such as the identity of individual male actors, of the infants they interact with, or of third parties (as in the case of triadic male-infant-male interactions). What at first appears to be a large inventory of data about male-infant interactions among primates, is, under close examination, a *melange* of descriptive reports revealing little about the social and biological foundations of paternalism among nonhuman primates.

One of the goals of this volume is to remedy the deficiency of quantitative data. Chapters 1 through 10, empirical studies, present the results of some recent studies of male-infant interactions among marmosets, macaques, and baboons. In marked contrast with most previous reports, the cornerstone of each chapter is the presentation of quantified data on many facets of the sociometric distribution of male-infant interactions in the species under study. Individually and collectively, these ten chapters provide the most detailed, quantified, and comprehensive view of primate paternalism available today, especially as this phenomenon occurs among the polygynously mating Cercopithecoidea. If there is a weakness in this section, it is perhaps the lack of new data on the monogamously mating primates—the marmosets/tamarins, calliceboids, and hylobatids—among whose members males are noted for being heavily involved in paternal behavior. During my solicitation of contributions for this volume, I was indeed surprised to find little new research available on the monogamously mating primates, and there is a need for further, quantitative studies, comparable to those reported in Chapters 2 through 10, on natural populations of such primates.

Chapters 11 through 13 comprise a section on theoretical considerations. In these chapters the authors integrate the empirical occurrence of male-infant interactions into a broader conceptual arena of current evolutionary biological theory. Much of the data necessary to "test" certain theoretical hypotheses of both the function and proximate/ultimate causation of primate paternalism are still unavailable (e.g. cytogenetically defined degrees of relatedness between male actor and infant recipient). These papers are not presented as definitive tests of such hypotheses. Rather, they provide collectively both general and specific theoretical frameworks within which the evolution of primate paternalism may be approached, analyzed, and interpreted. In addition, they identify problem areas for the future research needed if more definitive data sets are to be collected for a more thorough, prospective understanding of these parenting systems.

Chapters 14 through 16, which comprise a section of reviews, represent a compendium of the occurrence of paternalistic behavior among primates. They are presented to provide a fairly complete and practical literature resource, assembling data scattered throughout a varied and disparate literature. I believe this section represents the definitive summary of the occurrence of male-infant interactions among primates. I hope it is useful because of its completeness and unique because of its inclusion (in Chapter 16) of an overview of father-infant relations among humans, which provides a comparative perspective generally unavailable in the usual primate literature.

Just a note about semantics. The term "paternalism" and its derivatives are used *only* as *convenient* rubrics under which a wide variety of structurally (and probably functionally) diverse interactions can be accomodated easily; they are not intended to imply genetic or social relationships of unknown derivation.

As with any publishing undertaking, there are numerous individuals without whose talents and hard work such enterprises would never become reality. This volume is no exception to that rule. I wish to warmly thank the editors at Van Nostrand Reinhold, Susan Munger and especially Eric Rosen, for their tireless support. The production staff for this volume did a superior job, and I am particularly indebted to Bernice Pettinato for her selfless, untiring, and thoroughly professional guidance on this project. I wish to acknowledge warmly the assistance on the author index provided by two of my technicians, Myra Gwin and James Weed. Lastly, I want to express my gratitude and appreciation to all who contributed to this volume; publication has taken longer than any of us anticipated when the project first began, and I have greatly appreciated the cooperation and understanding shown by all the authors during the development and production of this volume. Their collective generosity made the editor's job that much easier, and I thank each and every one for his or her support.

DAVID MILTON TAUB

Contributors

Curt Busse
Yerkes Regional Primate Research Center, Emory University, Atlanta, Georgia 30332

Mary Scott-Cebul
Zoological Society of Philadelphia, 34th and Gerard Avenue, Philadelphia, Pennsylvania 19104

Diana L. Davis
Tangram Rehabilitation Network, P. O. Box 1051, San Marcos, Texas 78667

Gisela M. Epple
Monell Chemical Senses Center, University of Pennsylvania, 3500 Market Street, Philadelphia, Pennsylvania 19104

Alejandro Estrada
Institute of Biology, National University of Mexico, Veracruz, Mexico

Steven J. C. Gaulin
Department of Anthropology, 3H01 Forbes Quadrangle, The University of Pittsburgh, Pittsburgh, Pennsylvania 15260

Harold Gouzoules
Field Research Station, Rockefeller University, Tyrrel Road, Millbrook, New York 12545

William J. Hamilton III
Department of Environmental Studies, University of California, Davis, California 95616

Jeffrey A. Kurland
Department of Anthropology, 416 Carpenter Building, The Pennsylvania State University, University Park, Pennsylvania 16802

Michael E. Lamb
Departments of Psychology, Pediatrics, and Psychiatry, University of Utah, Salt Lake City, Utah 84112

Douglas B. Meikle
School of Interdisciplinary Studies, Miami University of Ohio, Oxford, Ohio 45056

Patricia G. Peffer-Smith
Yerkes Regional Primate Research Center, Emory University, Atlanta, Georgia 30322

A. G. Pook
26 Duke Street, Cheltenham GL52 6BP, England

William K. Redican
792 Ashbury Street, San Francisco, California 94117

Euclid O. Smith
Yerkes Regional Primate Research Center, Department of Anthropology and Department of Biology, Emory University, Atlanta, Georgia 30332

David M. Stein
Department of Psychiatry, John A. Burns School of Medicine, University of Hawaii at Manoa, 1356 Luistana #408, Honolulu, Hawaii 96813

Shirley C. Strum
Department of Anthropology, University of California-San Diego, La Jolla, California 92093

David Milton Taub
Yemassee Primate Center, P. O. Box 557, Yemassee, South Carolina 29945 and Department of Psychiatry and Behavioral Sciences, Medical University of South Carolina, Charleston, South Carolina 29425

Stephen H. Vessey
Department of Biological Sciences, Bowling Green State University, Bowling Green, Ohio 43403

Jerry Vogt
Department of Psychology, St. John's University, Collegeville, Minnesota 56321

Primate Paternalism

Two Barbary macaque subadult males ("BL" and "Ro") caretaking an infant ("iLN") in their possession. Photo by D. M. Taub from a study of wild Barbary macaques in Morocco (see Chapter 2).

1

Father-Offspring Relationships in Laboratory Families of Saddle-Back Tamarins *(Saguinus fuscicollis)*

Mary Scott Cebul
Zoological Society of Philadelphia
Philadelphia, Pennsylvania

Gisela Epple
Monell Chemical Senses Center
Philadelphia, Pennsylvania

INTRODUCTION

To date, most primate studies of social relationships during ontogeny have been carried out on Old World monkeys and apes. However, the Callitrichidae, the South American marmosets and tamarins, show a number of specializations that make them interesting alternative subjects for research in this area. Unlike most Old World simians, the callitrichids are monogamous and live in small groups (Epple, 1975a; Kleiman, 1978). Laboratory studies and a few field observations suggest that most callitrichid groups contain one breeding female, her dependent young (usually twins), and an adult male with whom she has established a permanent pair bond. In addition, other older individuals may be present. Often in some species these late juveniles and adults may emigrate from their own group and temporarily join others (Epple, 1975a; Izawa, 1979; Castro and Soini, 1977; Terborg, in press). Such temporary group members have been observed in both subspecies of *Saguinus oedipus* (Dawson, 1977; Neyman, 1977), but in other species, such as *Saguinus fuscicollis,* the groups may be more stable (Terborg, in press).

In the laboratory and in the wild, each group contains only one breeding female, even though several adult females, related or not related to each other, may be present (Epple, 1975a). In laboratory groups, it is the dominant, pair-bonded female who is reproductively active and some recent studies have suggested that she may be able to suppress ovarian cyclicity in subdominant females (Hearn and Lunn, 1978; Katz and Epple, 1980). In laboratory families of *S. fuscicollis,* all reproduction ceases when one of the bonded mates is lost (Epple, unpublished), a fact suggesting that this species observes a behavioral incest taboo. The existence of such a taboo makes it likely that in families of this species, all offspring of the breeding female are sired by her bonded partner. If another nonrelated male is present, however, paternity of the offspring is not certain, since the female may also mate with the nonbonded male (Epple, 1972).

While in all callitrichid groups studied so far only one female reproduces, the whole group usually participates in caring for her twin offspring. This fact makes the marmosets and tamarins interesting subjects for studies investigating the relationships of infants both with their fathers and with other members of their natal group. Even the earliest reports of the successful breeding of these small primates in captivity stress the fact that the father performs much of the infant care, aided by other members of the group, while contact between the mother and her young infants may be limited to nursing (Fitzgerald, 1935; Franz, 1963; Lucas et al., 1927, 1937; Rabb and Rowell, 1960; Roth 1960; Wendt, 1964). Because of the strong involvement of the father (when compared to Old World primates), some summarizing reports on reproduction of marmosets and tamarins give the impression that the father performs most, if not all, of the care (Crandall, 1951; Hill, 1957; Napier and Napier 1967; Mitchell and Brandt, 1972; Zukowsky, 1940). More recent studies, however, have reported considerable variability in the extent to which members of callitrichid groups are involved in the care and socialization of infants, both across species as well as across individuals of the same species (Box, 1977; Ingram, 1977; Epple, 1975b, Hoage, 1977; Vogt et al., 1978; Wolters, 1978). In the present report, we examine the relationships between father or the dominant, pair-bonded male and the offspring of the breeding female in laboratory families and other groups of saddle-back tamarins in three areas: (1) carrying of dependent infants during the first month of life; (2) provisioning of juveniles; and (3) socialization.

THE ROLE OF THE FATHER AS A CARRIER OF DEPENDENT INFANTS

During their first 30 days of life, saddle-back tamarins are almost always carried by a member of their group, riding on the back of their carrier except for periods of nursing when they cling to the ventral fur of the mother. Epple (1975b) recorded the amount of time each group member was involved in carrying babies in seven breeding groups of tamarins (note that one group VII and VIIIA [Table 1-1] had the same mother but different breeding males; consequently, for computational purposes the study population was considered to be seven groups). Each group was observed for a series of 30-minute periods between day 1 and 43 of the infant's life. However, the observations did not all cover the same days in the lives of the infants (for details see Epple, 1975a).

Each 30-minute period was divided into 120 intervals, each lasting 15 seconds. For each of the 15-second intervals, every group member received a score of 1 if it carried one infant and a score of 2 if it carried two infants. Thus, the maximum score per interval each member of a group with twin babies could obtain was 2. The mothers received a score for carrying their infants only when the infants were not nursed, but carried on the back of the mother. If infants were in the nursing position, their mothers received a score of 1 or 2 for nursing, but no score for carrying. If a mother was carrying one baby and nursing the second one, her behavior was scored accordingly. Mothers received a score of 1 or 2 in the category "being in possession of infants" when they were either carrying or nursing 1 or 2 babies. From these scores we computed the percentage of scores obtained for nursing only and that obtained for carrying infants (Table 1-3).

Mean scores for carrying infants per 30 minutes were computed for each individual and mean nursing scores for the mothers. Statistical tests follow Siegel (1956).

Table 1-1 shows the carrying patterns for each group by presenting every individual's percentage of the group's mean total score for carrying infants.

Table 1-1. Individual Percentages of Each Group's Mean Total Score For Carrying Infants (The Sex and Age of Group Members Other Than Mother and Father are Given In Parentheses).

Group	Mother	Father	Siblings (Juveniles)	Nonrelated Group Members
I	31.45	68.55[a]		
II	69.06	30.30[b]		0.64(♀ juv.)
III	28.53	39.26	32.15(♂)	0.06(♂ ad.)
IV	35.30	38.01	8.54(♂), 18.15(♂)	
V	1.04	95.63[b]	2.24(♂), 1.09(♂)	
VI	3.79	59.47	25.33(♂), 11.41(♀)	
VII	54.77[c]	29.87		15.36[c](♂ ad.)
VII-A	38.26[c]	58.74[c]	3.00(♂)	

[a]This father had been castrated prior to the birth of his infant.
[b]In these groups, the fathers had been removed and new adult males had been introducted prior to the birth of the infants.
[c]This signifies the same mother in groups VII and VII-A and the same male studied as the subdominant male of group VII and the breeding male of group VII-A.

It also depicts the group structure. Note that in groups II and V, the dominant, pair-bonded males were not the fathers of the infants. The natural fathers had been removed and the new adult males had established a pair bond with the pregnant female. Groups II, III, and VII each had a group member who was not related to the breeding pair and its offspring.

The table shows that there was not one among the seven groups in which one of the groups members carried the infants exclusively. As a group, the dominant males tended to carry infants most frequently, followed by mothers and other group members. However, due to high variability in the performance of all animals, there was no statistical difference between the carrying scores of the pair-bonded males and the mothers. Mothers and fathers, however, carried more frequently than subdominant group members (dominant males $p = 0.002$ and mothers $p = 0.02$ for groups carrying twins; Mann-Whitney U-test).

In every group, all members shared in carrying of infants. However, there was considerable variability in the extent to which each member carried infants, both within and between groups. In five out of seven groups, the fathers, or the dominant males who assumed the father role, tended to carry the infants most frequently. However, paternal performance varied strongly among the dominant males. The least active father performed only 29.87% of all carrying in his group, while the most active male obtained 95.63% of all scores (Table 1-1). His high score was obtained in spite of the fact that he was not even the father of the infants but was placed with the pregnant mother and her two sons shortly before parturition. It should be noted that this male, who has bred in our colony with several females for over 10 years, consistently performs much of the carrying regardless of his individual mate or the size of his family. The paternal care provided by the castrated male, who performed most of the carrying in his group ($p = 0.013$, Mann Whitney U-test), was in line with those of other males.

Table 1-2. Nursing and Carrying Expressed as Percentage of Total Scores Obtained for Being in Possession of Infants.

Mother in Group	% Nursing	% Carrying
I	23.93	76.07
II	39.53	60.47
III	42.45	57.55
IV	22.17	77.83
V	82.41	17.59
VI	90.96	9.04
VII	18.84	81.16
VII-A	12.67	87.33

The scores obtained by the mother for carrying their infants outside of nursing time also varied greatly (Table 1-2). The most active mother performed 69.06% of all carrying in her group, the least active one 1.04% (Table 1-1). In group VII-A, for instance, the mother nursed only during 12.6% of all intervals during which she was in possession of infants, carrying them on her back during the remaining 87.33%. In group VI, on the other hand, the mother nursed during 90.96% and carried during only 9.04% of all intervals in which she was holding infants. Thus, five of the seven mothers took over a considerable part of the carrying duties, while two groups more or less limited contact with their offspring to nursing.

Besides the parents, other members of the groups also shared in carrying the infants. Again, the carrying scores of siblings and non-related group members show a high degree of variability, which may reflect the individual's motivation to carry as well as the parents' willingness to allow carrying (Table 1-1). Thus, for instance, the low scores of the nonrelated members of groups II and III reflect the fact that the parents did not allow these animals to carry the babies (Epple, 1975b).

Carrying performance also varied among the older offspring of the breeding female, although parents were much more tolerant toward carrying attempts by their own offspring, and all of them participated in carrying.

Since group structure and kinship relations within groups varied so much, it is difficult to assess the contributions of each variable to the data. It is clear that dominant males pair-bonded to the mother assume a good portion of the carrying, regardless of whether or not they are the biological fathers of the infants. In pairs who do not have older offspring living with them, the father may be the predominant carrier. This fact may have given rise to some of the generalizations as to the exclusive role of the father in infant carrying in the Callitrichidae. Some earlier authors (Crandall, 1951; Hill, 1957; Marik, 1931; Sanderson, 1957; Zukowski, 1940) have pointed out that callitrichid mothers concern themselves only with nursing infants and rarely carry them. In *S. fuscicollis* as well as other species, however, the father shares the carrying duties with the mother, and helpers may take over much of the carrying when they are present. Empirical data for some of the large families in our laboratory (numbering 10 to 12 animals) suggest that sometimes sub-adult siblings, particularly males, may perform much or even most of the carrying. More recent studies on larger groups of other callitrichid species also demonstrate participation of all group members in infant carrying, but again show considerable interindividual and interspecies variability (Box, 1977; Christen, 1974; Ingram, 1977; Hoage, 1977; Rothe, 1973; Vogt et al., 1978; Wolters, 1978).

The degree to which father, mother, and other caregivers carry infants may also change as a function of the age of the infants. In *S. fuscicollis,* both parents carried the babies more frequently than did other group members during the first 10 days (Epple, 1975b). Wolters, (1978) found that in three out of four families of *S. oedipus oedipus,* the mother was the predominant carrier during the first week postpartum, while in the fourth family, the mother made a very small contribution to carrying. During the following weeks, participation of the father and other group members in infant carrying increased. However, considerable interindividual variability was evident in this species also. In *Leontopithecus rosalia rosalia,* mothers were found to be the primary carriers during the first three weeks of the infants' lives, while fathers become primary carriers from week four on. Juvenile siblings served as tertiary carriers (Hoage, 1977).

THE ROLE OF THE FATHER AS A FOOD PROVIDER

During the first months of life, much of the solid food consumed by callitrichid infants is provided by other members of the family rather than obtained by the youngsters through foraging on their own. At approximately four weeks of age, the infants first begin to sample solid food. Most frequently the first solid food is taken from the mouth or out of the hands of the animal who carries the infant, and it is given up to the infant without resistance. As the infants grow older, their intake of solid food increases, but they persist in obtaining much of their food from other members of the group rather than foraging for it (Epple, 1967; Hoage, 1977; and others).

The following data from studies by Cavalliere (unpublished) and Cebul (1980) document the frequency of food stealing in family groups of captive *S. fuscicollis* and examine the extent to which the father acts as a provider of solid food during the first year of life.

In the first study (Cavalliere, unpublished[1]), attempts to steal food from other members of the group were recorded in a series of 15-minute observation sessions on five groups[2], each having a single juvenile who was the subject of the study. Each observation period was divided into intervals of 15 seconds and the subject received a score of 1 for each interval it attempted to steal food from another group member. Table 1-3 shows the total attempts at food stealing directed at each group member by the young subject and gives the percentage of attempts during which food was obtained in parentheses. From the table it is clear that, similar to their participation in the carrying of small infants, mother, father and siblings provide food for the growing youngsters. No consistent relationship was found between the youngster's

[1]Independent study project at the Monell Chemical Senses Center.

[2]The groups were not identical with those in which infant carrying was recorded.

Table 1-3. Total Attempts at Food Stealing Seen During the Study and Percentage of Attempts Resulting in Food (Given in Parentheses).

Group	Mother	Father	Siblings	
I [21-31 ♀*]	37 (51%)	37 (60%)		
II [16-26 ♂*]	76 (67%)	40 (70%)		
III [17-27 ♂*]	7 (0%)	15 (0%)	23 (5%) (adult brother)	23 (17%) (juvenile brother)
IV [17-27 ♀*]	2 (0%)	38 (90%)	8 (50%) (juvenile sister)	
V [21-31 ♂*]	27 (15%)	1 (0%)		

*Sex and age of the infant in weeks covered by observations given in brackets.

attempts to steal food and the type of group member at which these attempts were directed nor between the type of group member and the success rate of stealing attempts.

In the second study (Cebul, 1980), juvenile subjects living in their family groups were observed in 15-minute sessions during the course of which every instance of food stealing was recorded, as was the identity of the other group member involved. Data were collected during the first year of life. Figure 1-1 illustrates the dramatic drop in active food stealing (referred to here as foodsteal *a*, or active) by juvenile callithricids over the course of the first year and the simultaneous rise in frequency of seeking food independently. *AP* is an abbreviation for Age Period and data are presented for male and female juvenile subjects of Age Period 1 (8-12 weeks), Age Period 2 (13-24 weeks), Age Period 3 (25-36 weeks) and Age Period 4 (37-52 weeks).

It is interesting to note that during AP 2, juveniles began to seek food independently as often as they stole food from other group members. Juvenile subjects had food items stolen from them (foodsteal *p*, or passive) during every Age Period, and by AP 4, they stole at approximately the same low rate that others were stealing from them.

Figure 1-2 is a series of four histograms showing the distribution of active and passive food stealing by juvenile subjects to mother, father, twin, and older and younger siblings (Cebul, 1980) during each of the four Age Periods. These data show a definite time-dependent trend. The youngest juveniles took food mostly from older siblings and their mothers. It was not until AP 2 that youngsters exchanged food frequently with their twins (either active or passive foodsteal). During AP 2 and 3, the majority of food stealing occurred among the siblings. Juveniles actively took food from both parents, but they were not called upon to share food with their parents (passive

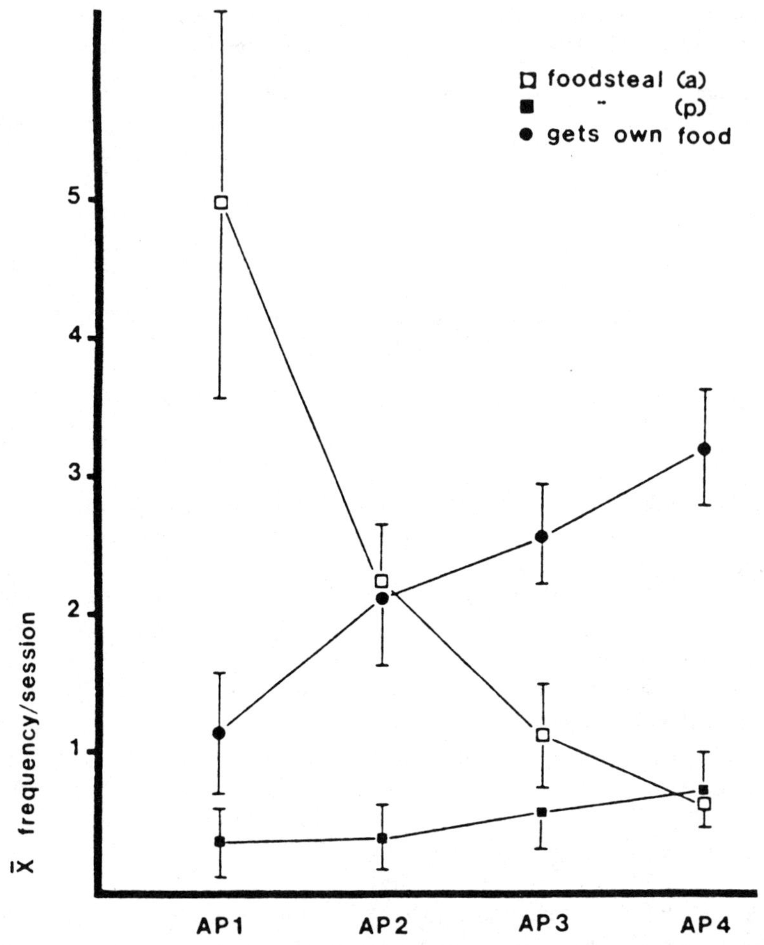

Fig. 1-1. Age changes in food-getting behaviors: active *(a)* and passive *(p)*.

foodsteal), as they were by their siblings. By AP 4, the majority of food stealing was among siblings, with a relatively small percentage of active food stealing still directed toward the parents—mostly the mother. Just as they stole from their older siblings when they were younger, these AP 4 juveniles experienced much food stealing from their new younger brothers and sisters and, in effect, became "providers."

Although Cavalliere's data (unpublished) fail to indicate any trends in food stealing between juveniles and other group members, it is clear from both his and especially Cebul's (1980) data that food stealing is distributed across all group members. Further, the data show that fathers do not play a

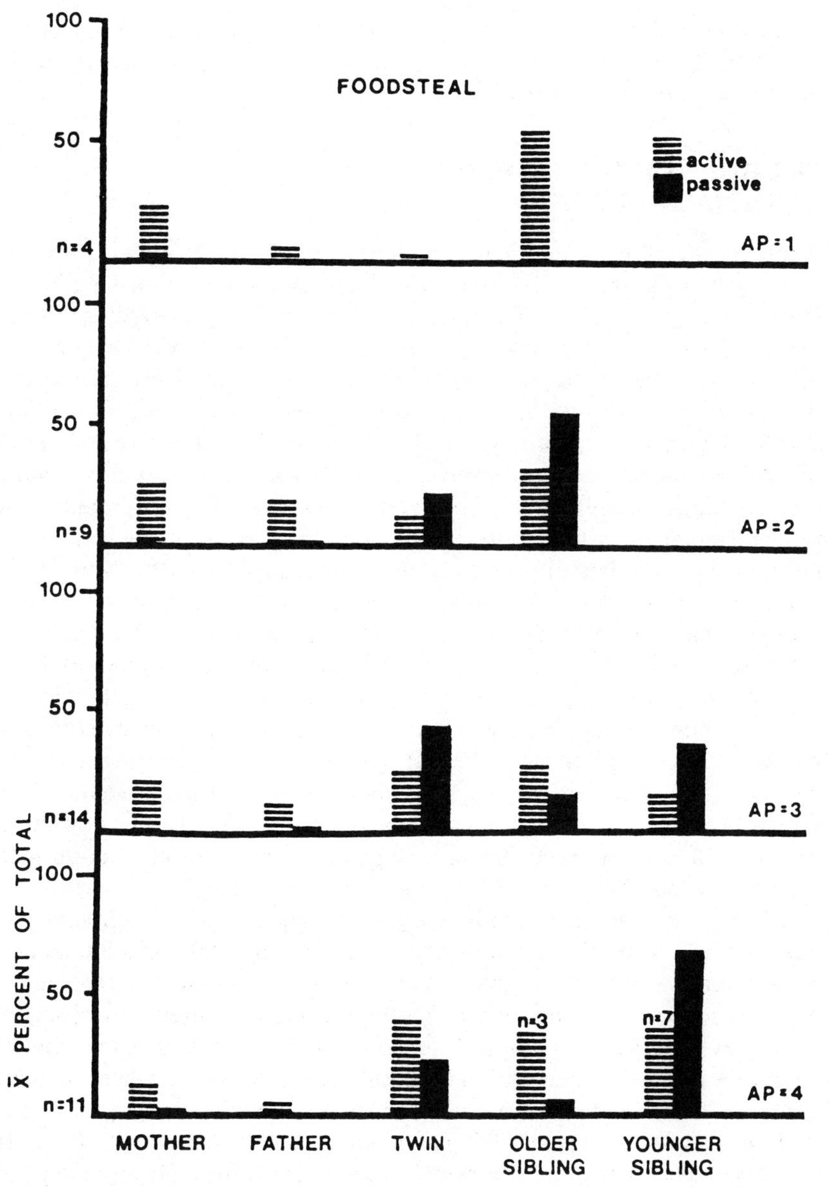

Fig. 1-2. Distribution of food stealing across group members at each age period (active and passive).

dominant role in this behavior pattern with their offspring. The mothers and other siblings of juvenile subjects in Cebul's (1980) study were far more involved in food stealing than were fathers.

THE ROLE OF THE FATHER IN SOCIALIZATION

Another view of male interaction patterns with their offspring is provided by the following data on *S. fuscicollis.* These data are taken from Cebul's (1980) study of the social development of 24 juvenile twins (11 males and 13 females) growing up in 12 different family groups. Each family group consisted of an adult male-female bonded pair (the subject's mother and father) and their offspring of various ages, ranging from neonates to sub-adults. In the course of 15-minute recording sessions, each instance of approximately 30 types of behavioral interactions between a subject and another group member was recorded. A general finding of the study was that in family groups of six to eight members with multiple sets of offspring, parents, and especially fathers, play a relatively minor role (as measured by observed social interaction) in the socialization of young. In fact, there was no behavior category that was performed preferentially with fathers.

Figure 1-3 shows the mean percent of total social interactions with each class of interacting group member—mother, father, twin, and older or younger sibling (sex of sibling was not controlled) over the course of the time blocks or Age Periods (AP 1, AP 2, AP 3, AP 4) comprising the first year. The curves in this figure are based on combined male and female data. Total social interaction is a rough measure and simply refers to the sum of all instances of each social behavior category performed in interaction with each family member.

It is clear that fathers were not interacting frequently with their youngest offspring; moreover, throughout the year the percent total social interaction with fathers was lower than that of any other family member. Interestingly, the curve for social interaction with mothers dropped off precipitously in AP 3 to a level similar to fathers. Interaction with twins and older siblings remained relatively high throughout the year. In AP 3, when subjects averaged 30 weeks of age and usually a new set of babies was present in the study groups, the pattern of sibling interactions is seen to repeat itself. Study subjects had become older siblings and showed the same high percentage of interaction with their baby siblings that was shown toward them as babies of AP 1 by their older siblings.

Figure 1-4 shows juvenile sex differences in the distribution of total social interaction within family groups. Histograms are drawn for male and female

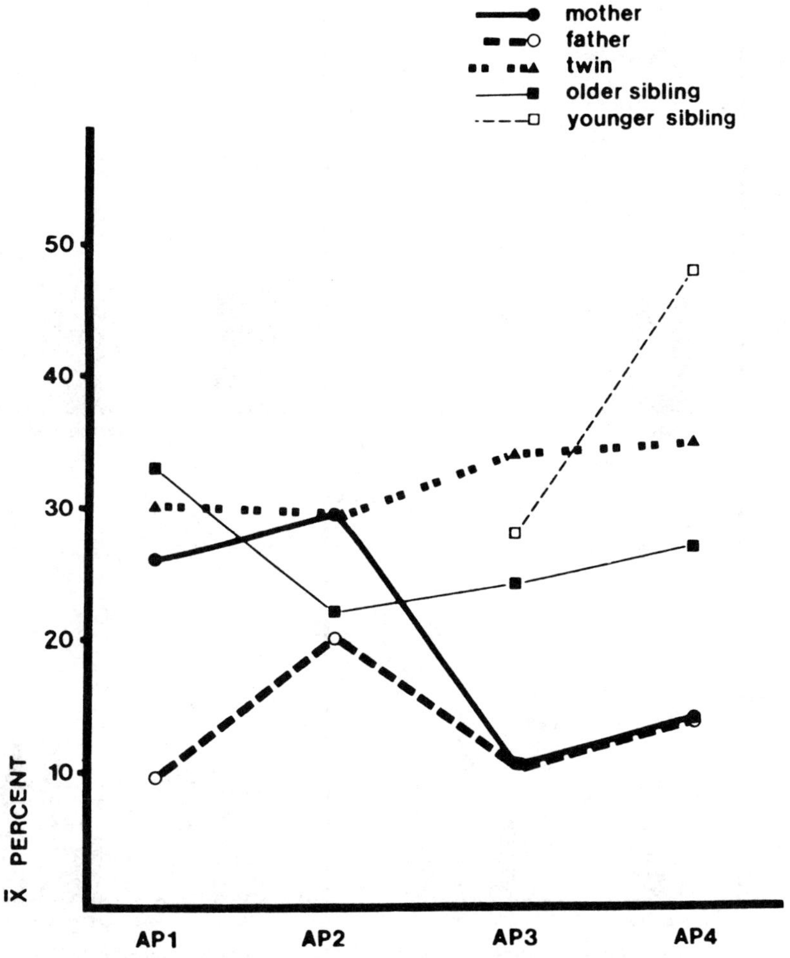

Fig. 1-3. Age changes in percent total social interaction with each type of group member.

juveniles in AP 2, 3, and 4, its columns representative of the percent total social interaction observed with each particular class of group member. Because of the small number of subjects, AP 1 female values are not presented. The horizontal line labeled *E* is the percentage of interaction expected were the social interactions to be evenly distributed among all group members (*E* =100/# of group members other than subject). In this way, it becomes clear which group members the juveniles of each sex were interacting with less than, and more than, expected and at what age. Columns

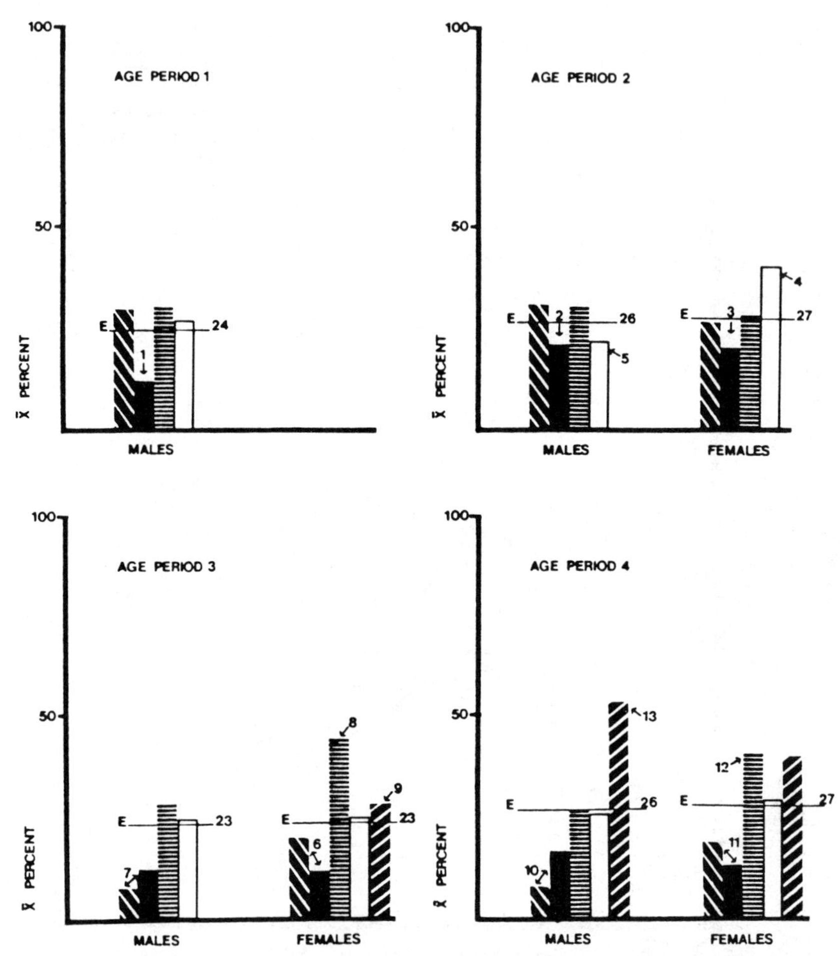

Fig. 1-4. Age changes and sex differences in distribution of total social interaction across group members. First bar (diagonal stripe, left to right) is interaction with mother; second bar (solid) is interaction with father; third bar (horizontal stripe) is interaction with twin; fourth bar (open) is interaction with older sibling; fifth bar (diagonal stripe, right to left) is interaction with younger sibling. Arabic numbered arrows drawn to bar indicate changes in the direction of the relationship which are referred to and explained in the text.

for older and younger siblings show weighted mean percentages such that data were averaged only for those subjects who lived in family groups that included that particular type of member.

For both males and females in all age periods, the relative percent total social interaction with fathers remained well below E—the father was not a preferred social interactant at any age. However, an interesting sex difference developed in AP 3, (this is not shown in Figure 1-3). In AP 3, social interaction with parents fell in both males and females, but the decrease in female interaction with mothers (6) was only very slight compared to fathers. There was a dramatic fall in male interaction with mothers and fathers from AP 2 to 3 (7) and though both percentages were below *E,* males maintained a greater percentage interaction with their fathers rather than their mothers. Females seemed to interact with their mothers not only more than with their fathers (6), but also much more than did same-aged males (7). AP 3 females also remained more active with their twins (8) than did AP 3 males. In addition, females directed a high percentage of social behavior toward their younger siblings (9), a tendency totally lacking in AP 3 males.

The relatively low levels of social interaction with parents versus siblings remained essentially the same through AP 4 in both sexes. In addition, the differential sex trend continued in that males still showed a tendency to interact more with their fathers (10) and females more with their mothers (11).

Juvenile relationships with their parents in AP 3 and AP 4 also indicated that male and female tamarins interacted quite similarly with their fathers through the year; i.e., although there were changes in behavioral content and percent of interaction, the sexes showed the same types of changes over time.

By AP 4, the greatest differences in relationships occurred between females and their mothers and females and their fathers. Figures 1-5, 1-6, and 1-7 are histograms representing male and female relationships with parents for AP 4. They describe relative behavioral content of relationships and provide a comparison of the frequency of types of interactions displayed by juvenile males and females toward each parent relative to those with other group members. Each column is labeled with the name of a behavioral category, and either an *(A)* or a *(P)* to indicate whether that category is the active form (juvenile initiated) or the passive form (juvenile received). Some of the categories are nondirectional and therefore have no *(A)* or *(P).* Column height is the mean percentage of the total frequency of a given behavior performed by juveniles toward either mother or father. Therefore, the column heights indicate the inclination of the juvenile and interacting animal to engage in each type of behavior. For example, in Figure 1-5, juvenile males performed an average of 19% of their total huddling with their fathers. This is slightly lower than what would be expected (*E*) given an even distribution of that behavior across all family members ($E = 100$/mean number of group members other than subject).

The profiles in Figures 1-5 (relationship of males to fathers) and 1-6 (relationship of females to fathers) are quite similar, with percentages generally

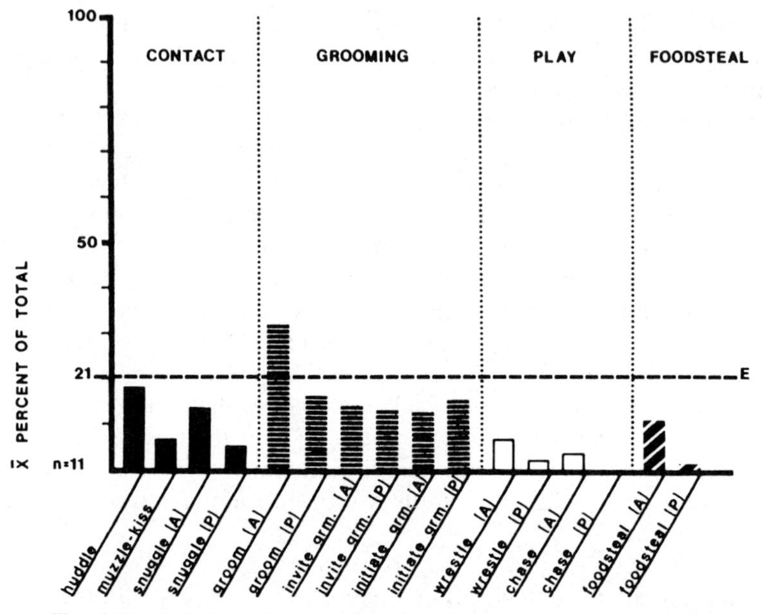

Fig. 1-5. Relationships of males to their fathers in age period 4.

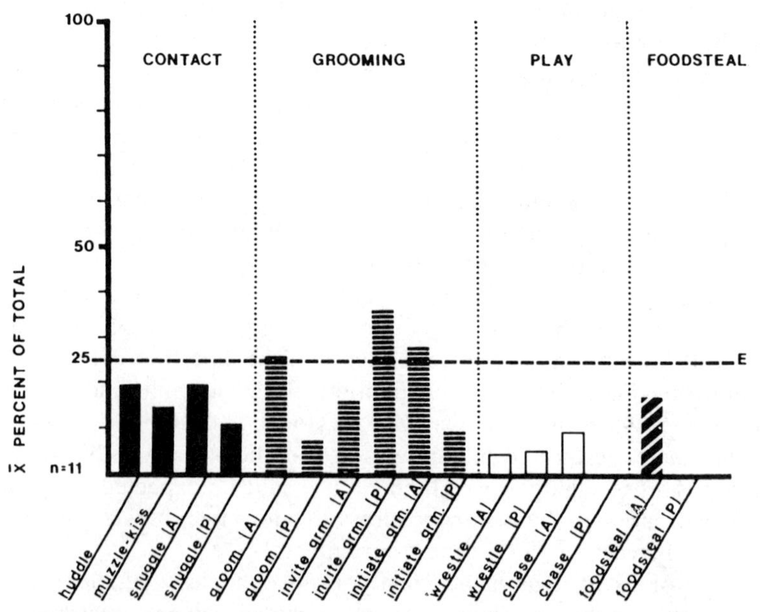

Fig. 1-6. Relationships of females to their fathers in age period 4.

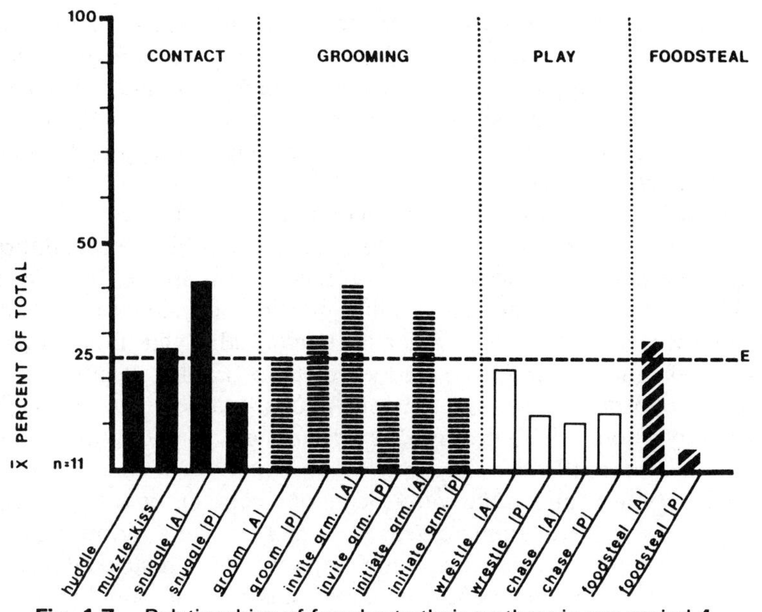

Fig. 1-7. Relationships of females to their mothers in age period 4.

lower than *E* across all categories. As opposed to earlier Age Periods, by AP 4 the frequency of juvenile interactions with fathers in all behaviors fell to very low levels, just as was shown in Figure 1-4. Grooming behaviors were the only behaviors whose percentage with fathers reached or surpassed *E* (Figures 1-5 and 1-6), although these fell well below the level shown toward other siblings (not shown in Figures 1-5, or 1-6).

Figure 1-7 illustrates further the fact that juveniles interact differently with each parent. Although juvenile females interacted far less with parents than with siblings, they nevertheless still interacted less with their fathers than with their mothers. It would seem that the father's most direct interactive role with offspring is exercised very early on and although he is quite active, this role (carrying, food sharing) is not exclusive to him but is very much a group activity.

DISCUSSION

There is no question that the role of the dominant pair-bonded callithricid male differs radically from that of the dominant male primate living in a multi-male group. Male callithricids invest much more of their energy into the care and socialization of their offspring than do most other Haplorhini, particularly when no helpers are present. Perhaps it is worthwhile to discuss this heavy investment in terms of the particular survival strategies of the Callithricidae, which appear to result in maximizing reproductive success, while coming to terms with small body size (chapter 13). Callithricid paternal behavior may be seen as just one part of these strategies, which resulted in a well-adapted primate social system evolved to meet the rigors of life in a particular niche of the neotropics.

In order to put the role of the pair-bonded male in proper context and perspective, we must understand the function and importance of callithricid family groups, for it is the social group that affords species survival and individual reproductive success. Callithricids are considered to be monogamous, living in small bands in which the bonded adult pair stays associated with its offspring for extended periods of time (Epple, 1975a). In some species, group members (especially sub-adults) may emigrate from their natal family to join neighboring groups for varying periods of time (Dawson, 1978 and Neyman, 1978 for *S. oedipus*).

In the laboratory and in the wild, each group contains only one breeding female (Dawson, 1978; Epple, 1975a; Izawa, 1978; Terborg, in press). In laboratory families and in groups containing several nonrelated adults, only the dominant female bears offspring, and it has been proposed previously that she actively inhibits reproduction in her submissive adult female group mates (Epple, 1975a). Among nonrelated common marmosets *(Callithrix jacchus),* only the dominant female exhibits ovarian cyclicity (Hearn, 1977), and in saddle-back tamarin families, most adult daughters do not cycle as long as they cohabit with their parents. In *S. fuscicollis,* this reproductive inhibition is short lived, since removal from the family and pairing with an adult male results in immediate onset of cyclicity (Katz and Epple, 1980).

Marmosets and tamarins mature quickly, normally reaching reproductive age when about 18 months old (Abbott and Hearn, 1978; Epple and Katz, 1980). It appears, however, that the rate of maturation may be accelerated by stimuli from adults of the opposite sex. In the course of a previous experiment, in which juvenile (six-month-old) *S. fuscicollis* were permanently mated with either adult or juvenile partners, young males and females pair-bonded to adults reproduced months earlier than subjects paired with partners of their own age (Epple and Katz, 1980). The earliest conceptions under these conditions occurred in two seven-month-old females living with adult males. These results suggest that adults facilitate early reproductive success in nonrelated, oppositely sexed youngsters, an effect that might involve an acceleration of puberty, as it is known to occur in some rodents (Bronson, 1979).

While subordinate females in the group cannot invest in reproduction as long as they do not maintain the dominant position, they do invest (as do the group males) in rearing the dominant female's offspring, who are their own siblings. The studies reviewed above show that all members of the group participate in the carrying and provisioning of the dominant female's infants; group members who are not offspring of the breeding pair do so to a lesser degree than do kin. Depending on group size and structure, older siblings, rather than the father, may be the major caregivers (Epple, 1975b; Hoage, 1977; Terborg, in press; Wolters, 1978). Help in the care of her infants reduces the energy demand on the mother, who would otherwise be burdened by the constant transport of twins whose combined body weight at birth may amount to 15-20% of her own and who grow quite rapidly. Moreover, helpers also relieve the pair-bonded male, who may perform most of the carrying in the absence of helpers. In some species (e.g., *S. oedipus, S. fuscicollis),* the pair-bonded male may perform most of the contact aggression during territorial defense (French and Snowdon, 1981; Terborg, in press) and it may be of advantage to him not to be overburdened with infant care as well. There are also advantages to the helpers. Rosenblum (1981) has recently pointed out that in small primates in which neonatal weight represents a high percentage of maternal weight, neonatal mortality is high if maternal care is not completely adequate very soon after birth. In callithricids, earlier experience in infant care as a helper assures high quality of maternal care by primiparas and, consequently, good infant survival. Callithricids who have been deprived of this experience fail to develop adequate parental behavior and lose successive sets of infants due to neglect or infanticide (Epple, 1975b; Hoage, 1977).

Marmosets and tamarins are small and vulnerable to predators. Therefore, a long delay in the production of offspring after a group has suffered the loss of its breeding female, or after a new group containing a nulliparous female has been formed, would be a disadvantage. The facts that the onset of ovarian cyclicity in females who have been released from the inhibitory influences of their mothers is very fast and that nonrelated adults may even accelerate puberty in juveniles of the opposite sex may be interpreted as a result of selective pressure to optimize the reproductive rate under these conditions.

Members of the family Callithricidae are also characterized by small body size, claws on all digits except the hallux, dental specializations, and the production of heterozygous twins, features that set them apart from the Cebidae and other Haplorhini. Although these features have been interpreted as primitive traits by many authors, others have interpreted them as derived (cf., Hershkovitz, 1977). Ford (1980), for example, interprets callithricids as "phyletic dwarfs," suggesting that the development of claws, changes in dentition, and reproductive twinning have evolved as a consequence of reduction in body size through time (see also chapter 13). It may be that accelerated maturation, the inhibition of reproduction in nondominant females, as well as twinning are also concomitants of a reduced body size evolved to maximize reproductive efficiency. These aspects, together with a monogamous mating system, a tight group affiliation, and the assurance of help in rearing offspring by reproductively inhibited helpers, work to produce and bring to maturity the greatest number of offspring per breeding female. The callithricid male role is therefore part of a unique social system with complex physiological underpinnings that we are just beginning to understand.

REFERENCES

Abbott, D. H., and Hearn, J. P. 1978. Physical, hormonal and behavioral aspects of sexual development in the marmoset monkey, *Callithrix jacchus*. *J. Reprod. Fert.* 53: 155-66.

Box, H. O., 1977. Quantitative data on the carrying of young captive monkeys *(Callithrix jacchus)* by other members of their family groups. *Primates* 18: 475-84.

Bronson, F. H. 1979. The reproductive ecology of the house mouse. *Quart. Rev. Biol.* 54: 265-99.

Castro, R., and Soini, P. 1977. Field studies on *Saguinus mystax* and other callitrichids in Amazonian Peru. In D. G. Kleiman (ed.), *The Biology and Conservation of the Callitrichidae,* pp. 73-78. Washington, D.C.: Smithsonian Institution Press.

Cebul, M. S., 1980. The development of social behavior in captive *Saguinus fuscicollis.* Ph.D. dissertation, Yale University, New Haven.

Christen, A. 1974. Fortpflanzungsbiologie und Verhalten bei *Cebuella pygmaea* und *Tamarin tamarin. Fortschr. Verhaltensforsch. Z. Tierpsychol.,* suppl. no. 14, pp. 1-78.

Crandall, L. S. 1951. Those forest sprites called marmosets. *Animal Kingdom,* 54: 178-84.

Dawson, G. A. 1977. Composition and stability of social groups of the tamarin, *Saguinus oedipus geoffroyi,* in Panama: Ecological and behavioural implications. In D. G. Kleiman (ed.), *The Biology and Conservation of the Callitrichidae,* pp. 23-37. Washington, D.C.: Smithsonian Institution Press.

Epple, G. 1967. Vergleichende Untersuchungen über Sexual- und Sozialverhalten der Krallenaffen (Hapalidae). *Folia Primat.* 7: 37-65.

———. 1972. Social behavior of laboratory groups of *Saguinus fuscicollis.* In D. D. Bridgewater (ed.), *Saving the Lion Marmoset,* pp. 50-58. Wheeling, WV:Wild Animal Propagation Trust.

———. 1975a. The behavior of marmoset monkeys *(Callithrichidae).* In L. A. Rosenblum (ed.), *Primate Behavior,* vol. 4, pp. 195-239. New York: Academic Press.

———. 1975b. Parental behavior in *Saguinus fuscicollis* ssp. (Callitrichidae). *Folia Primat.* 24: 221-38.

Epple, G., and Katz, Y. 1980. Social influences on first reproductive success and related behaviors in the saddle-back tamarin *(Saguinus fuscicollis,* Callitrichidae). *Int. J. Primatol.* 1: 171-83.

Fitzgerald, A. 1935. Rearing marmosets in captivity. *J. Mamm.* 16: 181-88.

Ford, S. M. 1980. Callitrichids as phyletic dwarfs and the place of the Callitrichidae in Platyrrhini. *Primates* 21: 31-43.

Franz, J. 1963. Beobachtungen an einer Lowenaffchen-Aufzucht. *Zool. Garten* 28: 115-20.

French, J. A., and Snowdon, C. T. 1981. Sexual dimorphism in response to unfamiliar intruders in the tamarin *Saguinus oedipus. Anim. Behav.* 29: 822-29.

Hearn, J. P. 1977. The endocrinology of reproduction in the common marmoset, *Callithrix jacchus.* In D. G. Kleiman (ed.), *The Biology and Conservation of the Callitrichidae,* pp. 163-71. Washington, D.C.: Smithsonian Institution Press.

Hearn, J. P., and Lunn, S. G. 1975. The reproductive biology of the marmoset monkey, *Callithrix jacchus. Lab. Anim. Handb.* 6: 191-202.

Hershkovitz, P. 1977. *Living New World Primates (Platyrrhini),* vol. 1. Chicago: University of Chicago Press.

Hill, W. C. O. 1975. *Primates: Comparative Anatomy and Taxonomy,* vol. 3, *Pithecoidea, Hapalidae and Callimiconidae.* Edinburgh: University Press.

Hoage, R. J. 1977. Parental care in *Leontopithecus rosalia rosalia:* Sex and age differences in carrying behavior and the role of prior experience. In D. G. Kleiman (ed.), *Biology and Conservation of the Callitrichidae,* pp. 293-305. Washington, D.C.: Smithsonian Institution Press.

Ingram, J. C. 1977. Interactions between parents and infants and the development of independence in the common marmoset *(Callithrix jacchus). Anim. Behav.* 25: 811-27.

Izawa, K. 1978. A field study of the ecology and behavior of the black-mantle tamarin *(Saguinus nigricollis). Primates* 19: 241-74.

Katz, Y., and Epple, G. 1980. Social influence on urinary estradiol cyclicity of female *Saguinus fuscicollis* (Callitrichidae). *Proceedings of the 8th International Congress of Primotologists Antropologia Contemporaneo,* p. 219, Florence.

Kleiman, D., 1977. Monogamy in mammals, *Quart. Rev. Biol.* 52: 39-69.

Lucas, N. S.; Hume, M.; and Henderson-Smith, H. 1927. On the breeding of the common marmoset *(Hapale jacchus)* in captivity when irradiated with ultraviolet rays. *Proc. Zool. Soc. Lond.* 30: 447-51.

Lucas, N. S.; Hume, E. M.; and Henderson-Smith, H. 1937. On the breeding of the common marmoset *(Hapale jacchus)* in captivity when irradiated with ultraviolet rays. II. A ten years' family history. *Proc Zool. Soc. Lond. Ser. A.* 107: 205-11.

Marik, M. 1931. Beobachtungen zur Fortpflanzungsbiologie der Uistiti *(Callithrix jacchus L.). Zool. Garten* 4: 347-49.

Mitchell, G., and Brandt., E. M. 1972. Paternal behavior in primates. In F. E. Poirier (ed.), *Primate Socialization,* pp. 173-206. New York: Random House.

Napier, J. R., and Napier, P. H. 1967. *A Handbook of Living Primates.* New York: Academic Press.

Neyman, P. F. 1977. Aspects of the ecological and social organization of free-ranging cotton-top tamarins *(Saguinus oedipus)* and the conservation status of the species. In D. G. Kleiman (ed.), *The Biology and Conservation of the Callitrichidae,* pp. 39-71. Washington, D.C.: Smithsonian Institution Press.

Rabb, G. B., and Rowell, J. E., 1960. Notes on reproduction in captive marmosets. *J. Mamm.* 41: 401.

Rosenblum, L. A. 1981. Parity and rearing factors in squirrel monkey reproduction. *Fourth Meeting of the American Society of Primatologists,* San Antonio, TX.

Roth, H. H. 1960. Beobachtungen an *Tamarin spec. Zool. Garten* 25: 166-81.

Rothe, H. 1973. Beobachtungen zur Geburt beim Weissbüscheläffchen *(Callithrix jacchus* Erxleben, 1777). *Folia Primat.* 19: 257-85.

Sanderson, I. T. 1957. *The Monkey Kingdom.* New York: Hanover House.

Siegel, S. 1956. *Nonparametric Statistics for the Behavioral Sciences.* New York: McGraw-Hill.

Terborg, F. In press. *The Behavioral Ecology of Five New World Primates.* Princeton: Princeton University Press.

Vogt, J. L.; Carlson, H.; and Menzel, E. 1978. Social behavior of a marmoset monkey group. I. Parental care and infant development. *Primates* 19: 715-26.

Wendt, H. 1964. Erfolgreiche Zucht des Baumwollköpfchens oder Pincheäffchens *Leontocebus (Oedipomidas) oedipus* (Linné, 1758), in Gefangenschaft. *Säugetierkdl. Mittlg.* 12: 49-52.

Wolters, H. J. 1978. Some aspects of role taking behaviour in captive family groups of the cotton-top tamarin *Saguinus oedipus oedipus.* In H. Rothe; H. J. Wolters; and J. P. Hearn (eds.), *Biology and Behaviour of Marmosets,* pp. 259-78. Gottingen: Eigenverlag Rothe.

Zukowsky, L. 1940. Zur Haltung und Pflege einiger Neuweltaffenarten. *Zool. Garten* 12: 92-110.

2

Male Caretaking Behavior Among Wild Barbary Macaques *(Macaca sylvanus)*

David Milton Taub
Yemassee Primate Center
Yemassee, South Carolina
and
Department of Psychiatry and
Behavioral Sciences
Medical University of South Carolina
Charleston, South Carolina

INTRODUCTION AND BACKGROUND

Intrauterine gestation and the physiological adaptations of females for neonatal nourishment that characterize the class Mammalia have had considerable consequences on the evolution of social organization and mating patterns (Trivers, 1972; Brown, 1975; Maynard-Smith, 1977; Whittenberger and Tilson, 1980). In contrast to birds, for example, rearing of the mammalian young is principally a female specialty, and the male's role in the care and rearing of offspring has been greatly reduced or has disappeared altogether. With the marked exception of the monogamously mating primates (see chapters 1 and 14), male primates have not been thought to participate significantly in the care of the young. The primary supportive role of males has been viewed as that of a protector of the entire social unit, of which the neonate is but one member.

As field and laboratory research on primates has increased in recent years, it has become clear that a diversity of interactions occurs between males and infants (see reviews by Mitchell, 1969; Mitchell and Brandt, 1972; Hrdy, 1976; Redican, 1976; Redican and Taub, 1981; chapters 14 and 15, this volume). From these data it would appear that there is a strong, perhaps causal, relationship between the mating system and the occurrence of male caretaking among most primates. Thus among some species of New World monkeys (mainly marmosets and tamarins) and members of the Hylobatidae, males are extensively involved with neonates for an extended period during the neonates' early lives. The occurrence of such male care is apparently restricted to those species that are monogamous in social structure and mating patterns, those that actively defend a finite territory, and those that are exclusively arboreal in habitat (Redican and Taub, 1981).

Male care of infants is infrequent among multi-male/female social groups with polygamous mating. Male-infant associations that do occur among such primate groups do not appear to be as frequent or as intense as those observed among the monogamously mating primates. Indeed, the principal male-infant interaction reported among polygamously mating multi-male/female social groups appears to be an "exploitative" or "usery" relationship between the "caregiver" and the infant. The Barbary macaques of North Africa (Taub, 1977) may be an exception to this general pattern.

Studying a zoo group, Lahiri and Southwick (1966) first reported that Barbary macaque males showed intensive interest in and interacted extensively with infants. In the controlled and manipulated colony of Barbary macaques resident on Gibraltar, first MacRoberts (1970) and later Burton (1972) observed similar behavior between males and infants. Observations on a semifree-ranging, provisioned colony in France (Merz, 1978) have shown that these Barbary macaques too exhibit a heightened interest in and extensive involvement with infants, to the point of males carrying dead infants. When this species was first studied in the wild condition a decade ago, Deag and Crook (1971) confirmed that under natural conditions the interactions between males and infants were elaborate, and they were immediately impressed with the magnitude, intensity, and multiplicity of such relationships. Male care activities were not artifacts of captivity or provisioning. Taub's (1978, 1980(a), 1980(b)) subsequent study of wild Moroccan Barbary macaques confirmed these observations, establishing without doubt that intensive and elaborate interactions between males and infants were typical and characteristic of this, the only African macaque.

This paper reports further finding from a 15-month field study, amplifying and clarifying many aspects of the sociometry of male caretaking behavior among wild Barbary macaques.

METHODS AND MATERIALS

Detailed descriptions of methodologies used in this study can be found in Taub, 1978 and 1980b.

Field Site and Study Group

This study was conducted in Morocco, North Africa, at a field site adjacent to the forestry post of Ain Kahla, 5° 13′ W by 33° 15′ N, 30 km. SW of the city of Azrou. This area lies in the center of the Middle Atlas mountain range (elevation averages about 2000 m.) and has extensive snowfalls each winter from December through April. The forests of this region are dominated by two species of trees, the Atlantic or Moroccan cedar *(Cedrus atlantica* Endl. Carriere) and the evergreen oak *(Quercus ilex),* and forests composed of mixtures of these trees appear to be primary habitat for Barbary macaques (Taub, 1977).

A study group, "S," was located and habituated to close approaches during the fall months of 1973, and the behavioral study was done on this group. After the birth of all infants of the 1974 birth season, "S" group contained 39 animals; the males and infants during this study were seven adult males (Mo, RD, CM, WN, V, NM, and BB), four sub-adult males (4SA, Ro, BL, and BS), five juvenile males (LJM, MJ, Ba, DOF, and NJ), and five infants (iLP, iBu, iLN, iRSE, and iPP).

Protocols

During the initial phase of the study (1973), only the ad libitum data collection technique was used (190 observation hours). During 1974, data were collected by the ad libitum, scan sample, and focal female techniques (559 observation hours). Qualitative descriptions of male-infant interactions are derived from data collected in both 1973 and 1974, whereas quantitative analyses were done using only the data collected in 1974, during which time 2,231 episodes of male-infant interactions were recorded.

Terminology

The male-infant interactions of Barbary macaques are classified into two general categories based on their structure (Taub, 1978, 1980a):

1. Dyadic male-infant interactions, which corresponds to Deag and Crook's (1971) "male care" category, include all interactions between a single male and an infant.
2. Triadic male-infant interactions are those that involve two males and an infant; this term can be modified to describe interactions involving more than two males: e.g., quadradic male-infant interaction, tetradic male-infant interaction, etc.

QUALITATIVE SUMMARY OF THE DIVERSITY OF MALE CARETAKING BEHAVIOR

Mitchell and Brandt (1972) have described eight types or levels of interactions that may exist between males and infants. Barbary macaque males regularly and predictable show at least seven of these types—touching, carrying, approaching, retrieving, grooming/playing, protecting, caring for— and perhaps also the eighth, adoption. Interactions between a Barbary macaque male and an infant may take many individual forms, but the principal types include the specific interactions and associations summarized below.

Carrying. The most frequently observed type of male-infant association was males carrying infants (Table 2-1), which is identical in form and structure to that used by females. Males carried infants in both the dorsal and ventral positions (Figs. 2-1 and 2-2), but the choice of carriage position appeared to be both age and context dependent. Males carried infants in all social and nonsocial contexts, over a diversity of terrain, on the ground and in the trees.

Holding. This type of association included: males sitting with infants clinging to their backs; the infant sitting in the male's lap or clinging to his ventrum; the male cuddling or cradling the infant in his lap or to his ventrum; or the male huddling with the infant in his lap. Barbary macaque male-infant interactions of this type are isomorphic in structure with those displayed by mothers to their infants (see Redican and Mitchell, 1973, for contrast with rhesus macaques) (Figs. 2-3 and 2-4).

Grooming and Manipulation. Males often groomed infants as the infants sat with their mothers. During grooming bouts, or independent of them, males also nuzzled, licked, smelled, or mouthed the infants, tactilely manipulated or licked the genitalia of the infant, clutched or cuddled the infants tightly against their chests, or lip-smacked/teeth-chattered (Van Hoof, 1969) at or directly onto the infants.

Soliciting Approaches. Barbary macaque males solicit approaches from infants in several ways. A male may (1) stand facing the infant and vigorously assume the teeth-chatter face; (2) stand oriented in a presenting position with rump toward and lowered to the infant (i.e., legs flexed), with or without the teeth-chatter face; (3) stand next to the infant and lower the shoulders (i.e., flex the forearms) or the hindquarters to facilitate mounting by the infant. Infants usually responded to these solicitation gestures by approaching and climbing onto the male's back. Infants also mounted males without any overt signals of solicitation from the males.

Fig. 2-1. Juvenile male carries infant in ventral carriage position.

Fig. 2-2. Adult male carries infant in dorsal or "jockey" carriage position.

Fig. 2-3. Adult male sits with infant in his lap.

Fig. 2-4. Adult male sits with infant protected in lap, in a manner indentical to that shown by mothers to their infants.

Playing. The most common play companions of infants tended to be other infants and yearlings of both sexes. Sometimes juvenile, adult, and subadult males played with infants by wrestling, grappling, and mouthing them. The most common play episodes with adult and subadult males tended to be unidirectional; i.e., the infants climbed all over and hopped on and around the impassive male, who was solicitous and patient of this activity, but did not otherwise interact directly with the infant.

Monitoring. On occasion a male might sit in close proximity to an infant and watch it closely. During this time the male would not engage in other activities, but would focus his attention on the infant. Thus a male might stand under a tree and intently watch an infant just above him for a lengthy period. Monitoring behavior by a male usually terminated when the male retrieved the infant by soliciting the infant's approach, or when the male simply walked off and left the infant.

Retrieving and protecting. Males approached infants directly and pulled them to themselves. Such retrievals occurred in a variety of contexts: e.g., males would run and gather up infants in response to external threats (alarm barks). Males were also very responsive to stress vocalizations given by infants, and ran to and retrieved them in response to these vocalizations, even if no external threat could be identified. At other times, retrievals occurred in affiliative social encounters. Males were also observed to walk up to unattended infants, in the absence of any signals the observer could identify, pick them up, and walk off with them.

Male protection of infants has been observed among many nonhuman primates. There may be a qualitative difference, however, in the retrieval-protective behavior of male Barbary macaques and other primates. Redican and Mitchell (1973, 1974; Mitchell, 1976, pers. comm.) emphasize that for rhesus macaques, when an infant is threatened, a male responds first to the external threat and only afterward does he respond to the infant, if at all. In contrast, Barbary macaque males respond first to the infant during a crisis (Taub, 1978).

Triadic male-infant interactions. Among the Barbary macaque, there is a ritualized interaction that occurs between two or more males and an infant, a behavior previously termed "agonistic buffering" by Deag and Crook (1971; see Taub, 1980a, for a different interpretation of this phenomenon). In its most elementary form, triadic male-infant interactions involve one male carrying an infant to another, after which both males sit together for a few seconds, displaying a series of exaggerated and stereotypic behaviors to each other and to the infant.

Barbary macaque males begin to interact with an infant within days after the infant is born (Deag, 1974). Male associations with the infant are frequent and intensive throughout the infant's first year of life, although the frequencies of association decline steadily as the infant matures. This decline is partly the result of increased neuromotor competence, enabling the infant to play a more active, independent role in determining social interactions. Additionally, increasing age is associated with a preference for establishing play groups with other infants and yearlings. These patterned changes among infants coincide with a shift in male interest toward estrous females during the mating season, which occurs when infants are from five to seven months of age. Interactions with infants thus reach their lowest point just before the onset of the birth season, which contrasts sharply with Japanese macaques (Itani, 1959), among whom the most intensive interactions occur between males and yearlings as the new birth season approaches. Among Barbary macaques, male interest focuses on new infants, and little interest is shown toward the previous year's infants, now yearlings. Indeed, no adult male was ever seen carrying a yearling, although sometimes subadult and juvenile males did so.

RESULTS: THE SOCIOMETRY OF MALE-INFANT INTERACTIONS

During this phase of the study, 2,231 interactions between individually known adult, subadult and juvenile males, and infants were recorded (Table 2-1): 1, 587 (71%) were of the dyadic type, 577 (26%) were triadic interactions, while 67 (3%) were of the quadradic or greater type. The analyses presented below concentrate on the dyadic type of male-infant association.

Table 2-1. Distribution Of Types Of Male-Infant Interactions.

Type Of Male-Infant Interaction	Number of Cases Obtained[a]	Percent of All Cases
1. Infant sits on male's dorsum	475	(22)
2. Infant sits on male's ventrum	137	(6)
3. Male walks/stands with infant dorsal	602	(28)
4. Male walks/stands with infant ventral	13	(.6)
5. Male grooms infant	112	(5)
6. Male and infant play together	22	(1)
7. Male monitors infant	33	(1)
8. Male retrieves infant	158	(7)
9. Male approaches and touches infant	35	(2)
All Dyadic Male-Infant Interactions	1587	(73)
10. Triadic male-infant interactions	577	(27)
All Male-Infant Interactions[b]	2164	(100)

[a]Includes observations on adult, subadult, and juvenile males.
[b]Excludes 67 cases of quadradic or greater male-infant interactions.

Distribution of the Types of Male-Infant Caretaking Interactions

Table 2-1 shows the distribution of the various types of male-infant interactions, while Tables 2-2, 2-3, and 2-4 delineate these interactions for adult, subadult, and juvenile male age classes, respectively (excludes quadradic interactions). As can be seen, there is a significant difference in the frequency distribution of the different types of interactions. Three types predominate: males sitting with infants on their backs accounted for 22% of all interactions; males walking/standing with infants on their backs accounted for an additional 28%; while triadic interactions as a class accounted for 27%. Retrieving infants (7%) and grooming them (5%) were the next most frequent behaviors occurring between males and infants, but they occurred only about a third as often as the most common interactions. The remaining behaviors occurred relatively less frequently. When one considers dyadic interactions only, males carrying infants accounted for 39% (602 + 13 ÷ 1,587), with another 39% being accounted for by males sitting with infants (475 + 137 ÷ 1,587), or a total of 78% of all dyadic interactions were some form of males carrying or holding infants. Among all males, the distribution of the types of male-infant interactions was fairly concordant; that is, there were no significant differences in the types of male-infant interactions most frequently engaged in by the different age classes of males ($W = 0.979$, $\chi^2 = 96.92$, $df = 9$, $p < .001$).

Distribution of Male-Infant Behaviors Among Infants

Table 2-5 depicts the distribution of the different types of male-infant interactions among the five infants of "S" group. One of the most striking features of this distribution is that male infant iPP never interacted with any adult or subadult male, nor was it ever involved in a triadic interaction. Given that all males are involved regularly with infants in general, it is unknown why iPP attracted no interest from males; there was no obvious feature of this infant or its mother, PP, to suggest why males would not interact with it. But iPP did interact on rare occasions with juvenile males DOF and NJ, and interacted frequently with the other infants and with yearlings of both sexes. I believe it is not coincidental that iPP was the only case of infant mortality, disappearing from the group when it was about 4 to 4½ months of age. Although this is not conclusive evidence, it suggests strongly that for this species in this habitat, caretaking behavior by adult and subadult males is crucial to a neonate's ability to survive its first year of life.

Infants received care from several different males (Table 2-8), but they did not receive equivalent amounts of attention from these males (Table 2-6). Moreover, there were significant differences among the four infants (analysis

Table 2-2. Distribution Of Types Of Male-Infant Interactions Among Adult Males.

Type Of Male-Infant Interaction	Adult Males							All Adult Males
	Mo	RD	CM	WN	V	NM	BB	
1. Infant sits on male's dorsum	19	17	33	20	41	29	24	183
2. Infant sits on male's ventrum	9	13	17	17	24	7	3	90
3. Male walks/stands infant dorsal	36	31	43	26	73	36	28	273
4. Male walks/stands infant ventral	1	1	2	0	1	2	1	8
5. Male grooms infant	18	14	6	12	8	17	3	78
6. Male and infant play together	0	0	5	2	0	0	1	8
7. Male monitors infant	3	2	2	1	5	3	2	18
8. Male retrieves infant	4	10	17	13	14	9	9	76
9. Male approaches and touches infant	4	2	7	5	0	2	0	20
10. Triadic male-infant interaction	29	36	49	23	36	54	24	251
Sum All Behaviors	123	126	181	119	202	159	95	1005
Percent Adult Males[a]	(12)	(13)	(18)	(12)	(20)	(16)	(9)	
Percent All Males[b]	(6)	(6)	(8)	(5)	(9)	(7)	(4)	(46)

[a]Figured as a percent of the distribution among adult males only.
[b]Figured as a percent of the distribution among all males.

Table 2-3. Distribution Of Types Of Male-Infant Interactions Among Subadult Males.

Type Of Male-Infant Interaction	Subadult Males				All Subadult Males
	4SA	Ro	BL	BS	
1. Infant sits on male's dorsum	93	86	27	47	253
2. Infant sits on male's ventrum	13	21	9	3	46
3. Male walks/stands infant dorsal	100	79	29	40	248
4. Male walks/stands infant ventral	1	2	1	1	5
5. Male grooms infant	10	14	6	0	30
6. Male and infant play together	0	1	2	0	3
7. Male monitors infant	10	4	0	0	14
8. Male retrieves infant	28	15	16	8	67
9. Male approaches and touches infant	3	2	3	0	8
10. Triadic male-infant interaction	112	67	55	50	284
Sum All Behaviors	370	291	148	149	958
Percent Subadult Males[a]	(39)	(30)	(16)	(16)	
Percent All Males[b]	(17)	(13)	(7)	(7)	(44)

[a]Figured as a percent of the distribution among subadult males only.
[b]Figured as a percent of the distribution among all males.

Table 2-4. Distribution Of Types Of Male-Infant Interactions Among Juvenile Males.

Type Of Male-Infant Interaction	Juvenile Males					All Juvenile Males
	LJM	MJ	Ba	DOF	NJ	
1. Infant sits on male's dorsum	0	15	12	3	9	39
2. Infant sits on male's ventrum	0	0	0	0	1	1
3. Male walks/stands infant dorsal	0	28	34	0	19	81
4. Male walks/stands infant ventral	0	0	0	0	0	0
5. Male grooms infant	0	1	2	1	0	4
6. Male and infant play together	0	5	2	1	3	11
7. Male monitors infant	0	1	0	0	0	1
8. Male retrieves infant	0	6	6	1	2	15
9. Male approaches and touches infant	0	4	2	0	1	7
10. Triadic male-infant interaction	0	17	15	0	10	42
Sum All Behaviors	0	77	73	6	45	201
Percent Juvenile Males[a]	(0)	(38)	(36)	(3)	(22)	
Percent All Males[b]	(0)	(4)	(3)	(.2)	(2)	(9)

[a]Figured as a percent of the distribution among juvenile males.
[b]Figured as a percent of the distribution among all males.

excluded iPP) in both the number of dyadic or caretaking interactions received ($X^2 = 343.24$, $df = 3$, $p < .001$) and the total frequency of male-infant interactions received (i.e., dyadic plus triadic) ($X^2 = 515.62$, $df = 3$, $p < = .001$). There were no significant differences among the infants in the rankings of the *kinds* of interactions they received ($W = 0.924$, $X^2 = 33.26$, $df = 9$, $p < .001$), however, which indicates that the inter-infant differences came from differences in the quantity of attention received from the different males interested in them and not in the quality of the behaviors themselves.

The magnitude of these inter-infant differences in the amount of care received can be seen qualitatively in Figure 2-5. Interactions with iLN accounted for 43% of the all-male total, while iBu received 28% of all recorded male attention. Both iRSE and iLP were associated with the least: 15% and 13% each, respectively. The age and sex of the infant did not appear to be critical in determining the frequency of male attention received. For example, iPP was the youngest and a male; iRSE, a male, was younger than male infant iLN; but iBu, a female, was older than iRSE, but younger than female infant iLP. Neither could maternal social rank account for these differences: LP ranked first among the females, followed by Bu, LN, RSE, and PP, respectively (Taub, 1978). Later it will be suggested that kinship may be important in determining how frequently a male associates with an infant, but the proximate mechanisms responsible for the unequal distribution of male interest shown toward infants remain unknown. What is clear from these data, however, is that infants do not receive equivalent amounts of care from males during the first year of their lives.

A corollary of the fact that infants received unequal amounts of male attention is that individual males are not equally involved with infants. The different frequencies and relative percentages of each sexually mature male's caretaking activity are presented in Table 2-2 and 2-3. Considering all male-infant interactions for the adult and subadult males, the null hypothesis —that there are no inter-male differences in caretaking activity—is rejected ($\chi^2 = 364.12$, $df = 10$, $p < .001$), showing that there are striking differences among individual males in the extent of their involvement with infants. Similar differences are seen when only dyadic interactions are considered ($\chi^2 = 294.69$, $df = 10$, $p < .001$). Moreover, there are significant differences between the adult and subadult male subgroups in the extent of their caretaking behaviors; subadult males are involved with infants significantly more often ($t = 2.29$, $df = 9$, $p < .05$). This may be appreciated more qualitatively: the four subadult males accounted for 44% of all the male-infant interactions (or an average 11% per subject), while seven adult males accounted for 46% (or about 6.5% on average per subject.)

Table 2-5. Distribution Of Types Of Male-Infant Interactions Toward The Infants Of "S" Group.

	Infants											
	iLP		iBu		iLN		IRSE		iPP		All Infants	
Type Of male-Infant Interaction	*n*	%[a]	*n*	%	*n*	%	*n*	%	*n*	%	*n*	%[b]
1. Infant sits on male's dorsum	61	13	139	29	207	44	75	16	3	.1	475	22
2. Infant sits on male's ventrum	15	11	43	31	55	40	24	18	0	0	137	6
3. Male walks/stands infant dorsal	87	14	175	29	246	41	94	16	0	0	602	28
4. Male walks/stands infant ventral	1	8	4	31	4	31	4	31	0	0	13	.6
5. Male grooms infant	17	15	35	31	51	46	9	8	0	0	112	5
6. Male and infant play together	4	18	4	18	11	50	2	9	2	9	22	1
7. Male monitors infant	8	24	14	42	8	24	3	9	0	0	33	1
8. Male retrieves infant	24	15	40	25	72	46	21	13	1	.6	158	7
9. Male approaches and touches infant	0	0	10	29	18	51	7	20	0	0	35	2
10. Triadic male-infant interaction	60	10	160	27	262	46	95	16	0	0	577	27
Sum All Behaviors Per Infant	277		614		934		334		5		2164	
Percent All Infants[c]	(13)		(28)		(43)		(15)		(0.2)			

[a]Figured as a percent of each infant's contribution to each behavioral category; i.e. iLP accounted for 13% of all "infant sits on male's dorsum" observed.
[b]Figured as a percent of all male-infant interactions, $n = 2164$.
[c]Figured as a percent of each infant's contribution to all male-infant interactions, $n = 2164$.

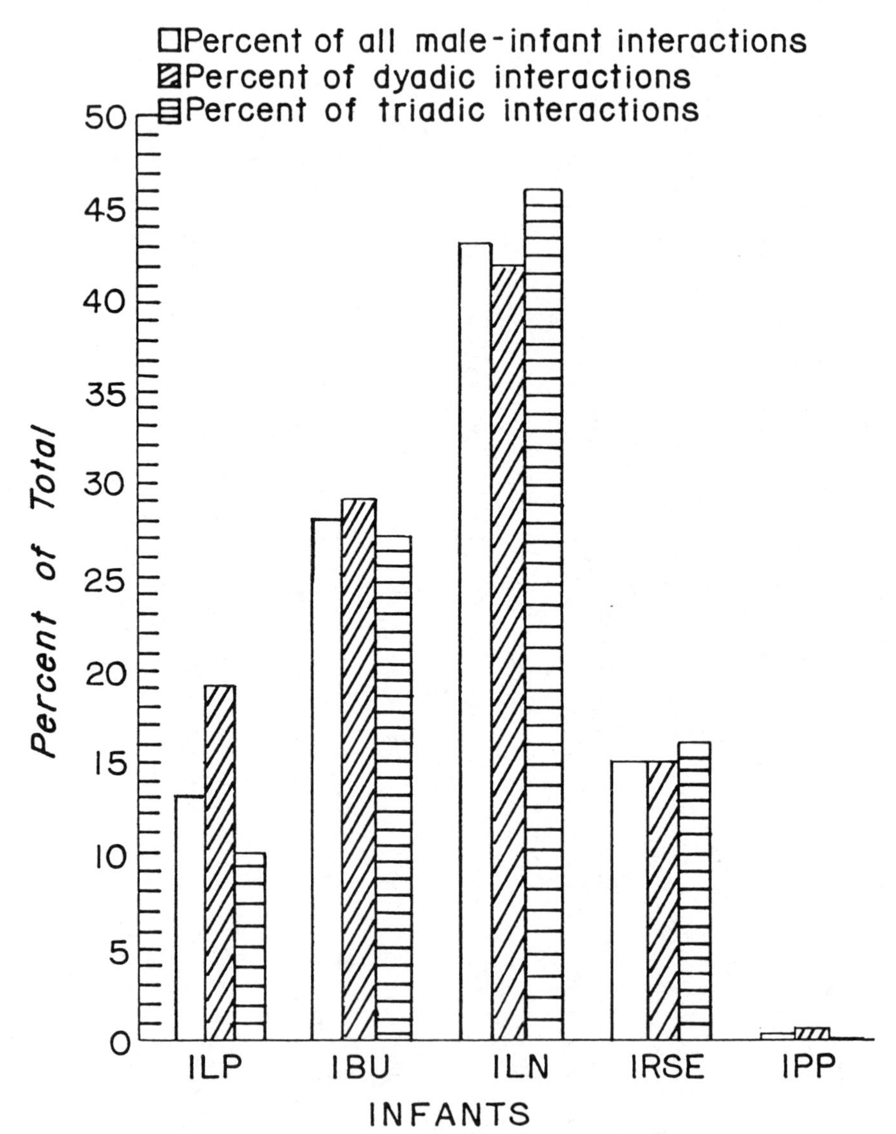

Fig. 2-5. Percentage distribution of all male-infant interactions among "S" group infants.

The qualitative differences among the males in their caretaking activities can be seen in Figure 2.6. Considering all the caretaking activities combined (Tables 2-2 and 2-3), the largest portion of all interactions accounted for by a single male was 17% (subadult male, 4SA). The second most active was

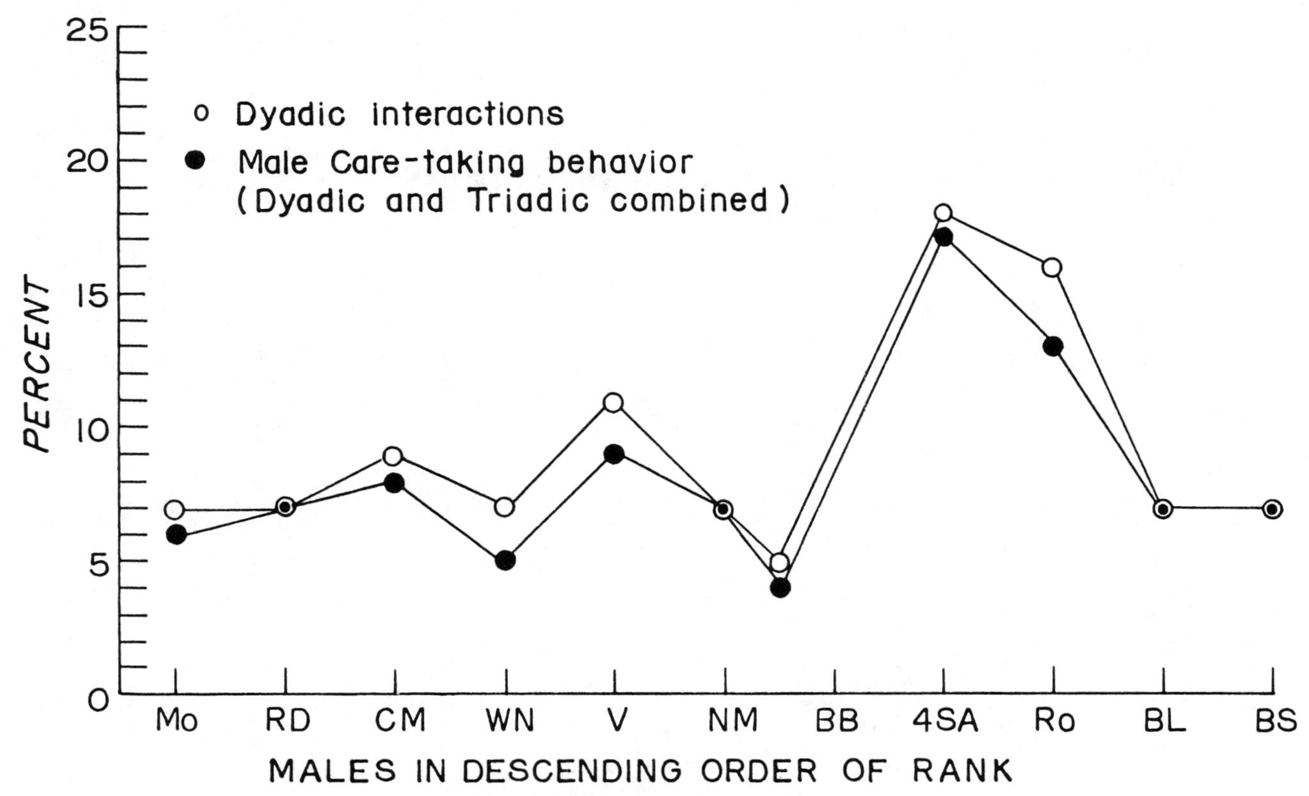

Fig. 2-6. Distribution of male caretaking behavior among "S" group adult and subadult males.

subadult male Ro, with 13% of the total. Adult male V was third with 9%, followed by CM with 8%. Three males, NM, BL, and BS, clustered with 7% each, followed by RD and Mo with 6% each. WN with 5% and BB with 4% of all male-infant behaviors were the least actively involved with infants of all the sexually mature males. Although the percentages differ, the same ranking of male interest obtains when considering only dyadic caretaking behaviors (Fig. 2-6).

Distribution of Male-Infant Behaviors by Individual Males and Individual Infants

Males did not interact equally with each infant in the group. The distribution of all male-infant behaviors by each male versus each infant receiving care is given in Table 2-6. The most remarkable feature of this distribution is that a male's attention to infants is not distributed randomly toward them. On the contrary, a male tends to restrict his interactions to one or two infants, essentially ignoring the others. Thus 11 of the 16 (69%) adult, subadult, and juvenile males directed a majority (53% +) of their associations toward just a single infant. All 16 males directed from 62% to 100% of their attention toward only two infants. The specificity by which a male directs his interests to a particular infant can be seen clearly from Table 2-6 and 2-7. These data establish unequivocally that among Barbary macaques, a male's caretaking activity tends to be organized around one or two particular infants.

A male's "primary infant" may be defined as the infant to which the male directs the majority of his caretaking activity. The "secondary infant" is the one that receives between one-half and one-third as much attention as the male shows to his primary infant. Any infant that received less than one-third as much attention from a male as his primary infant received was considered to be "ignored" (these indices were used because no male distributed his attention equally to more than three infants—and then only one male, CM (in Table 2-6), did so—which suggests that three infants are probably the upper limits of a male's effective capability to contribute to an infant's care.

Each male's primary and secondary infant and those he ignored (excluding iPP, whom all males ignored) are shown in Table 2-7. Nine of the 11 adult and subadult males of "S" group (82%) had a single primary infant; one male (9%) had two, while one male (9%) had three. Of the nine males that had only a single primary infant, five of them (56%) had no secondary infants, as they essentially ignored all other infants. Three of these nine males with just a single primary infant (33%) had only a single secondary infant. One of them had two secondary infants. Thus 9 of the 11 sexually mature males (82%) of the group interacted extensively and almost exclusively with one or two

Table 2-6. Distribution Of Each Male's Percentage Of Total Interactions with Each Infant.

	iLP		IBu		iLN		iRSE		iPP	
Males	Number	Percent	Number	Percent	Number	Percent	Number	Percent	Number	Percent
Adult Males										
Mo	20	(16)	13	(11)	87	(71)	3	(2)	0	(0)
RD	4	(3)	16	(13)	51	(40)	55	(44)	0	(0)
CM	21	(12)	48	(27)	53	(29)	59	(33)	0	(0)
WN	63	(53)	18	(15)	25	(21)	13	(11)	0	(0)
V	38	(19)	96	(47)	43	(21)	25	(13)	0	(0)
NM	19	(12)	104	(65)	30	(19)	6	(4)	0	(0)
BB	65	(68)	13	(14)	15	(16)	2	(2)	0	(0)
Subadult Males										
4SA	2	(.5)	276	(75)	91	(25)	1	(.2)	0	(0)
Ro	2	(.6)	11	(4)	273	(93)	5	(2)	0	(0)
BL	2	(1)	8	(5)	130	(88)	8	(5)	0	(0)
BS	1	(1)	4	(3)	53	(36)	91	(60)	0	(0)
Juvenile Males										
LJM	0	(0)	0	(0)	0	(0)	0	(0)	0	(0)
MJ	34	(44)	1	(1)	19	(25)	22	(29)	0	(0)
Ba	3	(4)	3	(4)	54	(74)	13	(18)	0	(0)
DOF	2	(33)	0	(0)	0	(0)	0	(0)	4	(77)
NJ	1	(2)	1	(2)	15	(33)	27	(60)	1	(2)

Table 2-7. Specificity of a Male's Interactions with "S" Group Infants.

Males	Primary Infant[a]			Secondary Infant[b]			Infant Ignored[c]		
	Infant	Number	Percent	Infant	Number	Percent	Infant	Number	Percent
Mo	iLN	87	(71)	none			iLP	20	(16)
(n = 123)[d]							iBu	13	(11)
							iRSE	3	(2)
								36	(29)
RD	iRSE	55	(44)	none			iBu	16	(13)
(n = 126)	iLN	51	(40)				iLP	4	(3)
		106	(84)					20	(16)
CM	iRSE	59	(33)	iLP	21	(12)	none		
(n = 182)	iLN	53	(29)						
	iBu	48	(27)						
		160	(88)						
WN	iLP	63	(53)	iLN	25	(21)	iBu	18	(15)
(n = 119)							iRSE	13	(11)
								31	(26)
V	iBu	96	(47)	iLN	43	(21)	iRSE	25	(12)
(n = 202)				iLP	38	(19)			
					81	(40)			
NM	iBu	104	(65)	none			iLN	30	(19)
(n = 159)							iLP	19	(12)
							iRSE	6	(4)
								55	(35)

BB	iLP	65	(68)	none			iLN	15	(16)
(n = 95)							iBu	13	(14)
							iRSE	2	(2)
								30	(32)
4SA	iBu	276	(75)	iLN	91	(25)	iLP	2	(.5)
(n = 370)							iRSE	1	(.2)
								3	(.7)
Ro	iLN	273	(93)	none			iBu	11	(4)
(n = 291)							iRSE	5	(2)
							iLP	2	(.6)
								18	(6.6)
BL	iLN	130	(88)	none			iRSE	8	(5)
(n = 148)							iBu	8	(5)
							iLP	2	(1)
								18	(11)
BS	iRSE	91	(60)	iLN	53	(36)	iBu	4	(3)
(n = 149)							iLP	1	(1)
								5	(4)

[a]Defined as the infant that received the majority of the male's attention.
[b]Defined as the infant that received one-half to one-third fewer interactions than the primary infant received from that male.
[c]Defined as the infant that received less than one-third the amount of the primary infant's interactions with that male.
[d]Total number of male-infant interactions recorded for this male.

infants. These data establish that Barbary macaque males are exceedingly discriminating in their choice of which infants to interact with.

When the frequency of care given to the primary infant is compared to that given the secondary infant, each male contributed significantly more to the primary infant (all Chi-square tests were significant at the $p < .01$ level). Likewise all primary infants received significantly more attention from a male than the "ignored" infants (all Chi-square tests were significant at $p < .001$ level). However, in the case of two males, V and WN, there were no statistical differences between the frequency of interactions with their secondary infants versus that received by the "ignored" infants (Table 2-7).

It should be noted that with one exception (BS, 60%), each subadult male directed a higher percentage of all his caretaking behavior to a single primary infant than did any adult male: 75% for the *least* for subadult males (except BS) versus 71% the *highest* for adult males (Table 2-7). The average percentage of caretaking behavior directed toward a single primary infant for all subadult males was 79% while that for all adult males was 54%. Thus not only are subadult males more heavily involved in caretaking in general, they also may be able to discriminate more finely than can adult males which infants to interact with.

While males tended to restrict the number of infants they interacted with to one or two, infants, on the other hand, received care from many different males, although the absolute number of males caring for each infant varied considerably among them (Table 2-8). For example, iLN was the primary or secondary infant for nine males (82% of the sexually mature males of "S" group), whereas iRSE was the primary infant for three males (27%) but was not a secondary infant for any male.

That portion of a male's caretaking given to a particular infant was not necessarily consistent with that male's contribution to all infants relative to all other males. Depending on the focus of analysis, e.g., how much care was contributed by each male relative to all males, or how much care an infant received relative to either all infants or to all males giving care, different proportional distributions of the male's caretaking activities should be considered. Several examples will illustrate this dichotomy: (1) the percentage of each individual male's attention shown to his primary infant ranged from a low of 27% (for male CM) to a high of 93% (for male Ro); on the other hand, viewed as a proportion of all caretaking recorded, CM accounted for 8% of the total and Ro for 17%, respectively. (2) For male BB, 68% of his interactions with infants was directed toward iLP (65 episodes, Table 2-7), but of all the interactions received by iLP (n = 227), BB's portion accounted for 23% of iLP's total (Table 2-9), and similarly, for male Mo, 71% of his interactions were with iLN (87 episodes, Table 2-7), but this accounted for 9% of the care iLN received (Table 2-9). (3) That two males might have similar relative percentages of caretaking infants did not mean that they were equally involved with infants; e.g., 47% of male V's interactions with infants were accounted for by 96 behavioral episodes, but 44% of RD's came from 55 interactions (Table 2-6), thus three percentage points diffference in the male's *relative* involvement with an infant amounted to almost twice as many actual interactions with them. These differential percentage distributions have derived from the important fact that there are significant differences among males in the interest shown to infants in general, and to specific infants in particular. Put another way, subadult male 4SA had the third highest number of interactions with iLN (91 episodes, Table 2-6) but for 4SA, iLN was a secondary infant (Table 2-7); whereas male CM had 53 episodes with iLN (which ranked 6th for iLN's total, Table 2-6) but for male CM, iLN was a primary infant (29% of all CM's interactions with infants, Tables 2-6 and 2-7).

Curiously, the largest juvenile male, LJM, was never observed to interact with an infant during the 1973-74 study (Taub, 1978). During a short 1977 project, however, LJM, by then clearly a subadult male, was observed to be extensively involved with infants in all manner of caretaking behavior (Kurland and Taub, in prep.).

Multiple Male Interest in Individual Infants

It has been shown that each infant received differential attention from several different males in the group. How then is that infant's caretaking activity divided among the males attending to it? Table 2-9 shows the distribution of caretaking behavior each male devoted to his primary and secondary infants. Thus for three infants, iBu, iLN and iRSE, a majority (61% +) of the total care each received came only from those males for which it was the primary infant; iLP received 46% of its care from such males. All four infants received from 67% to 86% of all their care from those males for which the infant was both the primary and secondary infant (only iLP and iLN). The amount of care an infant received from the several males attending it was not provided equally by them. For example, 8% of iBu's care came from male CM, while 45% of it came from subadult male 4SA. Also

Table 2-8. Multiple Male Interest in Individual Infants.

Infant	Males for Which Infant Was Primary Infant	Males for Which Infant Was Secondary Infant	Total *n*	Total % Total
iLP	WN, BB	CM, V	4	36
iBu	CM, V, NM, 4SA	none	4	36
iLN	MO, RD, CM, Ro, BL	WM, V, 4SA, BS	9	82
iRSE	RD, CM, BS	none	3	27

Table 2-9. Distribution of Infants' Caretaking Behavior by Major Male Contributors.

	Interactions Received as Primary Infant			Interactions Received as Secondary Infant		
Infants	Number	Percent Total Received	Male Contributing Care	Number	Percent Total Received	Male Contributing Care
iLP	65	(23)	BB	38	(14)	V
($n = 277$)[a]	63	(23)	WN	21	(8)	CM
	128	(46)		59	(21)	
iBu	276	(45)	4SA			
($n = 612$)	104	(17)	NM	none		
	96	(16)	V			
	48	(8)	CM			
	524	(85)				
iLN	273	(29)	Ro	91	(10)	4SA
($n = 939$)	130	(14)	BL	53	(6)	BS
	87	(9)	Mo	43	(5)	V
	53	(6)	CM	25	(3)	WN
	51	(5)	RD			
	594	(63)		214	(23)	
iRSE	91	(28)	BS			
($n = 330$)	59	(18)	CM	none		
	55	(17)	RD			
	205	(61)				

[a]Total number of interactions in which this infant was involved.

4SA's contribution to iLN, as his secondary infant ($n = 91$), was absolutely greater than either Mo and RD's contribution to iLN as their primary infant ($n = 87$ and $n = 51$, respectively). Considering the distribution of caretaking behavior among only the males for which that infant was the primary infant, the following patterns are seen: for infants iLP and iRSE, each male contributed approximately equivalent amounts of care to them; but for infants iBu and iLN, there are significant differences in the amounts of care contributed by each male attending primarily to them ($\chi^2 = 227.98$, $df = 10$, $p < .001$ and $\chi^2 = 279.33$, $df = 10$, $p < .001$, respectively). These inter-male differences become more pronounced when the data for primary and secondary infants are considered collectively.

Even though these data establish that infants received care from many males—in some cases in significantly different amounts from each—it should be recalled that the care received by each infant varied significantly among them. Thus even when all the care an infant received from males is summed, some infants still received significantly more attention from males than did others.

Social Status and Male-Infant Interactions

Deag and Crook (1971) have suggested that the triadic type of male-infant interaction (the so-called agonistic buffering) is functionally related to the dominance rank of males. The analysis here of the relationship between a male's social status and his caretaking involvement with infants will, therefore, exclude the data on triadic interactions (see Taub, 1980a, for a detailed discussion of the triadic male-infant interaction). Social rank among the males does not appear to be a significant variable in the pattern of male-infant interactions, or in the amount of care each infant received (Fig. 2-7). Analysis by Spearman's rank correlation shows no linear relation between a male's position in the social hierarchy and his dyadic involvements with infants ($r_s = .22$, n.s.), or in all of his caretaking activities ($r_s = .32$, n.s.). Moreover, there were no significant differences between high-ranking (Mo, RD, CM, and WN) and low-ranking males in the frequency of male-infant activities each was involved in ($t = 1.12$, $df = 9$, $p < .20$). On the other hand, subadult males as a group were significantly more involved with infants than were adult males. It thus appears that age (i.e., subadulthood) rather than rank (subadult males were all low ranking) is the critical variable, hence rank per se is probably a spurious correlate of the distribution of infant caretaking. It will be recalled that there were significant individual male differences in the extent to which they were involved with infants; it appears to be the case that the primary basis for differential male involvement with infants is not a male's social rank.

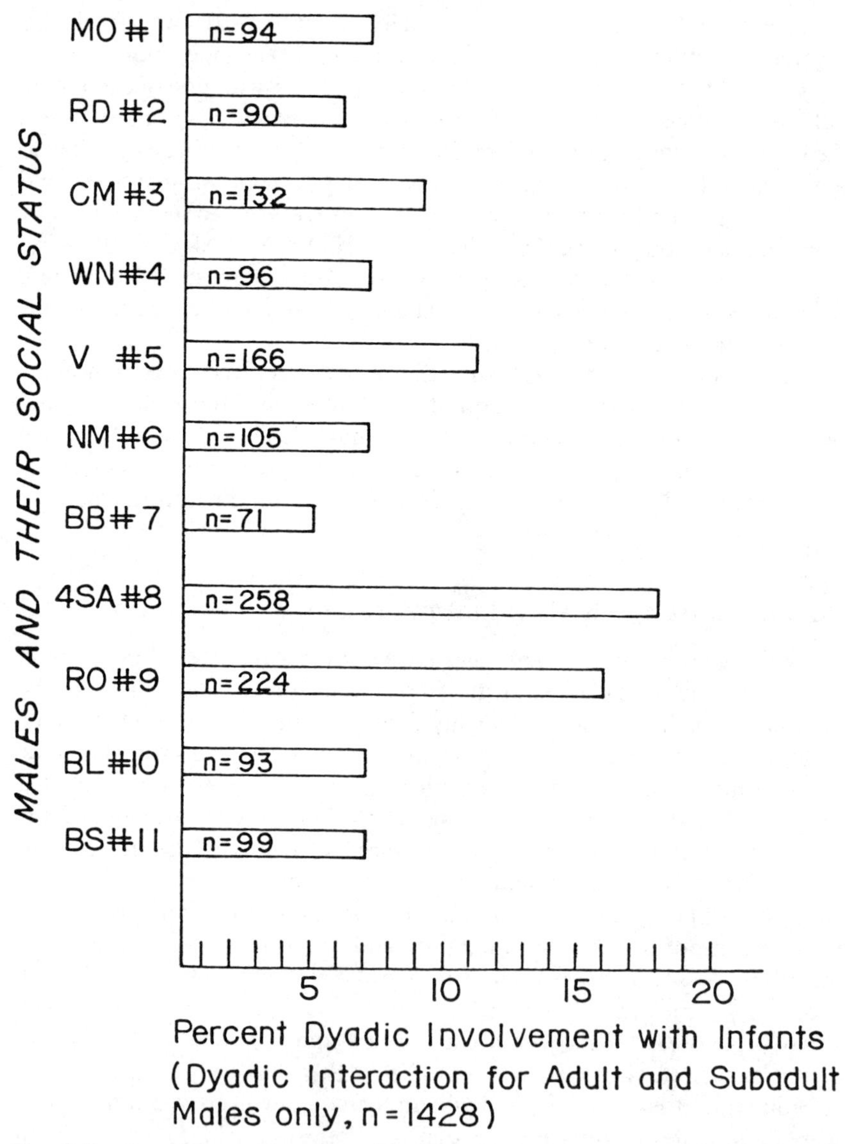

Fig. 2-7. Comparison of male social status by degree of involvement in dyadic caretaking.

The mothers of infants iLP, iBu, iLN and iRSE were the four highest ranking of the adult females. There was no linear correlation, however, between individual maternal rank and the amount of care the infant received ($r_s = .40$, n.s.). Qualitatively, the infant of the highest-ranking female, LP, received the least attention from males (excepting iPP). On the other hand, female PP, whose infant never interacted with adult or subadult males, was the lowest ranking adult female. Could it be the case that males respond to infants on the basis of the general rank of the infant's mother? This question is complicated due to the fact that during the 1973-74 study, the four highest-ranking females had infants, but only one of the lower-ranking females did. During the short 1977 project, all eleven adult females had infants, and although quantitative data were not collected on this question, all infants were observed to interact with males. Thus it is known that the infants of low-ranking females received some attention from adult and subadult males. Yet it may still be the case that the offspring of all lower-ranking females collectively received less care than the offspring of all high-ranking females. While it appears certain that male social rank and male caretaking activities are poorly correlated, the question of the relationship of maternal rank to male caretaking activities remains unresolved for the present time.

Other Aspects of Male-Infant Interactions

It has been suggested (Deag, 1974) that Barbary macaque males' interaction with infants are functionally related to social cohesion and dominance relations. Three general, mutually exclusive, behavioral categories relating to these functions were used to describe a male's behavior prior to his dyadic interactions with infants: (1) male involved in nonsocial activities; (2) male involved in affiliative social behavior; (3) male involved in aggressive interactions. Table 2-10 shows the distribution of 1,048 episodes in which the

Table 2-10. Behavior of the Male Prior to Dyadic Interactions with Infants.

Males	Nonsocial Behavior	Affiliative Social Behavior	Aggressive Behavior
Mo	51	6	4
RD	66	2	6
CM	86	11	2
WN	63	10	3
V	88	9	15
NM	81	6	4
BB	34	3	3
4SA	172	12	12
Ro	138	10	2
BL	72	10	6
BS	51	2	8
Totals	902 (86%)	81 (7%)	65 (6%)

Table 2-11. Behavior of Males During Dyadic Interactions with Infants.

Males	Nonsocial Behavior	Affiliative Social Behavior	Aggressive Behavior
Mo	71	18	4
RD	80	11	0
CM	113	19	2
WN	72	22	2
V	143	14	9
NM	94	10	2
BB	62	4	5
4SA	228	20	14
Ro	198	22	7
BL	72	12	8
BS	82	8	9
Totals	1215 (85%)	160 (11%)	62 (4%)

male's behavior prior to an involvement with an infant was known. As can be seen, the male's predominant activity was nonsocial, as this activity accounted for 86% of all cases. Males were involved about equally in both affiliative and aggressive encounters before interacting with an infant, but they were involved in these kinds of activities considerably and significantly less often than they were in nonsocial behavior.

These same three behavioral categories were also used to describe a male's behavior during the course of his association with an infant. It is clear from the data presented in Table 2-11 that during the interval when males have infants in their care, they are almost exclusively involved in nonsocial activities. (It was my subjective impression that males were frequently close to one another when in the company of infants, but since proximity data were not collected systematically, this impression cannot be tested. Deag (1974) found that this was true during his study.) Significantly, males seldom interacted in aggressive kinds of encounters while they had infants with them, as only 4% of all cases involved combative aggression. In 221 cases, however, a male that had an infant in his possession was threatened by or gave submissive gestures toward males, suggesting that agonistic behavior does occur when infants are present (also noted by Deag, 1974).

DISCUSSION

From the data just presented on the sociometry of the wild Barbary macaque's male-infant interactional system, a number of significant features stand out; these combine to form an integrated configuration defining an unusual, if not unique, caretaking system among polygynously mating, multi-male/

female organized nonhuman primates. At the very least, the patterns of dyadic interactions between Barbary macaque males and infants contrasts sharply and fundamentally with reports of male-infant interactions among all other such Old World primates (see chapter 15).

First, and probably ultimately the most important benefit to the infant, is that caretaking behavior from males may be crucial for infant survival. The only infant (iPP) that did not receive any caretaking behavior from either adult or subadult males was the only case of infant mortality. Intuitively, it seems reasonable to assume that receiving daily care from several different, sexually mature males is quite beneficial to an infant's chances for survival in the early part of its life.

The variety of discrete patterns comprising this dyadic caretaking system is large and rich in its complexity, eclipsed by the mother-infant interactional system only by virtue of the cumulative duration of all behaviors, and of suckling/nursing interactions. Thus during the first year of life, infants of both sexes are regularly (i.e. daily) carried, held, cuddled, groomed, played with, monitored, protected, retrieved, and otherwise interacted with by males of all ages and dominance ranks. Of 2,164 dyadic and triadic interactions recorded, three-quarters were between a single male and a single infant. The infant was the sole focus of the male's attention and the recipient of his directed behavior. This is clearly evident from the activity patterns exhibited while a male cared for an infant. For example, during 85% of all dyadic caretaking episodes, the male giving care was involved in nonsocial activities, activities that did not involve troop members other than the infant receiving care. These dyadic behaviors do not differ in their physical composition from analogous behaviors between mother and infant. Nevertheless, there is a differential distribution in frequency in the types of dyadic caretaking behaviors directed toward infants: holding and carrying behaviors predominate, accounting for over three-quarters of all interactions recorded. There was, however, no difference among individual males in how they distributed these different caretaking behaviors toward infants; i.e., from each male giving it care, all infants received a predominant amount of holding and carrying behavior, even though there were significant differences among individual males in the frequency with which they interacted with infants. Thus with the exception of nursing behavior and the possible case of adoption (there were no orphans in the study group during either study period), males' dyadic caretaking interactions encompassed every conceivable form, including all mother-infant types of caretaking interactions.

While male caretaking may be critical for infant survival, and males interact regularly with them in a complex variety of ways, individual infants did receive significantly different amounts of care from the males attending to them. Neither maternal rank, sex, nor age of the infant could account for the different amounts of care received by each infant. Thus at one end of the distribution, iLN received 43% of all caretaking behavior recorded, while iLP received 13% and iPP received zero percent (from adult and subadult males). Infants also receive care from different numbers of males. Thus iLN was the "primary infant" for five males and the "secondary infant" for four others—nine of the eleven males of this group showed significant amounts of interest in this infant. At the other extreme, iRSE had three males for which it was the "primary infant" but was not a "secondary infant" for any male. The differential distribution of care directed toward each of the infants is probably a function of several interdependent factors: (1) total number of males attending to the particular infant; (2) identity of the individual male contributing care (for there is a significant difference among individual males in the amount of care each contributes to infants in general and to specific infants in particular); (3) how active a role the infant itself played in initiating and promulgating the interactions; (4) the probability to which a male might be able to ascertain his probable relatedness to the infants attended to. Unfortunately, the data sets that would allow ascertaining how and to what extent these factors account for the differential distributions seen in male care are not available from this study. It can be concluded with certainty from this study, however, that infants did receive care from different total numbers of males, and further, that they did receive significantly different total amounts (though not kinds) of care.

Some of the most interesting aspects of the Barbary macaques' caretaking system are revealed in the patterns of interactions between individual males and individual infants. While all males interacted with infants on a regular basis, there were, nevertheless, significant differences among all males in the degree to which they did so. Thus the most involved male, 4SA, accounted for 17% of all dyadic caretaking behavior recorded, while the least involved male, BB, accounted for 4%. Thus compared with one another, males exhibited substantial individual differences in the degree to which they were involved with infants. Also reflecting the cumulative individual male differences, subadult males, as a class, were significantly more involved with infants than were adult males: four subadult males accounted for 44% of all male caretaking behavior versus seven adult males who accounted for 46%.

Yet when each individual male's pattern of dyadic interactions with infants is considered by itself, a most striking, and perhaps the most significant, facet of this system emerges: Barbary macaque males predictably orient their caretaking attention only to specific infants. Interactions with infants by males do not take place on a random or indiscriminate basis. Most of the study group's males directed a majority of their caretaking behavior to a single infant (the so-called "primary infant"), while all of the males directed a minimum of 62% of their interactions to just two infants. Nine of eleven sexually mature males had but a single "primary infant." Five of these all but ignored other infants, while four of them did show some interest in a "secondary infant." Yet, even for these four males, their "primary infant" still received significantly more attention from them than did their "secondary infant." Only two males, RD and CM, divided their caretaking interests more or less equally between two and three "primary infants," respectively. In regard to the specificity for which males were attracted to particular infants, subadult males again contrasted with adult males as a group: the average amount of care directed to a single "primary infant" by subadult males was 79% compared to 54% for adult males. Thus even though most all males oriented to particular infants, subadult males showed a more exaggerated specificity than did adult males. While a male might be relatively uninvolved with infants when compared to other males in the group, the caretaking activities that he was involved in were nonetheless directed to a specific individual. The fact that males focused their interests in and interacted with such a delineated and specific few infants bears dramatic witness to the fact that, by whatever means and through whatever proximate and/or ultimate mechanisms, Barbary macaque males are exceedingly discriminating in their choice of which infant to caretake.

Much has been made of the relationship of social rank and the Barbary macaque males' interactions with infants. This interest has focused almost exclusively on the triadic rather than the dyadic form of the male-infant interaction. My interpretation of the triadic interaction in this species has been presented elsewhere (Taub, 1980a) and will not be dealt with here. It is sufficient for this discussion to point out that males prefer to interact triadically with the same infants that they prefer for a dyadic interaction. Relative to dyadic infant caretaking, social rank did not appear to be a critical factor. Thus there were no significant correlations among individual males, nor in high-versus low-ranking males, in the degree to which they were involved with infants. Again subadult males contrasted with adult males, but I believe this reflects an age-dependent rather than a rank-dependent phenomenon.

Summarizing the salient patterns that have emerged from this study, we may conclude at the least the following about the Barbary macaque male's dyadic caretaking system. Barbary macaque males regularly and predictably interact with infants in a wide diversity of interactions, all of which are analogous to the mother-infant types of interactions. During these dyadic interactions, the infant is the sole focus of the caregiving male's attention and interaction; as such infants may be considered true caretaking investments. Males differ significantly in the amount of care they invest in each infant; and, concordantly, infants differ significantly from one another in the amount of care each receives and in the number of males giving them care. Each male exhibits a remarkable specificity in selecting a particular one or two infants to care for, essentially ignoring all other infants. Male investment in infants, then, is widespread, common, and substantial, at least in terms of frequency if not in fact in both duration and quality. Male investment may be critical for infant survival. Striking and unusual in itself, this system of caretaking among Barbary macaques is all the more remarkable in that it contrasts so sharply in so many of its fundamental features with the reports of male-infant interactions among all other polygynously mating Old World primates (see chapter 15).

Why this system has come to develop as it has is, of course, ultimately the most intriguing question, but unfortunately also the one we are least able to answer on the basis of the accumulated data on this species. It is probable that there is no single, unidimensional causative agent or selective force that is responsible; and therefore there is no unifactorial explanatory model that will be likely to accommodate all the elements of this system of caretaking behavior operating at the present time. I would like to conclude this chapter with some speculations on this system which may be useful in suggesting future areas for investigation.

I have suggested elsewhere (Taub, 1980b) that there is a causal relationship between the Barbary macaque males' infant caretaking system and the females' peculiar form of "choice" behavior in selecting mating partners. Briefly, the system operates in the following way. An estrous female changes sexual partners at a rate averaging one new consort every 17 minutes. A female does not allow any male to sexually monopolize her because she terminates one consociation (i.e., brief consort association) after a single ejaculatory copulation to purposefully seek out another male in the group. This serial selection of different, sequential sexual partners continues unabated throughout the several days of each female's estrous period, when her perineal skin is at its maximum turgescence. The net result of this behavior is to allow each sexually mature male in the group to have sexual access to the estrous female during a time when conception is likely. However, males do differ significantly in the total numbers of copulations each achieves daily and cumulatively through estrus with each estrous female. Since this system ensures that each estrous female copulates daily with most of the males in the group, I have suggested that by selecting multiple, serial sexual partners, an estrous female gives each male some chance of siring her offspring: "Choosing to allow all males to copulate with her—indeed, by actively seeking out multiple males to insure that many of them, in fact, do so, and therefore offering them all a chance to sire offspring—may be the most effective means for females to induce some care from many males and thereby obtain a maximum cumulative investment in her offspring. Simply stated, by manipulating their sexual associations so that many males participate in mating, the females promote more total male care that is available for contribution to infants." (Taub, 1980b, p. 338-339). In this system, a male could be investing in an infant that was (1) his real progeny, (2) his probable progeny, (3) his relative other than progeny, or (4) unrelated.

Some facets of the Barbary macaque's infant caretaking system are probably a subset of kinship altruism; some of these males are probably interacting with their biological offspring. This situation could be understood and accommodated easily within a kinship selection framework (Trivers, 1972; chapter 11). Many investigators have assumed a causal relationship for males between frequency of copulation and the probability of inseminating a female (Trivers, 1972; Bertram, 1975). If this conjectural assumption is true for Barbary macaques, then the differential frequencies of copulations seen among males should equate to differential probabilities of siring offspring, hence differential male reproductive success. If the frequency of associating and copulating with females is a proximate mechanism whereby a male assesses his relative probabilities of siring an offspring with a particular female, then this could be one means whereby the remarkable specificity of interest in particular infants shown by these Barbary macaques (i.e., directionality and identification) could be explained. In terms of quantity of male care (i.e., frequency of association), the data from this study show that it is the subadult males, as a class, that are the most heavily involved with infants. But subadult males, while they do achieve copulations with estrous females, are those obtaining the lowest frequencies (Taub, 1980b). According to the conventional wisdom, males will invest in an infant in relation to the likelihood that they have sired that infant. This may not necessarily be so, however, especially if others who are also related to the infant (but not as the sire) also invest in it (see, for example, chapter 1, for some species of callithrichids). Then it would be to a male's advantage to modulate his investment in an infant, reducing it in concordance with the increases in investments from other, nonmaternal caregivers.

The converse assumption, that any copulation during estrus is as likely to inseminate a female as any other is at least as probable (perhaps more so, as work on rhesus macaques suggests that high-ranking males that do the majority of the copulating are not necessarily inseminating females in relation to that high frequency of sexual activity; Smith, 1981). If this assumption is correct, all Barbary macaque males would be equally likely to sire any female's progeny, and hence the probability of each male's reproductive success would be roughly equivalent. At any single point in time, real inter-male reproductive success would probably vary since only some males would sire offspring each mating season; nevertheless, over time, all male reproductive success would, on the average, be roughly equivalent. If so, then associating and copulating with females as the Barbary macaques do should not provide a proximate mechanism for a male to assess the differential likelihood that he had sired a particular infant, since all males would be equally likely to have done so. If male reproductive success were equivalent, then all males would, theoretically, invest more or less equally in all infants. But as we have seen, this is not characteristic of the Barbary macaques' caretaking system, because males differ substantially in their respective investment in infants. Nevertheless, these assumptions could lead to a condition that could partially explain why all sexually mature males do invest in infants, but probably could not explain either differential male investment and/or infant specificity.

I believe, however, that there is another explanation to account partially for why subadult males are so heavily involved in infant caregiving (note too that for many cercopithecoid primates, it is the subadult males that are most involved with infants): subadult Barbary macaque males are investing in their maternal half-siblings or the offspring (i.e., "nieces" or "nephews") of their mothers' female siblings (i.e., "aunts"). A proximate mechanism for recognizing or assessing relative relatedness is simple enough to envision: among macaques, younger siblings tend to be the most favored and frequent play partners, and hence a younger animal is more likely to have an immediate knowledge of its siblings and mother through close associations. Such close associations form the foundation of strong social bonds that last throughout adulthood, as seen, for example, in kin-based intercession in intra-group aggression (Kurland, 1977; Kaplan, 1978); mere association has been found to be the basis for the formation of strong social bonding (Zajonc, 1971). Barbary macaque males do not seem to migrate as readily from their natal groups (Taub, 1978) as is commonly reported for other macaque species. By the natural course of frequent and intense associations during the developmental process in the absence of mass emigration from the natal group, young Barbary macaque males (and females, too) have a convenient and effective means of assessing relatedness. Because of the intensity of associations among younger versus older animals, the younger the animal, the greater the probability of "certainty" estimates of relatedness. This association mechanism could explain the very high degree of specificity shown by subadult males toward a particular infant (their maternal half-siblings), as well as the high frequency of caretaking interactions (high certainty probability of relatedness). If this interpretation is correct, I believe it also provides an answer (or better, a partial answer) to the perplexing and controversial question of the function of the triadic interaction (i.e., "agonistic buffering") for this species.

It will be recalled that only certain males interact triadically with one another and that they do so through the common link of a shared interest in the *same* infant (Taub, 1980a). I have also suggested that the directionality of the distribution of the triadic interactions is dependent on *age* rather than on rank (Taub, 1980a). Let us extend the line of reasoning just advanced above (i.e., male assessment of relatedness by close associations with younger siblings): Males that are at present adults would have cared for what are at present subadult males when the latter were infants and the former were themselves subadult males. Hence the current adult males were able to assess their relatedness to the current subadult males previously through the caretaking complex; conversely the subadult males would recognize their special relationship with these adult males by having been cared for by them when they were infants. Current subadult males are also able to assess, by the association mechanism suggested above, their relationship to the current crop of infants, which are the ones that they caretake. Consequently, when a subadult male takes such an infant (to whom it is presumably related) to specific other adult males (also to whom he has some reasonable expectation of relatedness), through the aegis of the triadic interaction, the subadult male is, in effect, "telling " the adult male that they are both related to this particular infant. Thus the triadic interaction could provide the proximate mechanism whereby *adult* males also come to assess which infants they are related to (but *not* by virtue of being their fathers). Once the adult male "learned" which infant he had a biological relationship with through this triadic interactional process with certain subadult males, he could, in turn, use the triadic interaction to deliver the infant to the subadult male, thereby relieving himself of the burden of continued caretaking while at the same time ensuring that the infant would nonetheless receive caretaking from the subadult male (a male he could "trust" because they are both related to one another and to the infant): in a word, "enforced baby-sitting." This could explain the anomaly of (1) why high-ranking males take infants to low-ranking males (i.e., subadults) and (2) why high-ranking males without infants in their possession approach lower-ranking subadult males who do have infants in their possession, patterns that cannot be accommodated by the "agonistic buffering" hypothesis (Taub, 1980a).

With the current, rapid development of primate cytogenetics, the means for accurately assessing paternity is advancing to the state that in the future such speculative predictions can be tested and quantitatively evaluated.

ACKNOWLEDGMENTS

This research was financed in part by the following persons and institutions and I thank them all warmly for their generous support: Mrs. Pam Taub, The National Science Foundation (GB-37497); The Fauna Preservation Society of Great Britain (Oryx 100% Fund); The New York Zoological Society; The Rockefeller Foundation; Sigma Xi; and the Explorers' Club of New York.

I wish to thank the government of Morocco and, in particular, the National Forestry Service (Eaux et Foret) for granting permission to conduct this research, and to its director, M. Zaki, and forestry officials, M. Hessim and M. Benjalloun, for their support and assistance.

I would further like to acknowledge the encouragement and support I received from Dr. John M. Deag when I initially undertook this study. Drs. William A. Mason, William K. Redican, Peter S. Rodman, Donald G. Lindburg, Jeffrey A. Kurland, Sarah B. Hrdy, and Ronald L. Tilson all provided stimulating ideas about Barbary macaques and "paternalism" as these materials were prepared, and I thank them for sharing their creativity with me.

REFERENCES

Bertram, B. C. R. 1975. Social factors influencing reproduction in wild lions. *Journal of Zoology, London* 177: 463-82.

Brown, J. L. 1975. *The Evolution of Behavior.* New York: Norton.

Burton, F. D. 1972. The integration of biology and behavior in the socialization of *Macaca sylvana* of Gibraltar. In F. Poirier (ed.), *Primate Socialization,* pp. 26-62. New York: Random House.

Deag, J. M. 1974. A study of the social behaviour and ecology of the wild Barbary macaque, *Macaca sylvanus* L. Ph.D. dissertation, University of Bristol, England.

Deag, J. M., and Crook, J. H. 1971. Social behaviour and "agonistic buffering" in the wild Barbary macaque, *Macaca sylvana* L. *Folia Primatologica* 15: 183-200.

Hrdy, S. B. 1976. Care and exploitation of nonhuman primate infants by conspecifics other than the mother. *Advances in the Study of Behavior* 6: 101-58.

Itani, J. 1959. Paternal care in the wild Japanese monkey, *Macaca fuscata fuscata. Primates* 2: 61-93.

Kaplan, J. 1978. Fight interference and altruism in rhesus monkeys. *American Journal of Physical Anthropology* 49(2): 241-49.

Kurland, J. A. 1977. Kin selection in the Japanese monkey. *Contributions to Primatology,* vol. 12. Basel: S. Karger.

Kurland, J. A., and Taub, D. M. In preparation. Testing the Agonistic Buffering Hypothesis II: Does Agonistic Buffering Buffer Agonism?

Lahiri, R. K., and Southwick, C. H. 1966. Paternal care in *Macaca sylvana. Folia Primatologica* 4: 257-64.

MacRoberts, M. H. 1970. The social organization of the Barbary apes *(Macaca sylvana)* on Gibraltar. *American Journal of Physical Anthropology* 33: 83-100.

Maynard-Smith, J. 1977. Parental investment: A prospective analysis. *Animal Behaviour,* 25: 1-9.

Merz, E. 1978. Male-male interactions with dead infants in *Macaca sylvanus. Primates* 19(4): 749-54.

Mitchell, G. 1969. Paternalistic behavior in primates. *Psychological Bulletin* 71: 399-417.

Mitchell, G., and Brandt, E. 1972. Paternal behavior in primates. In F. Poirier (ed.) *Primate Socialization,* pp. 173-206. New York: Random House.

Redican, W. K. 1975. A longitudinal study of behavioral interactions between adult male and infant rhesus monkeys *(Macaca mulatta).* Ph.D. dissertation, University of California, Davis.

Redican, W. K. 1976. Adult male-infant interactions in nonhuman primates. In M. E. Lamb (ed.), *The Role of the Father in Child Development,* pp. 345-85. New York: Wiley.

Redican, W. K., and Mitchell, G. 1973. A longitudinal study of parental behavior in adult male rhesus monkeys. I. Observations on the first dyad. *Developmental Psychology* 8: 135-36.

Redican, W. K., and Mitchell, G. 1974. Play between adult male and infant rhesus monkeys. *American Zoologist* 14: 295-302.

Redican, W. K., and Taub, D. M. 1981. Adult male-infant interactions in nonhuman primates. In M. Lamb (ed.), *The Role of the Father in Child Development,* 2nd ed., pp. 203-58. New York: John Wiley and Sons.

Smith, D. G. 1981. The association between rank and reproductive success of male rhesus monkeys. *American Journal of Primatology* 1 (1): 83-90.

Taub, D. M. 1977. Geographic distribution and habitat diversity of the Barbary macaque, *Macaca sylvanus* L. *Folia Primatologica* 27: 108-33.

———. 1978. Aspects of the biology of the wild Barbary macaque (Primates, Cercopithecinae, *Macaca sylvanus* L. 1758): Biogeography, the mating system and male-infant associations. Ph.D. dissertation, University of California, Davis.

———. 1980a. Testing the "Agonistic buffering" hypothesis. I. The dynamics of participation in the triadic interaction. *Behavioral Ecology & Sociobiology* 6: 187-97.

———. 1980b. Female choice and mating strategies among wild Barbary macaques *(Macaca sylvanus* L.*)* In D. Lindburg (ed.), *The Macaques: Studies in Ecology Behavior and Evolution,* pp. 287-344. New York: Van Nostrand Reinhold.

Trivers, R. L. 1972. Parental investment and sexual selection. In B. Campbell (ed.), *Sexual Selection and the Descent of Man,* pp. 136-79. Chicago: Aldine Publishing Company.

Van Hooff, J. A. R. A. M. 1969. The facial displays of the Catarrhine monkeys and apes. In D. Morris (ed.), *Primate Ethology,* pp. 9-88. Garden City, NY: Doubleday & Company.

Whittenberger, J. F., and Tilson, R. L. 1980. The evolution of monogamy: Hypotheses and evidence. *Annual Review of Ecology and Systematics* 11: 197-232.

Zajonc, R. B. 1971. Attraction, affiliation and attachment. In J. F. Eisenberg and W. S. Dillon (eds.), *Man and Beast: Comparative Social Behavior,* pp. 141-80. Washington, D.C.: Smithsonian Institution Press.

3

Male-Infant Interactions Among Free-Ranging Stumptail Macaques

Alejandro Estrada
Institute of Biology
National University of México
Vera Cruz, Mexico

INTRODUCTION

The interest of males and of females other than an infant's mother in interactions with an infant has been reported quantitatively for primate species such as *Macaca mulatta* (Spencer-Booth, 1968, 1970; Hinde and Spencer-Booth 1967); *Macaca fuscata* (Itani, 1959, Alexander, 1970); *Macaca radiata* (Simonds, 1965); *Macaca sylvanus* (Deag and Crook, 1971; Burton, 1972; Taub, 1975); *Macaca sinica* (Dittus, 1977); *Papio anubis* (Ransom and Ransom, 1971; Ransom and Rowell, 1972); *Cercopithecus aethiops* (Lancaster, 1971); and others. In addition there are review papers on the display of friendly behavior by males (Mitchell and Brandt, 1972) and by males and females (Hrdy, 1976) toward infants.

In the case of *Macaca arctoides*, there is only one detailed quantitative study of care behavior directed by adult males to the troop's infants in captive conditions (Neely and Rhine, 1977). This study concentrated on the care behavior received by infants during their first 60 days of life. Another study, by Brandt et al. (1970), concentrated on care behavior displayed by juvenile and adult males toward infants. In this case, however, the ages of the infants were not specific and no details were given on the behavioral categories scored; the aim of the study was not to examine care behavior in detail but rather to compare the males of the four species of macaques in their general interest in infants.

In another study, Gouzoules (1975) provided quantitative data on the relationship between maternal rank and social contacts, submissive behaviors, manipulation (grooming and social investigation), harassment and genital manipulation received by infants from group companions in a captive stumptail group. Gouzoules found that maternal dominance status was a factor that shaped the nature of several of the recorded interactions.

In the case of free-ranging stumptail macaques, Estrada and Sandoval (1977) have presented preliminary quantitative data on patterns of male care for a troop of stumptails living on an island. In this study, the infants were a focal subgroup and the results indicated that infants under six months of age received more care behavior from males that did older infants; that male infants received more care contacts than did female infants; and that differences were found in the type of care contact received by infants from male adults and juveniles. No relationship was found between the infants' dominance rank and the amount of care they received from males. It was also found that the presence or absence of the mother influenced the males' interest in interacting with infants.

In the investigation reported here, interactions of male group members with troop's infants were sampled in the same stumptail free-ranging troop studied by Estrada and Sandoval (1977). The research reported here was aimed at sampling not only care behavior but also aggressive actions displayed by the same males toward the troop's infants; the purpose was to assess the contribution made by males in the infants' social experiences during the first year of their lives.

METHODS AND MATERIALS

Study Group

During the study period the troop consisted of two adult males, nine adult females, seven juvenile males, four juvenile females, and seven infants (two males and five females). Table 3-1 gives the identification codes as well as date of birth and age at the beginning and at the end of the study period for the members of the troop. The troop was released on the island of Totogochillo in lake Catemaco in Veracruz, Mexico in two stages. In August 1974, 20 stumptails were released on the island; 83 days later, an additional 12 animals were added (see Estrada and Estrada, 1976, and Estrada, 1977, for details).

Table 3-1. Age and Sex Composition of the Stumptail Troop.

Infants	Sex	Age at Start (Months)	Age at End (Months)
bla	F	10	15
ela	F	9	14
pie	F	6	11
ro	F	5	10
dj	M	4	9
br	F	3	8
ti	M	1	6

Juveniles	Sex	Birth Date	
na	F	July 19	1972
ne	F	July 14	1972
fo	F	August 7	1973
ma	F	March 21	1973
jd	M	January	1972
ch	M	April 13	1972
jo	M	November	1972
cha	M	February 26	1973
me	M	February 28	1973
pa	M	July 24	1973
pab	M	August	1973

Adult (All Feral Born)	Sex	Offspring Born to Each Female 1972	1973	1974	1975
Mac	M				
C	M				
J	F	ch	pa	ti	
Mar	F		cha	pie	
Fla	F	ne		ela	
Bl	F			br	
Can	F			dj	
F	F	jo	a*	ro	
R	F		me	ga*	
Bu	F		ma		
N	F				pri**

*These infants died before the start of the sampling period.
**Infant born toward the end of the study.

Observational Protocols

Observations in this troop were carried out daily for a period of 10 months (August 1974 to May 1975) after their release on the island. The data reported here were collected between December 1974 and May 1975. During this period, 551.25 hours of recorded observations were completed on interactions between all troop members and infants. Of these, 337.50 hours consisted of interactions between males (juveniles $\geq$ 1.5 yrs.) and infants ($\leq$ 1.5

yrs.). Data collection using the focal animal sampling technique (Altmann, 1974) was carried out in a systematic and regular order after the troop had achieved social stability (November 1974; see Estrada and Estrada, 1978, for details).

All the sexually immature animals (i.e., nonadults) were observed for focal samples of 45 minutes each. A total of 30 focal samples were completed for each nonadult in the troop. The order in which the individuals were observed was randomized prior to each observation session. All the adults were also observed using the same protocol except that 15 focal samples were completed for each. For both the adult and nonadult individuals, an attempt was made to obtain a similar number of focal samples in the morning (0700–1400) and in the afternoon (1400–1900). In addition, an effort was made to obtain at least one focal sample on each nonadult for each week of each month; for the adults, at least one focal sample on each individual every two weeks in each month.

Quantification of Interactions

To quantify interactions with the troop's infants, the following strategy was followed: In the focal samples of individual A, its interactions (both as an actor and as a recipient) with individual B were quantified; then in the focal samples of individual B, all its interactions, as an actor and as receiver, with individual A were quantified. The total frequency of interactions between A (e.g., a juvenile) and B, (e.g., an infant), both as actors and receivers, was the sum of their interactions with each other in their respective focal samples. This procedure was employed to quantify the interactions reported here of all the males with the troops' infants.

The Random Model

From the observed frequency data, expected frequencies for acts received and for acts displayed were calculated using Altmann and Altmann's method (1977) under the hypothesis that an individual's rate or frequency of behavior is independent of its age/sex class. This was done to determine if the frequency distribution of the observed behaviors could be accounted for in terms of the troop's composition (see also chapter 4 for use of this methodology).

The following formula, taken from Altmann and Altmann (1977), was used to calculate the expected frequencies: $Ex = Npx$, where N is the total number of behavior acts occurring and px the probability under the null hypothesis of a constant rate that the behavior will be exhibited by some members of the xth class. Because px equals the proportion mx/M, where

M is the total number of individuals in the class exhibiting the behavior, and mx is the number of individuals comprising the different classes receiving the behavior (in all cases shown in Table 3-2 through 3-6, these were two male and five female infants; all expected frequencies thus calculated are given in parenthesis in these tables), we have $Ex = Nmx/M$. The resulting differences between the expected frequencies and the observed frequencies were tested for significance ($p < .05$) using the Chi-square statistic (Siegel, 1971).

The dominance rank of individuals in the troop was determined on the basis of the directionality of agonistic encounters, displacement interactions, and submissive displays; in this stumptail troop, an infant's dominance status was dependent on its mother's rank (Estrada, 1977).

BEHAVIOR CATEGORIES

The behavior categories scored in this study are an amalgamation of behaviors recorded in stumptail macaques by Bertrand (1969), Rhine (1972), Neely and Rhine (1977), and Estrada and Sandoval (1977). The operational definitions of these behaviors are listed below:

Care Behavior

Play: includes play fighting and chasing and is characterized by frequent interruptions, deliberate positioning, repetitious incomplete behavior, the play face, and the "rrrrr" vocalization (given only in this type of interactions among stumptails.)

Touch-hand: simply touching the immature with the hand.

Touch-face: simply touching the immature with the face.

Genital manipulation: the individual handles the infant's genitals with its hands or mouth.

Carrying ventral: the infant clings to the ventrum of the individual while he walks around.

Carrying dorsal: the infant clings to the back of the individual while he walks around.

Bridging: the individual holding the infant forms a bridge with another individual who is also holding the infant.

Retrieving: an individual takes the infant from the mother or from someone else or from elsewhere.

Grooming: examination of the hair or skin of the infant using the hands and sometimes the mouth.

Sitting-touching: the individual sits in physical contact (no body clasping) with the infant.

Mounting: the individual graps the sides or hips of the infant, may seize its ankles with its feet, and/or may give pelvic thrusts.

Aggressive Treatment

Stare threat: predominantly a fixed stare toward the infant, causing the receiver to move away or to display a submissive facial expression, e.g., fear grimace or teeth chatter.

Open mouth threat: the eyes are wide open, looking fixedly at the infant, and the mouth is more or less open with an aperture appearing as an ellipsoid or round hole.

Chase: the performer displaying a threat face runs over a variable distance chasing another subject; the receiver usually runs screaming and behaving submissively.

Slap threat: a slap at the infant combined with a stare threat or an open mouth threat and threat grunts.

Nips: the individual may take a limb or a hand, biting them without causing serious injury; the receiver crouches or cringes and may also emit distress vocalizations.

Attack: biting combined with vigorous pulling of the infant's body causing the receiver to scream in distress, sometimes defecation and urination are present.

Joins

A "join" was defined as, and was scored when an individual approaches another stopping or walking alongside it within arm's reach for a minimum duration of five seconds. It was assumed that the performance of this action would give information as to who was attracted to whom and who was responsible for initiating other contacts under the assumption that the occurrence of social interactions is conditioned by the proximity of the individuals.

Figure 3-1 shows a series of drawings made from still photographs of some of the male infant interactions recorded in the stumptail troop at Totogochillo. It is hoped that these illustrations can become a small addition to the knowledge of the behavioral catalog for this macaque species.

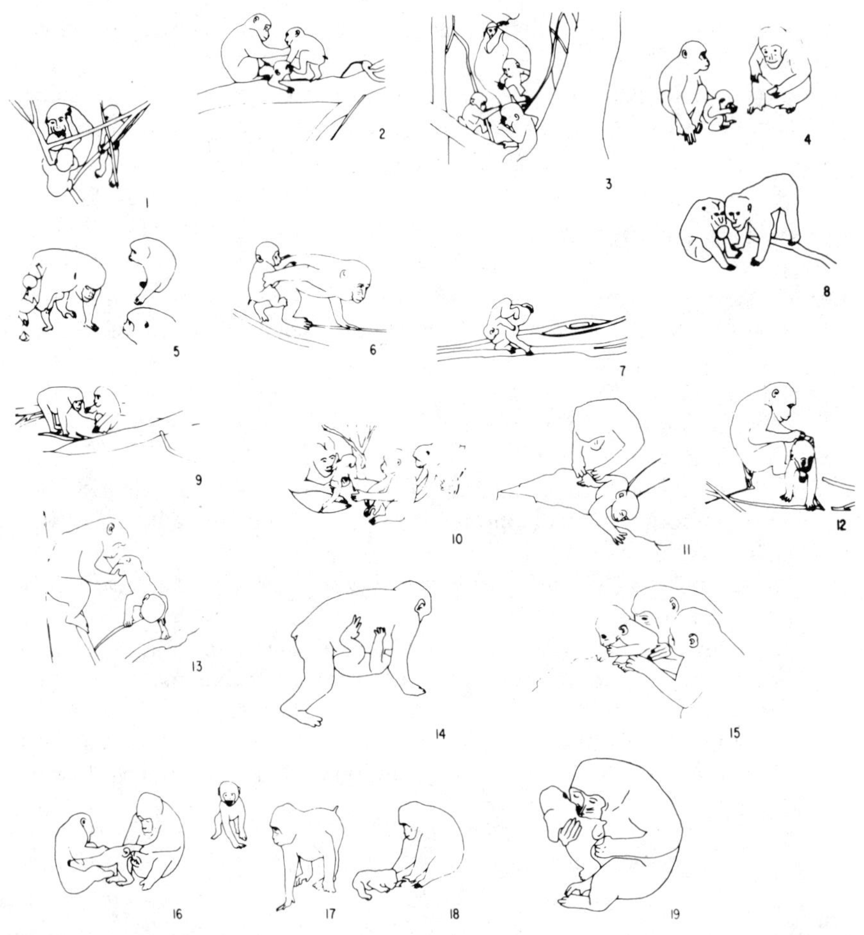

Fig. 3-1. Drawings made from still photographs of most of the contacts that took place between males and infants. *1* and *2* play; *3* and *4* sitting-touching; *5* retrieving; *6* carrying dorsal; *7* carrying dorsal but upside down; *8* touch-mouth; *9* touch-hand; *10* genital manipulation; *11* and *12* grooming; *13* genital manipulation; *14* carrying ventral; *15* touch-hand and touch-mouth; *16* bridging; *17* stare threat; *18* slap threat; and *19* nips.

RESULTS

Care Behavior Received by Infants

Distribution of Behaviors by Male Caretakers. Figure 3-2 shows the care contacts received by infants of both sexes from juvenile males and the two adult males. Considering the data for juvenile males first, the most

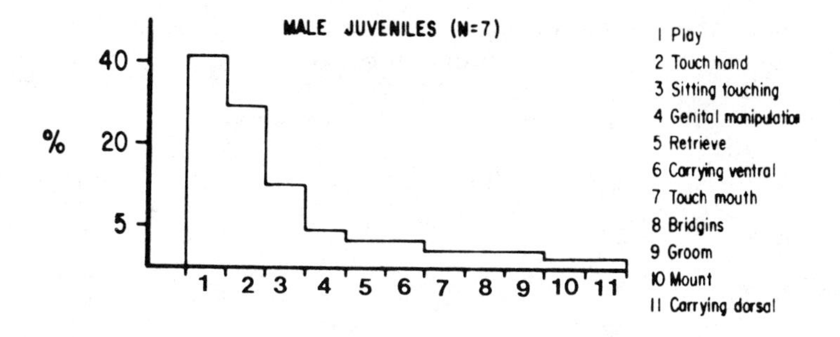

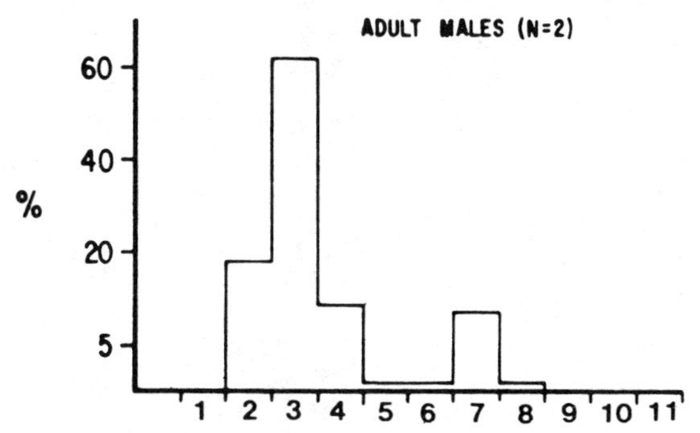

Fig. 3-2. Differential occurrence of each contact care directed at infants by the males in the troop.

frequently occurring behaviors were play, touch-hand, and sitting-touching (81%), while the next three most frequent behaviors together accounted for 13% of all juvenile male-to-infant contacts. The most frequent care contact that infants received from adult males, on the other hand, was sitting-touching (63%), followed by a clustering of three behaviors (touch-hand, 18%; genital manipulation, 10%; and touch-mouth, 7%). The remaining adult male-to-infant behaviors occurred relatively infrequently, or not at all (e.g., grooming, mounting, carrying dorsal).

Sex Discrimination. Tables 3-2 and 3-3 show the differential occurrence of care contacts received by male and female infants from juvenile and adult males, respectively. Let us consider the data for juvenile males first (Table 3-2). For infant males, the most frequent contacts received were touch-hand, play, and sitting together (72%), followed by genital manipulation, carrying ventral, and retrieving (17%). For the female infants, play behavior alone accounted for (72%) of the contacts, with sitting-touching (14%) and touch-hand (7%) a distant second and third most frequent. All other juvenile male-to-infant female behaviors together accounted for 7% of all

Table 3-2. Distribution of Juvenile Males' Care toward Male and Female Infants.

	Infant males ($N = 2$)		Infant females ($N = 5$)			
Behavior	Frequency	%[a]	Frequency	%	χ^2	p
Touch-hand	290 (82.90)[b]	45	35 (25.0)	7	521.37	.001
Play	125 (35.7)	19	348 (249.0)	72	262.73	.001
S. touch.	51 (14.6)	8	68 (48.6)	14	98.49	.001
G. manip.	44 (12.6)	7	3 (2.4)	1	78.59	.001
C. ventral	33 (9.43)	5	1 (.714)	.20	59.02	.001
Retrieving	32 (9.14)	5	5 (3.57)	1	57.74	.001
Touch-mouth	26 (7.43)	4	2 (1.43)	.40	46.63	.001
Bridging	20 (5.71)	3	1 (.714)	.20	35.87	.001
C. dorsal	10 (2.86)	2	1 (.714)	.20	17.93	.001
Grooming	9 (2.57)	1	8 (5.71)	2	17.0	.05
Mounting	1 (.286)	.15	12 (8.57)	2	3.15	n.s.
Total	641 (183.0)		484 (346.0)		1,201.29	.001

[a]Percentages were obtained from the total for each sex class.
[b]Numbers in parentheses indicate expected frequencies based on the differential N for infants of each sex (see description of method on pp. 59–60).

behaviors. Although juvenile males interacted with all infants, they displayed a marked preference for the two male infants. Only in mounting (a very infrequent behavior) was there no significant sexual differences in distribution. For almost all other behaviors, male infants received significantly more interactions from juvenile males than did female infants. However, juvenile males did sit-touching with infant females more often than they did with male infants.

Considering the data for adult males (Table 3-3), the most frequent care contact infant males received from them were sitting-touching (50%) and touch-hand (24%). Genital manipulation and touch-mouth accounted for 22%; and retrieving, bridging and carrying ventral accounted for 4% collectively. In contrast, infant females received only three types of care

Table 3-3. Distribution of Adult Males' Care toward Male and Female Infants.

	Infant males ($N = 2$)		Infant females ($N = 5$)			
Behavior	**Frequency**	**%[a]**	**Frequency**	**%**	χ^2	p
Touch-hand	25 (25.0)[b]	24	2 (5.0)	5	1.80	n.s.
Touch-mouth	8 (8.0)	8	2 (5.0)	5	1.80	n.s.
G. manip.	14 (14.0)	14	0 (0)	—	.99	n.s.
Retrieving	2 (2.0)	2	0 (0)	—	2.11	n.s.
C. dorsal	0	—	0	—	—	—
C. ventral	1 (1.0)	1	0 (0)	—	.22	n.s.
Bridgeing	1 (1.0)	1	0 (0)	—	.22	n.s.
Grooming	0	—	0	—	—	—
S. touching	51 (51.0)	50	41 (103.0)	90	37.32	.001
Play	0	—	0	—	—	—
Mounting	0	—	0	—	—	—
Total	102 (102.0)		45 (113.0)		40.92	.001

[a]Percentages were obtained from the total for each sex class.
[b]Numbers in parentheses indicate expected frequencies based on the differential *N* for infants of each sex (see description of method on pp. 59-60).

contact from adult males: sitting-touching (90%), touch-hand (5%), and touch-mouth (5%). As with the juvenile males, the male infants received significantly more care contacts from adult males than did the female infants. No sexual differences were found, however, for retrieving, carrying ventral, and bridging behaviors.

In comparing the distribution of infant care behaviors between the two male age classes, we find that although both interacted with infants in a friendly way, both the type and quality of the care displayed were different for each male age class; that is to say, infants received different tactile and social stimulation from adult and from juveniles males.

Dominance and Age Effects. No significant relationship was found between these variables and the amount of care infants received from male juveniles (r_s rank = 0; r_s age = .464, n.s.) and from adult males (r_s rank = .178, n.s.; r_s age = .607, n.s.).

Table 3-4. Care Contacts Directed to Siblings by Juvenile Males.

Infants		Caretakers			
		Kin	Not Kin		
Identification	Sex	Frequency	Frequency	χ	P
ti	Male	255 (87.78)	53 (219.91)[a]	422.57	.001
pie	Female	24 (9.51)	43 (57.41)	23.95	.001
ro	Female	67 (23.14)	98 (139.69)	93.39	.001

[a]Numbers in parentheses indicate expected frequencies based on the differential *N* for infants (see description of method on pp. 59–60).

Genealogical and Kinship Effects. Table 3-4 shows the amounts of care received by those infants that had juvenile brothers in the group. It can be observed that, except for *ti,* the other two infants, *pie* and *ro,* received more care behavior from nonkin than from their juvenile male siblings. Infant male *ti* received 83% of his scored male care contacts from his two brothers; infant female *pie* received 36% of her scored contacts from her brother *cha;* infant female *ro* received 40% of her scored contacts from her brother *jo.*

Male-to-Infant Aggression

Distribution by Male Caretakers. Figure 3-3 show the most frequent aggressive acts received by infants of both sexes from juvenile males. Slap threats accounted for 42% of the recorded aggressive acts, while open mouth threats, nips, and stare threats accounted for 46%. Attacks and chases accounted for 11%.

This figure also shows that stare threats were the most frequent (60%) type of punishment infants received from adult males. Slap threats were the next most common punishment (22%), followed by open mouth threats (10%). Nips and attacks were the least common and chase was never observed to occur.

Sex Discrimination. Table 3-5 shows that infant males received only two types of punishment from juvenile males: slap threats (67%) and nips (33%). In contrast, infant females received all types of aggressive treatment, and the most frequently occurring of these were slap threats (41%). Open mouth threats, nips, and stare threats collectively accounted for 47% of the punishment; attacks and chases accounted for 12%.

Table 3-5 shows that only the difference between the sexes in the slap threat category was statistically significant. The total in the table and its

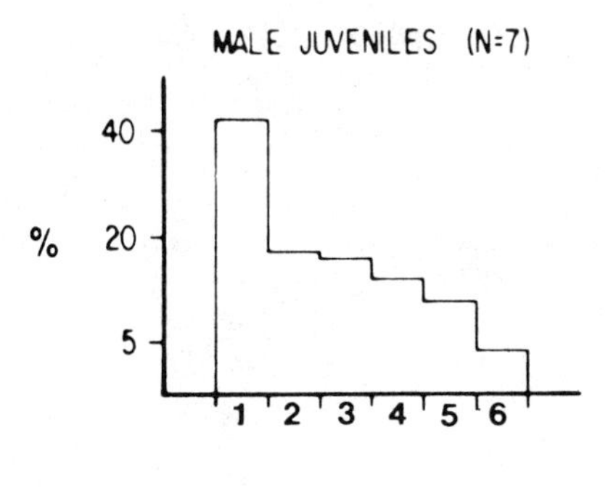

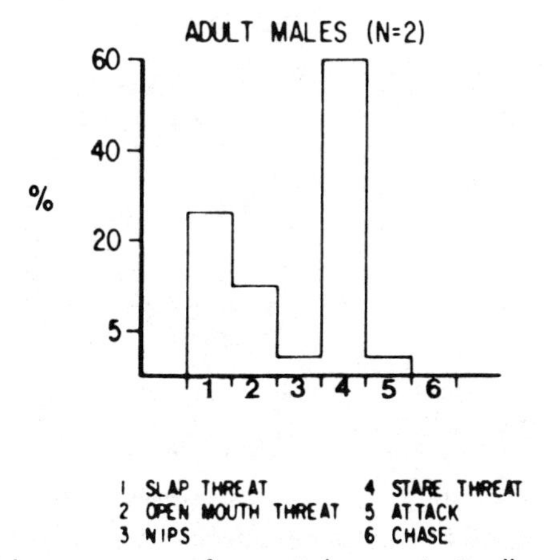

1 SLAP THREAT
2 OPEN MOUTH THREAT
3 NIPS
4 STARE THREAT
5 ATTACK
6 CHASE

Fig. 3-3. Differential occurrence of aggressive contacts directed to infants by males.

associated statistic indicate that female infants received significantly more aggressive acts than did male infants from juvenile males. These data also show that female infants received more contact and noncontact aggression (contact = slap threats + nips + attacks) than male infants did. These data indicate that juvenile males were not punishing infants randomly: the type of punishment varied according to the infant's sex.

Table 3-6 shows that infant males also received only two kinds of punishment from adult males: stare threats (76%) and slap threats (24%). Of the punishment received by infant females 52% consisted of stare threats; slap and open mouth threats combined accounted for 42% of the recorded punishment. Nips and attacks accounted for 4% and chases were never observed to occur.

Table 3-5. Distribution of Juvenile Males' Aggression toward Male and Female Infants.

	Infant Males ($N = 2$)		Infant Females ($N = 5$)			
Behavior	**Frequency**	%[a]	**Frequency**	%	χ^2	P
Slap threats	2 (.57)[b]	67	27 (19.3)	41	5.64	.001
Nips	1 (.28)	33	10 (7.14)	15	1.19	n.s.
Attacks	0 (0)	—	5 (3.57)	8	.83	n.s.
Stare threats	0 (0)	—	9 (6.42)	14	2.33	n.s.
Open mouth threats	0 (0)	—	12 (8.56)	18	3.50	n.s.
Chase	0 (0)	—	3 (2.40)	4	.20	n.s.
Total	3 (.85)		66 (47.1)		19.03	.001
Contact	3 (.85)		42 (30.0)		10.15	.001
Noncontact	0 (0)		24 (17.1)		2.50	.001

[a]Percentages were obtained from the total for each sex class.
[b]Numbers in parentheses indicate expected frequencies based on the differential N for infants of each sex (see description of method on p. 59-60).

The data-presented in Table 3-6 also show that adult males did differentially punish either male or female infants when total aggressive acts are considered. This is also true both for two individual categories of aggression and when aggression is grouped into contact and noncontact types.

Dominance and Age Effects. The amount of punishment infants received from juvenile males was not significantly associated with their mothers' dominance rank ($r_s = -.678$, n.s.), but it was positively associated with the infants' age ($r_s = .845$, $p < .05$).

The punishment received by infants from adult males was not associated with either their rank ($r_s = .189$, n.s.) or their age ($r_s = -.169$, n.s.)

Table 3-7 shows that when aggression is divided into the contact and noncontact types, we find that it is negatively associated with the infants' age. In addition, a positive association was found between the infants' rank and contact aggression received from adult males.

Genealogical and Kinship Effects. Our data show that only *cha* and *jo* aggressed their infant siblings. However, *cha* aggressed *pie* only once and

Table 3-6. Distribution of Adult Males' Aggression toward Male and Female Infants.

	Infant males ($N = 2$)		Infant females ($N = 5$)			
Behavior	**Frequency**	**%[a]**	**Frequency**	**%**	χ^2	***P***
Slap threats	4 (4.0[b])	24	11 (27.5)	27	9.90	.001
Stare threats	13 (13.0)	76	21 (52.5)	52	18.90	.001
Nips	0 (0)	—	1 (2.5)	3	.22	n.s
Open mouth threats	0 (0)	—	6 (15.0)	15	1.20	n.s
Attacks	0 (0)	—	1 (2.5)	3	.22	n.s.
Chases	0	—	0	—	—	—
Totals	17 (17.0)		40 (100.0)		36.00	.001
Contact	4 (4.0)		13 (32.5)		11.70	.001
Noncontact	13 (13.0)		27 (67.5)		24.30	.001

[a]Percentages were obtained from the totals for each sex.
[b]Numbers in parentheses indicate expected frequencies based on the differential *N* for infants of each sex (see description of method on p. 59-60).

Table 3-7. Relation between Contact-Noncontact Aggression and the Infants' Dominance Status and the Infants' Age.

Aggression	Infants' dominance		Infant's age	
Received From	**Contact**	**Noncontact**	**Contact**	**Noncontact**
Male juveniles	−.651 n.s	−.598 n.s.	−.812 <.05	−.794 <.05
Adult Males	+.776 <.05	−.473 n.s.	+.383 n.s.	+.178 n.s.

n.s. = not significant at .05 confidence level.

jo aggressed *ro* only four times. These frequencies are extremely low compared to the amount of care displayed by these same individuals and the other juvenile males toward their infant kin.

"Join" Interactions. Figure 3-4 shows that infants were more active in joining the males (i.e., seeking their proximity) in the troop than vice versa (first histogram for each male age category).

These data also show that juvenile males joined female infants significantly more often than they did male infants. Although the differences are not significant, a similar tendency is found for the adult males. Female infants joined juvenile and adult males significantly more often than infant males did.

In Figure 3-5, a comparison is made for join interactions with infants

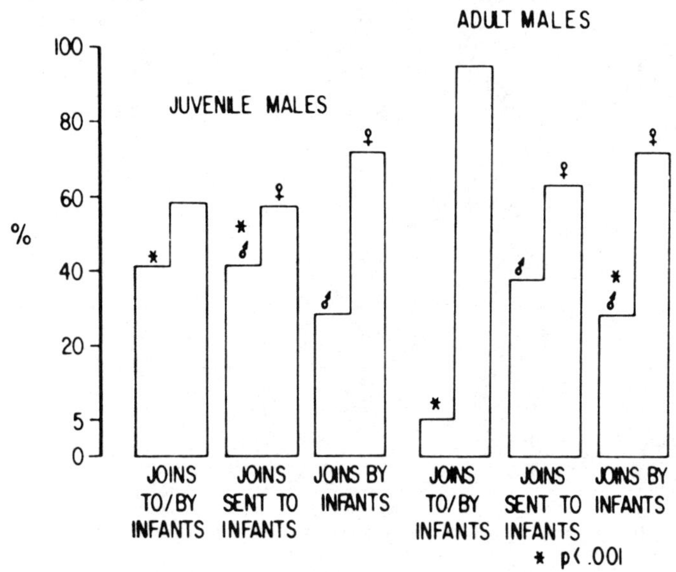

Fig. 3-4. Direction of join interactions between males and infants.

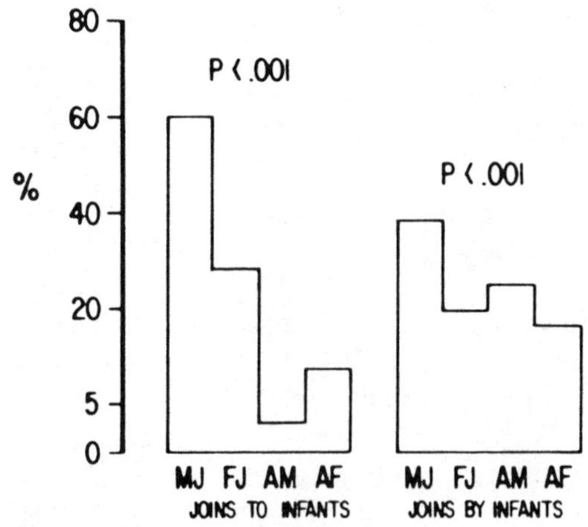

Fig. 3-5. Differential contribution of group companions to join interactions with infants.

between males and the juvenile and adult females in the troop. This figure shows that juvenile males were responsible for about 60% of the joins sent to infants by troop members, followed by juvenile females and adult females. Adult males joined infants the least. The statistical significance in the figure indicates that the differences between the age classes are not the result of the different numbers in each age/sex category in the troop.

Juvenile males received more joins from infants than the rest of the age/sex classes; adult males received more joins from infants than did the juvenile and adult females, an interesting datum in relation to the fact that adult males joined infants the least.

It is clear then that both juvenile males and infants were active in gaining proximity to each other, and were both responsible for initiating the care and aggressive contacts displayed toward the latter.

Caretaking Versus Aggression

Of the total interactions received by infants from males, 91% consisted of friendly acts and the remaining 9% of aggressive acts. Both adult males and juvenile males directed significantly more care than aggression to infants (Fig. 3-6). In addition, juvenile males were responsible for 52% of the care and 32% of the aggression displayed toward infants by all troop members (Estrada and Estrada, in prep. and Fig. 3-7). On the other hand, adult males contributed 7% of the total care displayed by troop members toward infants and 26% of the aggressive treatment displayed by troop members toward infants (Fig. 3-7). Using the data available on the juvenile and adult females of the troop ($\chi^2 = 16.24, p < .001$), infants received significantly more punishment from male than from female troop members. Also important is the fact that infants received 60% of their scored care behavior from male troop members and 40% from female troop members ($\chi^2 = 293.35, p < .001$); this shows that male troop members displayed more care toward infants than did females (the infants' mothers excluded).

The Males' Scores for Care and Aggression, Their Dominance Rank and Their Age. The data indicate that for the juvenile males, the display of care behavior toward infants was not significantly associated to their dominance rank ($r_s = .571$, n.s.), nor to their age ($r_s = .321$, n.s.), although the sign indicates positive tendencies for rank and age. However the display of aggressive treatment by juvenile males was found to be positively correlated with their dominance rank ($r_s = .714, p < .05$); age and aggression were not significantly associated ($r_s = -.467$, n.s.), although the sign of the coefficient may indicate a negative relationship.

A comparison between the two adult males indicated that both the alpha

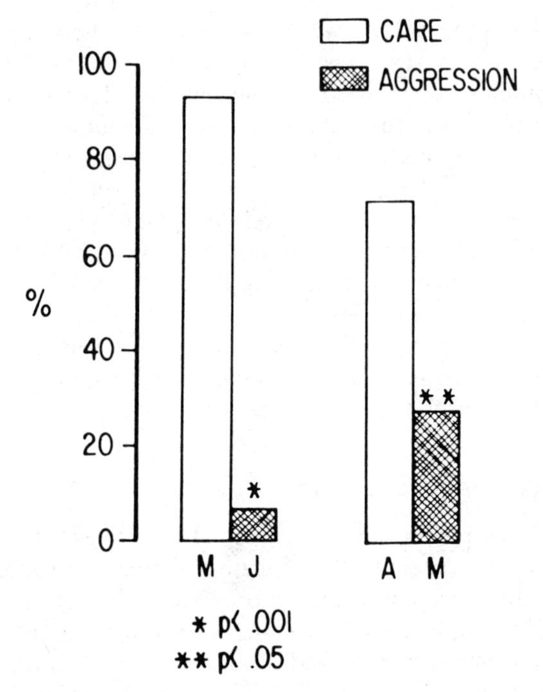

Fig. 3-6. Total aggression and care displayed toward infants by males.

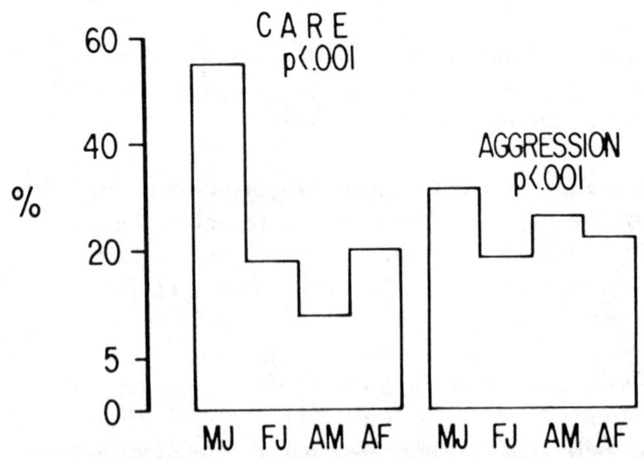

Fig. 3-7. Differential contribution of troop members to the care and aggressive contacts received by infants.

and the beta male displayed similar amounts of care behaviour toward infants ($\chi^2 = .22$, n.s.). In contrast, the alpha male displayed significantly more aggression toward infants than the beta male ($\chi^2 = 10.10, p < .05$).

Table 3-8 and 3-9 show the raw frequencies for care and aggressive behavior displayed toward individual infants by each male in the troop. It is

Table 3-8. Care Contacts Directed by Males toward Infants.

	Receiver									
Sender	Males		T	Females					T	Total
Adults	ti	dj		pie	ela	br	ro	bla		
Mac	48	4	52	2	—	18	4	1	25	77
C	45	5	50	5	3	1	2	9	20	70
Total	93	9	102	7	3	19	6	10	45	147
Juveniles										
ch	169[a]	99	267	20	6	12	7	5	50	317
pa	87	46	133	16	12	25	23	0	76	209
cha	24	36	60	24	38	14	25	9	110	170
jo	0	32	32	0	2	1	67	16	86	118
me	0	4	4	0	0	0	1	2	3	7
pab	4	76	80	9	49	8	42	42	150	230
jd	25	40	65	0	6	2	0	1	9	74
Total	308	333	641	69	113	62	165	75	484	1125

[a]Underscored numbers indicate a geneological relationship (siblings) between sender and receiver.

Table 3-9. Aggressive Contacts Directed by Males toward Infants.

	Receiver									
Sender	Males		T	Females					T	Total
Adults	ti	dj		pie	ela	br	ro	bla		
Mac	5	11	16	1	8	7	7	2	25	41
C	0	1	1	5	2	2	5	1	15	16
Total	5	12	17	6	10	9	12	3	40	57
Juveniles										
ch	0[a]	1	1	4	0	4	9	0	17	18
pa	0	1	1	3	7	2	1	5	18	19
cha	0	0	0	1	3	1	2	0	7	7
jo	0	1	1	0	0	0	4	3	7	8
me	0	0	0	0	0	0	0	4	4	4
pab	0	0	0	0	0	0	2	9	11	11
jd	0	0	0	0	0	0	0	2	2	2
Total	0	3	3	8	10	7	18	23	66	69

[a]Underscored numbers indicate a geneological relationship (siblings) between sender and receiver.

clear in both sets of data that there are marked preferences among the males for interacting with certain individual infants. These preferences are illustrated in a sociogram form in Figure 3-8 for care behavior, and in Figure 3-9 for aggressive treatment. In these sociograms, we have added the data available

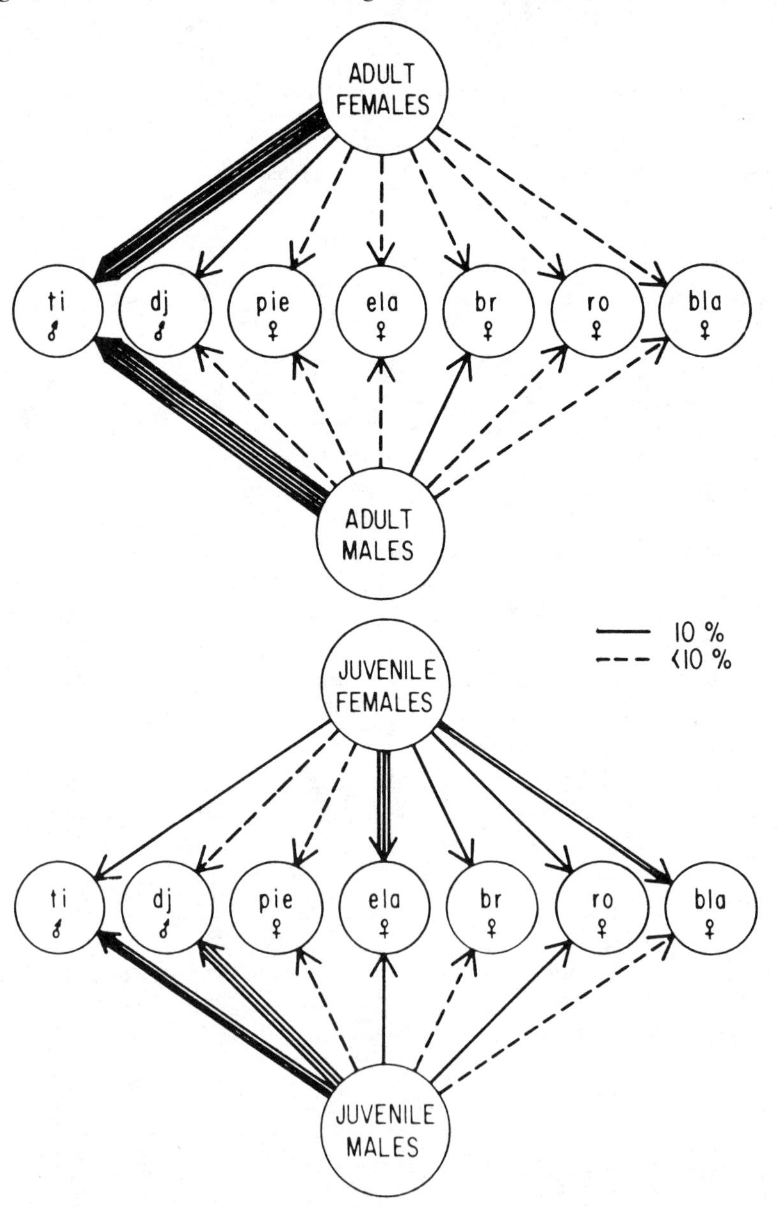

Fig. 3-8. Sociograms illustrating individual perferences by males and by females in care displayed toward infants.

on the juvenile and adult females (the infants' mothers excluded) for purposes of comparison.

Figure 3-8 shows the remarkable preference by adult and juvenile males for particular infants in the troop, especially for the two male infants *ti* and

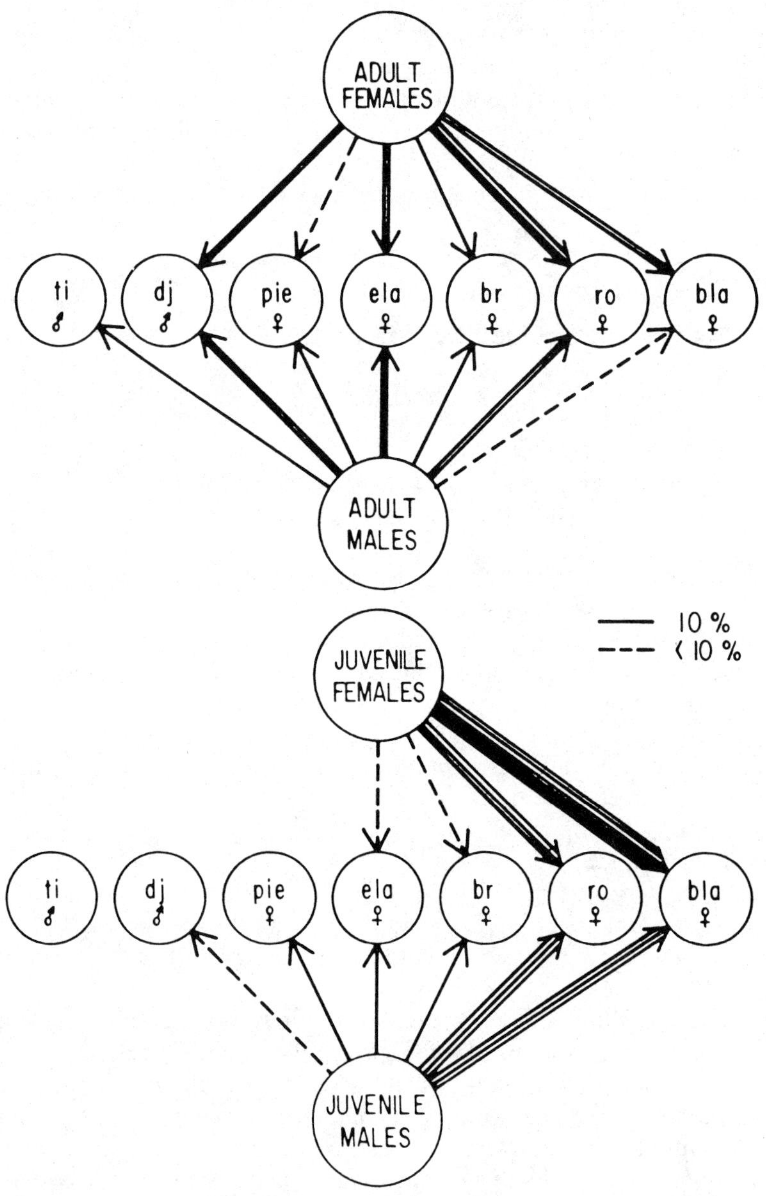

Fig. 3-9. Sociograms illustrating individual preferences by males and females in aggression displayed toward infants.

dj. Juvenile females were less selective than juvenile males in the attention they displayed toward infants.

Figure 3-9 show the individual preferences in aggressive treatment directed toward infants. Adult females (the infants' mothers excluded) directed most of their punishment to *ro, bla, dj,* and *ela,* but did not aggress *ti* at all. The adult males, on the other hand, concentrated their punishment on *dj, ela,* and *ro,* but displayed very little aggression toward *bla.*

Both the juvenile males and females were even more selective than the adults in the troop, as both classes concentrated their aggression on *ro* and *bla.* While juvenile females displayed no aggression toward the three highest-ranking infants, juvenile males aggressed all infants except *ti.*

DISCUSSION

Care Behavior Received by Infants

While juvenile and adult males interacted with infants in a friendly way, the type and quantity of the care displayed was different for each male age class. That is to say, infants received different tactile and social stimulation from adult and from juvenile males. There was also a marked sex preference (males preferred) by both male age groups in the display of infant contact behavior.

The marked preference by all males for infant males may be suggested to be the result of (1) the younger age of the male infants, (2) the males' conspicuous external genitalia, and (3) the special social position of male infants. However, we find that relative to age, the female infant *br* was a month younger than the infant male *dj,* and she received much less care than he did (see Table 3-8). It is possible that the infant male's relatively large penis and the difference between the scrotal sac and the small labia majora may act as important stimuli in releasing a differential response. Relative to social status, it is important to note that infant male *ti*'s mother *(J)* was the alpha female of the group, and it has been shown that in this troop, to gain access to dominant females, individuals directed friendly contacts to the females' youngest offspring (Estrada, 1977). Thus, the special social position of infant males, *ti*'s mother may help explain the great deal of attention he received from the males in the troop.

In the case of the other male infant *(dj),* the son of a midranking female *(Can),* it is important to recall some special events during its first months of life. *Can* arrived with *dj* on the island when he was about 1.5 months old. Shortly after their arrival with the second group of stumptails, *J,* the alpha female, for reasons unknown to the observer displayed a very strong interest in *dj,* even though there was a recently born infant female in the troop *(br). J* sought *Can*'s proximity very often to touch and manipulate *dj*'s body. Very soon, *ch* the oldest son of *J*, also started to pay much attention to *dj*, carrying him and providing him with friendly contacts; thereafter, other troop members also started to display the same type of interest in *dj*. This interest can be detected in the high scores for care received from juvenile males by this male infant (Table 3-8). In fact, *dj's* scores are higher than *ti's*. The reason for this situation may be that because of *ti's* younger age, he was more often with his mother, whereas *dj*, who was older, was more often away from his mother. In this context, Estrada and Sandoval (1977) have shown that in this troop the close proximity of the mother to her infant may inhibit the males' interest in interacting with the infant. However, if this is the case, why is it that only juvenile males exhibited a strong preference for *dj*, as shown by our data (Estrada and Estrada, in prep.) on the females' attention to the troop's infants? It could be that juvenile males were more easily tolerated by *ti's* mother and brothers when he was approached by them, and since they were interested in gaining social access to *ch* and *pa* (*ti's* juvenile brothers), they directed a great deal of attention to the infant (*dj*) much preferred by these two juvenile males.

In the case of adult males, their care contacts concentrated on the infant *ti* (Table 3-8), probably as a result of the close association between the alpha male and *J* and of the beta male's need to pacify the two most dominant adults in the troop.

Neely and Rhine (1977) found that in captive stumptails, infant males and females received about equal attention from adult males. They report that male infants were touched more frequently than were female infants by adult males and that female infants received more interactions in the remaining behavior categories scored. Brandt et al. (1970) found no sex preferences in care displayed by males toward infants in captive stumptails. Bertrand (1969) reports that only infant males established lasting ties with juvenile and adult males. In contrast, in other primate species a greater interaction of older males with female infants has been reported by Itani (1963), Kummer (1968), and Mitchell, and Brandt (1972).

Data on rhesus monkeys have shown that behavioral differences between the sexes can be observed early in life; e.g., male infants have been shown to be rougher in their play (Hansen, 1966). For the stumptail macaque, some sex differences have been reported by Estrada (1977) in behaviors such as joins, grooming, agonism, presents, and mounts, at each of three developmental stages (infant, juvenile, and adult [Estrada, unpublished data]). In this respect, it is pertinent to ask to what extent the differential treatment displayed by males toward infants of each sex are reflected in the differences in behavior and social treatment seen in later life stages for both sexes. For example, our data show that as infants and as juveniles, females are aggressed more often by adults of both sexes than males are (Estrada, unpublished data).

It is important to mention, however, that although no significant correlation was found with the infants' dominance rank, the great deal of attention that the alpha female commanded from others probably led them to display much attention to her infant. Care behavior displayed by males, then, was not distributed randomly, but rather it concentrated on those infants whose mothers were the highest ranking in the troop. It is thus important to note that 58% of the male care recorded was received by the two infants mentioned above (Table 3-8).

Our results are consistent with the preliminary information presented by Estrada and Sandoval (1977) for the same troop in that no association was found between care and dominance rank of the infant, but they differ in that these two authors found that infants under six months of age received significantly more care from males that did older infants. Dominant individuals were very successful at retrieving infants from low-ranking mothers, even though these mothers might show a resistance to such approaches. It seemed that mother tended to tolerate contacts with infants from males who had friendly bonds with her; with high ranking males, they tended to avoid such contacts or released the infant probably out of fear.

The information presented by Estrada and Sandoval (1977) and the data presented in this report for free-ranging stumptails indicate then that an infant's social position, or rather that of its mother, is an important variable in determining the amount of friendly attention an infant will receive from male group companions. Estrada and Estrada (in prep.) have also shown that the direction of the females' interest in infants is also affected by the same social circumstances. This has also been reported for baboons (Ransom and Rowell, 1972) and in captive stumptail groups (Gouzoules, 1975; Neely and Rhine, 1977).

In his study of captive stumptails, Gouzoules (1975) found positive significant correlations between maternal rank and the amount of social contact (touch, huddle, ventral clinging, dorsal clinging, jump on, and so forth) and grooming received from group companions by the infants during their first six months of life. However, Gouzoules does not specify the relative contribution of each type of group companion to the infants' social contact and grooming received scores.

With respect to age of the infant, our results may indicate a contradiction with Mitchell and Brandt (1972), who have suggested that an infant's attractiveness diminishes as it gets older due to changes in fur color, size, appearance, and behavior. This has been reported for vervet monkeys (Lancaster, 1971), baboons (DeVore, 1965), rhesus (Rowell et al., 1964), and other species (but see chapter 7). However, in our troop, due to the intricate web of interactions between individuals and the little tolerance of mothers to the approaches of others to their nursing infants, older infants, in general, tended to receive more care contacts than younger ones, particularly because they were able to leave their mother and join the males (see below).

As far as genealogical ties are concerned, the results indicate that infants do receive a substantial amount of care behaviour from their older male siblings. However, the remaining amount of infant care displayed by males was oriented to nonkin. The high percentage of infant care behavior directed to kin by male should be explained through the aegis of traditional kin selection theory (Hamilton, 1974a, 1964b) in the sense that older kin are helping the mother, to her genetic advantage, in the rearing of close kin and they could thus be, in the long run, augmenting the chances that their gene replicas will propagate in the group's gene pool. In addition, the performance of this behavior as adults for males could provide the individual with the opportunity to interact by chance with their own offspring, thus contributing indirectly to the socialization and well-being of their gene replicas.

In the context of genealogical ties, the difference in the size of each genealogy may have been influential in determining not only the recorded frequencies of male care displayed, but also the amount of social and tactile stimulation received by infants during their first year of life. For example, *J* had three offspring, two of which paid much attention to the youngest; *F* had only two, so that the youngest infant received care from only one sibling (a reduction of 50% in kin-related caregivers).

Another important observation is that interactions between juvenile males and their infant siblings were a highly pleasurable experience for the infant. Autonomic responses such as piloerection, stiffness of the body, urination, and defecation took place while the infant was being manipulated by older brothers. The juveniles reacted excitedly, with high-pitched vocalizations, rapid teeth-chatter, and lip-smacking. From these observed behaviors it is suggested that these regular interactions between males and their infant siblings tend to reinforce the existing emotional bond between kin. Bonds among siblings are long lasting and continue into adulthood (Estrada, unpublished data).

The great deal of attention displayed by males toward nonrelated infants seems to have been carried out not just as a general interest in infants but also as a means to gain social access (proximity and contacts) to the mother. Estrada (1977) and Estrada et al. (1977) have suggested that through this social strategy, troop members are able to buffer potential harm from the mother and her "friends," and thus the individual can remain in proximity or can interact with them. One case history from the study group illustrates this social access phenomenon. A male juvenile, *jd* used to be one of the most peripheral individuals in the troop, both spatially and socially, due to his very low rank and lack of relatives in the troop. He participated little in friendly social contacts, was frequently aggressed by others, and usually stayed behind the troop when the group was traveling. After the arrival of the second stumptail troop to the island and also after *ch* had started to display a strong interest in the infant *dj*, *jd* started to develop a special interest in this male infant, an interest that increased with time. Although all other juvenile males also attempted to interact with *dj*, only *ch*, his brother *pa*, and the son of the second-ranking female, *cha*, were successful. However, *jd* persisted in his attempts to be near *dj* and gradually established contacts with him. He was so persistent that eventually, in spite of many punishments received from *dj's* mother and friends, he became as successful as the high-ranking juvenile males in approaching and interacting with *dj*. Thus, as time passed, *jd* was able to remain more frequently and for longer periods of time in proximity to and was tolerated by the adults and nonadults that associated with *dj* and his mother (who did not tolerate his presence as easily before his infant care interactions). Also, after *ti* was born, *jd* also developed a very strong interest in him and through time was able to successfully remain in proximity to the highest-ranking individuals in the troop. As a consequence of these social changes and interactions with two high-ranking infants, *jd's* social network increased and he spent many hours playing and interacting not only with the infants but also with the infants' brothers and with other high-ranking juveniles.

Not only did *jd* pay much attention to *dj* and *ti* but all other juvenile males also did so, and in some cases they paid equal or more attention to nonkin infants. In this situation, it can be suggested that individuals are acting selfishly by interacting with high-ranking unrelated infants. In other words, they are attempting to improve their own social position at the cost of depriving their own siblings from caring behavior.

Male-to-Infant Aggression

Summarizing the results from this study, we find that in relation to infant-oriented aggression, (1) males exhibited a variety of aggressive acts but the most frequent type of punishments infants received were slap threats, stare threats, and open mouth threats; (2) adult and juvenile males provided infants of each sex with different types of aggressive experiences; and (3) not all interactions between males and infants were friendly, and, in fact, males played an active role in the infant's early aggressive experiences, ranging from mild (e.g., stare threats) to severe punishments (e.g., attacks).

Among the free-ranging stumptails of Totogochillo, the males, and in particular the juvenile males, actively discriminated between the male and the female infants in punishment behavior. It is possible, however, that male infants were punished less than female infants were because of their special social position rather than their sex per se.

In general, most of the aggressive acts directed to infants by males took place when the infants were away from their mothers; other incidences occurred without the mother's knowledge because in these instances the infant responded to the aggression with silent submissive displays such as teeth-chattering or lip-smacking. However, in those instances in which the infant responded with distress vocalizations to the aggression from males, the mother immediately moved to its aid. If the aggressor was subordinate to her, she immediately chased or attacked him. If the aggressor was dominant, just a few centimeters before touching him, she slightly slapped him and continued running so that the aggressor would turn toward her, thereby stopping the attack on the infant; if this was not successful, she positioned herself behind the aggressor and emitted distress barks, which also served, sometimes, to distract the aggressing male.

At the individual level, Table 3-9 shows that while infants received aggression from various males, *ti,* the infant son of the alpha female, was aggressed only by *Mac,* the alpha male. Even though *dj* had achieved a special position in the troop, he still received much aggression from males, particularly from *Mac.* Of the two infants, *ti,* the son of the alpha female, received significantly less aggression ($\chi^2 = 12.02, p < .001$) than the lower-ranking *dj* did.

From the juvenile males, *dj* received insignificant amounts of punishments, due, perhaps, to his close association with *ch* and *pa.* It was observed that on many occassions, *ch* supported *dj* in aggressive actions against some of the other juvenile males.

Juvenile males such as *jo, me, pab,* and *jd* aggressed only those infants that were lower ranking than they were. In contrast, *ch* and *pa* aggressed practically all infants. Although the data suggests a negative association ($r_s = -.800$, n.s.) between the infant female's dominance rank and aggression received from males, it is important to mention that aggression was heavily concentrated on *ro* and *bla.* Both of these female infants constantly approached the adult and juvenile males that responded with aggression.

Bertrand (1969) noted that two males in a captive group of stumptails occasionally attacked infants although they usually had few contacts with them. Gouzoules (1975) reports that infants in his captive group were threatened, slapped, dragged, bit, and chased by troop members, particularily after the second month of life. He also reports that when contrasted with the infant interactions of adult males, the interactions of subadult males were few and were generally agonistic in nature with the same infants. Also he found that the harassment received by an infant from troop members was negatively associated to the infant's mother's rank. In addition, a tabulation of the data presented by Gouzoules (1975, table 8, p. 412) in which the scores for harassment received by all infants ($N = 6$) were pooled together, yielded a negative association with the infants' age ($r_s = -1.0, p < .05$). These negative associations may be explained as resulting from the gradual changes in coat color in stumptail infants, who are cream-colored at birth but have changed to brown by the tenth to twelfth month of life. Hrdy (1976) has pointed out that one of the features that contributes to the infant's attractiveness is their distinctive natal coat, which also tends to inhibit aggression toward them or toward the infant holder, e.g., agonistic buffering (but see chapter 7).

As far as the effect of genealogical ties is concerned our data demonstrate that individuals interact mainly in a positive way with their siblings. Nevertheless, it is important to point out that the infant's mother was influential in inhibiting aggressive acts by juveniles toward the infants: mothers defended infants against older siblings and on occasion prevented (by moving away or by threatening the older offspring) the juvenile male's attempts to contact the infant.

Due to close association between the high-ranking mothers and the two adult males, these mothers' infants were more often in proximity to *Mac* and *C* and the infants made many attempts to join them, some of which resulted in the males' responding aggressively toward them. This may be the reason for the positive association found between contact aggression received by infants from adult males and their dominance rank.

In the context of social development, the adult male played an important role in an infant's early learning experiences. This can be seen in their aggressive contacts with the troop's infants. When an infant approached *Mac,* for example, if it displayed a behavioral sequence of approach then present, *Mac's* response invariably was a stare threat, an open mouth threat, or a slap threat, with the infant screeching and running away in distress. However, if the infant displayed the same sequence, but with a slight modification—approach, teeth-chatter, present—*Mac's* response was either manipulation of the infant's genitalia or tolerance of its proximity. After a few trials, the infant began to display the proper sequence on most of the occasions in which it approached *Mac*. Thus, the alpha male's positive and negative reinforcement of a particular sequence of signals seems to have taught the infant the appropriate way to approach him or more dominant monkeys. Adults were observed to display the same sequence, with either lip-smack or teeth-chatter, when approaching a more dominant individual.

Care Versus Aggression

When comparing care and aggression received by infants from males, the sociograms also indicate that in addition to the fact that when infants are grouped either by sex or age (Estrada and Sandoval, 1977; and this report) at the individual level they experience different types of friendly and aggressive social contacts from males and thus individual behavior differences in later life stages could be expected as a result of behavior modeling.

The sociograms (Figs. 3-8 and 3-9) show that while adults of both sexes displayed care behavior toward all infants, they focused attention on *ti,* the infant son of the alpha female. Additionally, this sociogram shows that while females paid more attention to *dj,* adult males preferred the infant female *br* after *ti.* The preference for *ti* and *br* by *Mac* was probably due to the fact that he consorted and had a special bond with *J.* He had interaction with her and even defended her and her offspring in aggressive actions against others. In the case of the other adult male, *C,* who also directed much care to *ti,* it is necessary to mention that he tended to associate a great deal with the low-ranking females *R* and *Bu* (Estrada, 1977). His special attention to *ti* may reflect his need to use these contacts to buffer potential agonism from *Mac* and *J,* who on some occasions attacked him. In this respect it is important to mention that *C's* friendly interactions with *ti* always took place when this infant was either with *J* or in close proximity (within arm's reach) to *J* and *Mac.*

In the case of the infant *br,* who received much attention from *Mac* (Table 3-8 and Fig. 3-8), it is important to mention that *Mac* also had a special preference for *br's* mother (Bl).

The fact that *Mac* displayed care behavior toward all infants and that he obviously was not in need of buffering agonism from others due to his undisputed dominant position may suggest that care behavior is a general response by males to stimuli such as fur color, size, etc., present in the infant. The group's social dynamics will determine on what infants caring behavior will be concentrated. Our data show that not just *Mac* but also *C* and the juvenile males paid attention to all infants, although in different degrees, thus supporting the conclusion made above.

Also very important is the active participation of infants in initiating caring interactions with males by approaching them and joining them. The join data showed that the joining by infants was not randomly distributed, but that they also showed a preference for particular males. For example, of all the joins to the two adult males, 63% of these were directed to the alpha male, in spite of the fact that *Mac* was much less tolerant of the infants' approaches than was the beta male. Similarily, 61% of the joins sent by infants to the juvenile males were directed to the three top-ranking juveniles (*ch, pa,* and *cha*). Thus, it seems that infants were very attracted toward the most dominant individuals in the troop.

The results also indicate that despite the fact that female infants were joined more often than male infants, they were, in turn, more active than infant males in joining the males in the troop, although they received less care and more aggressive treatment than infant males. It is possible that males joined female infants more as a result of the fact that the two male infants were a limited resource in the troop and as a result of a general interest by males in infants due to stimuli such as fur color, size, and behavior.

From the "join" data we can conclude that (1) infants were strongly attracted toward males and particularly toward adult males, where the greatest disparity is found between joins to infants versus joins by infants; (2) infants sought the proximity of males; and (3) many care and aggressive actions directed to infants by the males probably were initiated as a result of the infants' approaches to the males.

From the point of view of adaptive value, Hrdy (1976) refers to "flamboyant infants" as being strikingly different in color from adults and being quite perceptible at a distance to members of other species (including predators) as well as to conspecifics. Stumptail macaque infants could be categorized as having flamboyant natal coats that contrast with the juvenile and adult coloration. Since the creamy white fur color makes stumptail infant highly conspicuous, they may thus benefit by attracting group members. We have seen that stumptail infants tend to receive large amounts of attention from troop members when they are with or without their mother, which ensures that they are practically never without company.

A case history may help to illustrate this situation. On one occasion when the troop was traveling in the late afternoon toward their sleeping site, moving at the rearguard of the troop were some juveniles and some adult females with their infants. For reasons undetected by the observer, one of the infants with a creamy white fur color stayed behind the troop. Due to the dense foliage, the rest of the troop could not be seen, but it could be heard from where the infant was. The infant's mother was foraging with the main body of the troop and seemed unconcerned for the infant she had left behind. Then the infant "realized" that it could not locate its mother, it climbed up a tree, where it remained vocalizing in distress. As soon as this happened, a juvenile male (not related to the infant) left the troop and went toward the infant, climbing the tree and sitting in physical contact with it. Each time the infant emitted a distress call, the male seemed very nervous as he walked around the infant, giving him light touches with his hand and mouth (to comfort him?) The juvenile male remained with the infant until the mother retrieved it about 15 minutes later. Only then did the juvenile male move down to the ground and rejoin the troop. While no land predators exist on the island, there are avian predators in the area and these had been witnessed to attack the troop (Estrada and Estrada, 1976).

The intense attention infants receive from males (and females, see Estrada and Estrada, in prep.) ensures that the infants are seldom without the close company of an attentive individual. For example, infants in this troop received care contacts from males at a rate of 3.77 contacts per hour. This figure does not take into account the care received by infants from females other than their mothers (Estrada and Estrada, in prep.). The juvenile males were directing care contacts to infants at a rate of 3.57 per hour and were thus the individuals who made the major contribution to infant care received from males. As for aggressive contacts, infants received these at a rate of .19 per hour from juvenile males and .32 per hour from adult males—indicating that the adult males were less tolerant of the infants' approaches and participated more intensively than juvenile males did in the infants' early aggressive experiences. In contrast, the females directed care contacts to infants at a rate of 2.22 per hour and aggressive contacts at a rate of .24 per hour; the juvenile females directed care attention to infants at a rate of 1.46 contacts per hour and punishments at a rate of .17 contacts per hour.

The study has shown that all males, as well as the rest of the group companies, are active participants in the infants' early social experiences and also that the juvenile males (and females) are actively involved in the socialization of infants. In this context it is important to mention that infants were receiving care and aggressive contacts from troop members at a rate of 3.71 and .38 per hour, respectively. Most of the contacts were of a positive nature and some infants were treated differently as a result of the social circumstances given by the relative position of each individual in the group's dominance hierarchy.

The following salient features concerning the interactions between males and infants in wild stumptail macaques can be summarized from the data presented in this report:

1. Males are active participants in the infants' early social experiences, both aggressive and friendly.
2. Juvenile males show a strong interest in interacting with infants and participate more intensively than adult males do in the infants' early friendly experiences.
3. Males displayed preferences for particular infants in the direction of care and aggression due to the infant's social position, the infant's availability, given the degree of independence from its mother, and the infant's network of interactions with other troop members.
4. The preferences of males for particular infants were also strongly influenced by the degree of genetic relatedness between them, with the result that juvenile males invest substantial amounts of care in their infant siblings.
5. During their first year of life, infants actively participate in approaching

troop members and elicit care and aggressive responses from them due to stimuli such as fur color, appearance, behavior, and social position.

6. The attention commanded by stumptail infants from males (and from females) ensures that they are seldom without the close company of an older individual, with all the attendant benefits from these associations.

REFERENCES

Alexander, B. K. 1970. Parental behavior of adult male Japanese monkeys. *Behaviour* 36(4): 270-85.

Altmann, J. 1974. Observational study of behavior: sampling methods. *Behaviour* 49: 227-65.

Altmann, S., and Altmann, J. 1977. On the analysis of rate of behavior. *Animal Behavior* 25: 364-72.

Bertrand, M. 1969. The behavioural repertoire of the stumptail macaque. *Bibliotheca Primatologica.* no. 11. Basel: S. Karger.

Bernstein, I. S. 1975. Activity patterns in a gelada monkey group. *Folia Primatologica* 23: 50-71.

Brandt, E. M.; Irons, R.; and Mitchell, G. 1970. Paternalistic behavior in four species of macaques. *Brain, Behavior and Evolution* 3: 415-20.

Burton, F. D. 1972. The integration of biology and behavior in the socialization of *Macaca sylvana* of Gibraltar. In F. E. Poirier (ed.), *Primate Socialization,* pp. 29-62. New York: Random House.

Deag, J. M., and Crook, J. H. 1971. Social behavior and agonistic buffering in the wild barbary macaque, *Macaca sylvanus. Folia Primatologica* 15: 183-200.

DeVore, I. 1965. Male dominance and mating behavior in baboons. In F. Beach (ed.), *Sex and Behavior,* pp. 261-89. New York: Wiley.

Dittus, W. P. J. 1977. The social regulation of population density and age/sex distribution in the Toque monkey. *Animal Behavior* 63: 281-322.

Estrada, A. 1977. A study of the social relations in a free ranging troop of Stumptail macaques (*Macaca arctoides*). *Boletin de Estudios Medicos y Biológicos* 29(5-6): 313-94.

Estrada, A., and Estrada R. 1976. Establishment of a free ranging colony of Stumptail macaques: Relations to the ecology I. *Primates* 17(3): 337-55.

———. 1978. Changes in social structure and interactions after the introduction of a second group in a free ranging troop of stumptail macaques (*Macaca arctoides*): Social relations II. *Primates* 19(4): 665-80.

———. In preparation. Interactions of female with infants in a free ranging troop of stumptail macaques (*Macaca arctoides*).

Estrada, A.; Estrada R.; and Ervin, F. 1977. Establishment of a free ranging colony of stumptail macaques (*Macaca arctoides*): Social relations I. *Primates* 18(3): 647-76.

Estrada, A., and Sandoval, J. M. 1977. Social relations in a free ranging troop of stumptail macaques (*Macaca arctoides*): Male care behavior I. *Primates.* 18(4): 793-813.

Gouzoules, J. 1975. Maternal rank and early social interactions of infant stumptail macaques, *Macaca arctoides.* Primates 16(4): 405-18.

Hall, K. R. L., and DeVore, I. 1965. Baboon social behavior. In I. DeVore (ed.), *Primate Behavior: Field Studies of Monkeys and Apes,* pp. 53-110. New York: Holt Rinehart and Winston.

Hamilton, W. D. 1964a. The genetical evolution of social behavior. I. *Journal of Theoretical Biology* 7: 1-16.

———. 1964b. The genetical evolution of social behavior. II. *Journal of Theoretical Biology* 7: 17-52.

Hansen, E. W. 1966. The development of maternal and infant behavior in the rhesus monkey. Ph.D. dissertation, University of Wisconsin, Madison.

Hinde, R. A., and Spencer-Booth, Y. 1967. The behavior of socially living rhesus monkeys in their first two and a half years. *Animal Behavior* 15: 169-96.

Hrdy, S. B. 1976. Care and exploitation of nonhuman primates by conspecifics other than the mother. In *Advances in the Study of Behavior,* vol. 6, pp. 101-58. New York: Academic Press.

Itani, J. 1959. Paternal care in the wild Japanese monkey, *Macaca fuscata. Primates.* 2(1): 61-93.

———. 1963. Paternal care in wild Japanese monkey, *Macaca fuscata.* In C. H. Southwick (ed.), *Primate Social Behavior,* pp. 91-97. New York: Van Nostrand Reinhold.

Koford, C. B. 1963. Ranks of mothers and sons in bands of rhesus monkeys. *Science* 141: 356-57.

Kummer, H. 1968. *Social Organization of the Hamadryas Baboons.* Chicago: University of Chicago Press.

Kauffman, J. H. 1967. Social relations of adult males in a free ranging band of rhesus monkeys. In S. Altmann (ed.), *Social Communication among Primates,* pp. 73-98. Chicago: University of Chicago Press.

Lancaster, J. 1971. Play mothering: The relations between juvenile females and young infants among free-ranging vervet monkeys *(Cercopithecus aethiops). Folia Primatologica* 15: 161-87.

Mitchell, G. D. 1969. Paternalistic behavior in primates. *Psychological Bulletin* 71: 399-417.

Mitchell, G. D., and Brandt, E. M. 1972. Paternal behavior in primates. In F. E. Poirier (ed.), *Primate Socialization,* pp. 173-206. New York: Random House.

Neely, H. H., and Rhine, R. 1977. Social development of stumptail macaques *(Macaca arctoides):* Interactions with adult males during the infant's first 60 days of life. *Primates* 18: 589-600.

Ransom, T. W., and Ransom, B. S. 1971. Adult male infant relations among baboons *(Papio anubis). Folia Primatologica* 16: 179-95.

Ransom, T. W., and Rowell, T. 1972. Early social development of feral baboons. In F. E. Poirier (ed.), *Primate Socialization,* pp. 105-44. New York: Random House.

Rhine, R. J. 1972. Changes in the social structure of two groups of stumptail macaques *(Macaca arctoides). Primates* 13(2): 181-94.

Rowell, T.; Hinde, R. A.; and Spencer-Booth, Y. 1964. "Aunt" infant interactions in captive rhesus monkey. *Journal of Animal Behavior.* 12: 219-26.

Sade, D. 1967. Determinants of dominance in a group of free ranging rhesus monkeys. In S. Altmann (ed.), *Social Communication among Primates,* pp. 99-114. Chicago: University of Chicago Press.

Siegel, S. 1971. *Non Parametric Statistics for the Behavioral Sciences.* New York: McGraw-Hill.

Simonds, P. E. 1965. The bonnet macaque in south India. In I. DeVore (ed.), *Primate Behavior,* pp. 175-96. New York: Holt, Rinehart and Winston.

Spencer-Booth, Y. 1968. The behavior of group companions toward rhesus monkey infants *Animal Behavior* 16: 541-57.

———. 1970. The relationship between mammalian young and conspecifics other than the mother and peers: A review. In D. S. Lehrman; R. A. Hinde; and E. Shaw (eds.), *Advances in the Study of Behavior,* pp. 120-80. New York: Academic Press.

Taub, D. 1975. Paternalism in free-ranging Barbary macaques, *Macaca sylvanus. Proceedings of the American Association of Physical Anthropologists,* Denver, Co.

van Lawick-Goodall, J. 1967. Mother offspring relationships in free ranging chimpanzees. In D. Morris (ed.), *Primate Ethology,* pp. 287-346. New York: Doubleday.

Yamada, M. A. 1963. Study of blood relationships in a natural society of Japanese monkeys. *Primates* 4: 43-63.

4

Adult Male-Immature Interactions in Captive Stumptail Macaques *(Macaca arctoides)*

Euclid O. Smith
Yerkes Regional Primate Research Center
Department of Anthropology and Department of Biology
Emory University
Atlanta, Georgia

Patricia G. Peffer-Smith
Yerkes Regional Primate Research Center
Emory University
Atlanta, Georgia

INTRODUCTION

Alloparental care is limited to the most advanced animal societies and is most richly expressed among the primates (Wilson, 1975). There is considerable inter- and intraspecific variability in alloparental care among the primates, ranging from brief, sporadic encounters to prolonged association. As Wilson (1975) notes, alloparental care can be divided into two aspects: allomaternal and allopaternal. Although allomaternal behavior is far more common, allopaternal interactions do indeed occur. Because of the pervasiveness, complexity, and importance of mother-young interactions in understanding certain aspects of nonhuman primate sociality, only recently have researchers become interested in the role of males in the development and socialization of young. When compared to mother-young interactions in most primate societies, male-young interactions are subtle and infrequent and exhibit high interanimal variability. (See Hrdy, 1976; Mitchell, 1969,

1977; Redican and Taub, 1981; Spencer-Booth, 1970; and chapters 14 and 15 for reviews of male-infant interactions.)

Among one cercopithecoid genus, *Macaca*, considerable variability in male-immature interactions has been observed (see chapter 15). This variability ranges from the extensive male care system of Barbary macaques (Burton, 1972; Deag and Crook, 1971; Taub, 1980; Whiten and Rumsey, 1973) to the very restricted male care of free-ranging rhesus macaques (cf. Lindburg, 1971; Makwana, 1978; Southwick et al., 1965; Taylor et al., 1978; chapter 5).

Of particular interest for this paper were the interactions between adult male and immature stumptail macaques *(Macaca arctoides)*, since males of this species have been characterized as exhibiting high levels of paternalistic interactions with immatures (Brandt et al., 1970; Estrada and Sandoval, 1977; chapter 3). As such, it was an interesting species for investigating the parameters of male-immature interactions. Therefore, the specific objectives of this study were: (1) to determine the nature and types of adult male-immature interactions in a captive stumptail macaque group; (2) to examine the effects of certain variables on these interactions; and (3) to identify specific patterns of interactions among adult male-immature dyads.

In both free-ranging (Estrada and Sandoval, 1977) and captive (Hendy-Neely and Rhine, 1977) groups of stumptail macaques, adult males have been observed to carry, cradle, retrieve, groom, and touch infants. Rhine and Hendy-Neely (1978) noted that adult males and infants (individuals under 60 days of life) engaged in behaviors commonly seen in mother-young pairs. Adult males interacted significantly more often with infants than did immatures or adult females other than the mother. Bernstein (1970) noted considerable interest in infants among male stumptail macaques; Jones and Trollope (1968) also found that adult male stumptail macaques were very aware of infants and would even lip-smack to color pictures of infants. Conversely, infants also show interest in males and often seek their proximity (Bertrand, 1969; Estrada et al., 1977). On the other hand, Bertrand (1969) noted that zoo-reared male stumptail macaques may attack and bully infants. From these scattered reports of interactions between adult male and immature stumptail macaques, some of the salient features of the relationship might be identified, but more data are required to precisely delineate their nature.

METHODS

Study Group and Housing

The subjects of this study were a captive group of stumptail macaques *(Macaca arctoides)* housed at the Yerkes Regional Primate Research

Center Field Station near Lawrenceville, Georgia. The group was housed in a 28.4 × 32.7 m. outdoor enclosure attached to 4.4 × 12.2 m. indoor quarters via two metal tunnels (Fig. 4-1).

The group consisted initially of 36 animals: 4 adult males (classification based on full dentition, developed temporal musculature, and general physical conformation), 18 adult females (cycling, at approximately 4 years of age), 1

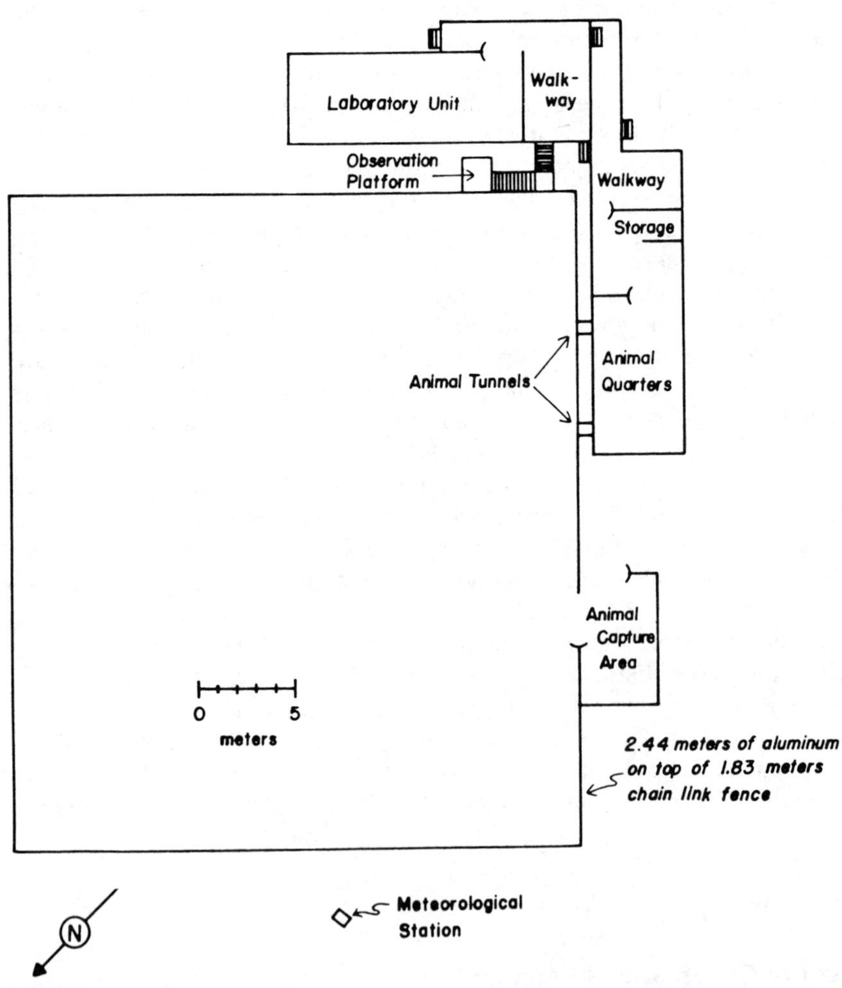

Fig. 4-1. Schematic view of the animal enclosure and associated structures.

subadult male (4 to 5 years old), 3 subadult females (3 to 4 years), 3 juvenile males (2 to 4 years), 2 juvenile females (2 to 3 years), 3 immature males, and 2 immature females (birth to 2 years). See Table 4-1 and Figure 4-2 for the genealogies and birth dates of individuals in the study group (3 deceased animals were not included).

During the course of the study, there were seven changes in the demography of the group, including: (1) two births, (2) permanent removal or death of three adult females, and (3) maturational changes of one male and two females. Methodological problems associated with maturation of group members are discussed later. The demography of the group and the associated changes during the study are detailed in Table 4-2.

Table 4-1. Birthdates of Group Members.

Males		Females	
Code	Birthdate	Code	Birthdate
06	1970	01	1961
10	2/04/70	02	4/29/66
13	9/16/72	03	1965
18	2/15/74	04	1965
24	8/11/75	05	1970
26	11/12/76	07	1970
28	12/24/76	08	1970
30	7/04/77	09	1970
33	3/04/78	11	6/06/72
35	8/05/78	12	8/28/72
37	1/22/79	14	12/14/72
38	6/12/80	15	1/01/73
39	7/09/80	16	1/22/73
		17	6/18/73
		19	9/19/74
		20	2/20/75
		21	3/06/75
		22	3/18/75
		23	7/13/75
		25	7/05/76
		27	12/16/76
		29	12/31/76
		31	9/09/77
		32	9/20/77
		34	3/22/78
		36	8/10/78

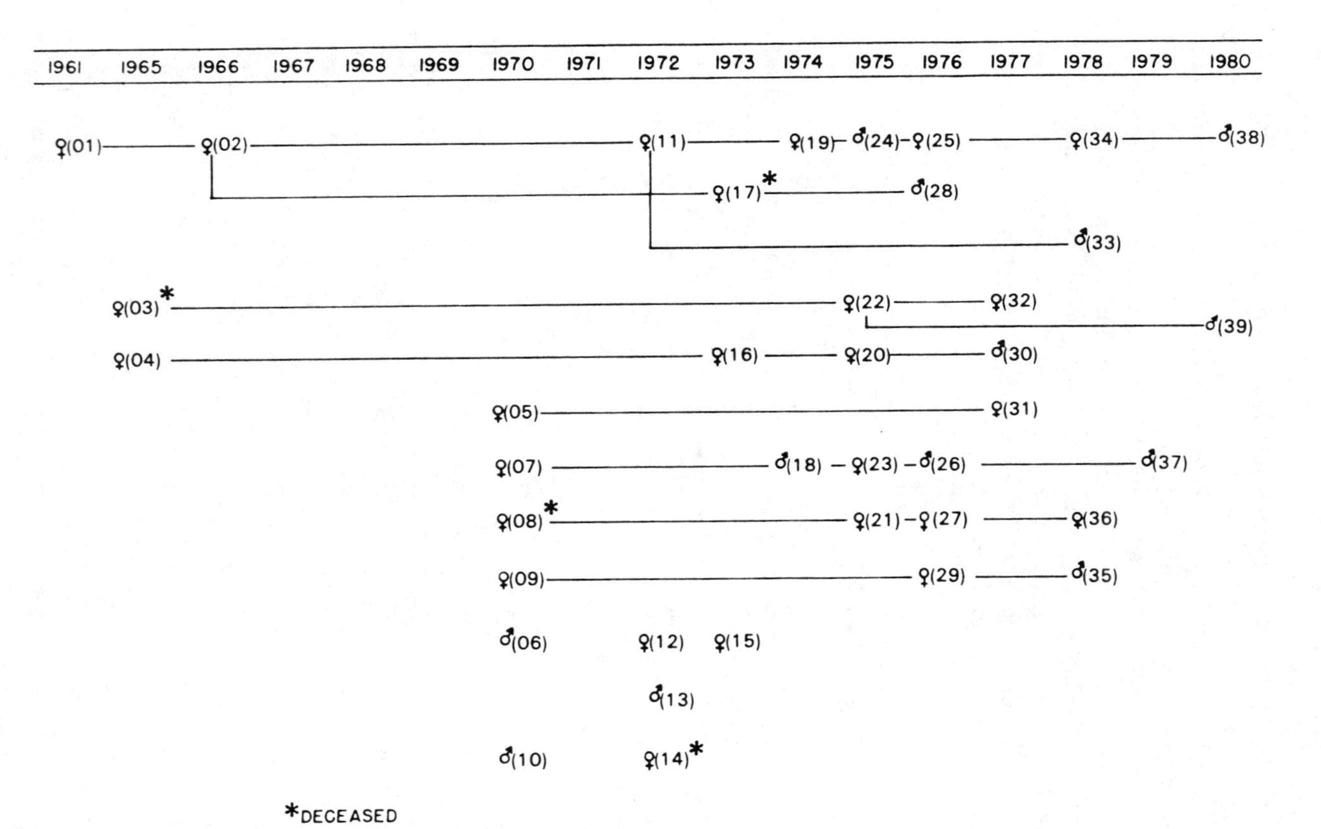

Fig. 4-2. Genealogical relationships of members of the study group.

Table 4-2. Demographic Changes in the Study Group.

Time Block	Dates	Adult Male	Adult Female	Subadult	Juvenile	Immature	Reason for Change
1	12/17/79-12/21/79	4	18	4	5	5	
2	12/22/79-3/3/80	4	17	4	5	5	Permanent removal, adult female 08
3	3/4/80-3/12/80	4	17	4	6	4	Maturation, 33
4	3/13/80-3/21/80	4	16	4	6	4	Death-adult female 03
5	3/22/80-6/11/80	4	16	4	7	3	Maturation, 34
6	6/12/80-7/4/80	4	16	4	7	4	Birth, 38
7	7/5/80-7/8/80	4	17	3	7	4	Maturation, 25
8	7/9/80-8/7/80	4	16	3	7	5	Birth, 39
							Permanent removal, adult female 17

Procedures

Data were collected from 17 December, 1979, to 7 August, 1980, using a microprocessor-based data collection device, the Datamyte 900 (see Smith and Begeman, 1980, for details). Data were collected on the interactions of the four adult males with other group members using the focal animal observation technique (Altmann, 1974); 173.5 hours of focal adult male data were obtained totaling 46,053 male-initiated interactions (Table 4-3). Data were collected at preselected time periods throughout the day in a randomized block design so that subjects were observed equally for all time periods over several days. Focal tests lasted 15 minutes per subject; observation sessions typically involved two to three tests in succession. Observations were made from a platform located 4.27 m. above the enclosure substrate (see Fig. 4-1). Complete visual access to the group during observation sessions was maintained by restricting the animals' movements to the outdoor enclosure.

A total of 120 different types of behaviors was scored during observation sessions. The behavioral inventory utilized was influenced by the work of Bertrand (1969) and Chevalier-Skolnikoff (1974), but was primarily developed during a previous study conducted on another stumptail macaque group at the study site (Peffer-Smith, 1978). For a complete discussion of the behavioral inventory, see Smith and Peffer-Smith (unpub. ms.).

Based on previous observations of stumptail macaques and for the purpose of developing a broader perspective on group interactions, the behavioral inventory was organized into seven functional and analytical classes of behavior (Rosenblum, 1978):

1. *Aggressive:* behaviors that cause actual physical injury or signal the potential for harm—also, behaviors that result in preferential access to incentives.
2. *Submissive:* behaviors that are a reaction to a real or perceived possibility of bodily injury.
3. *Affiliative:* any positive behavior that does not involve bodily harm or that is an attempt to safeguard another.
4. *General social:* any behavior that is social because it involves another individual, but carries no specific social messages (e.g., approach, watch, look at, move off from).
5. *Play:* vigorous, exaggerated but relaxed movements; structurally similar behaviors seen in aggressive and submissive behavior, but contact is less forceful, is silent, and the roles of the interactants are frequently reversed.
6. *Sexual:* behaviors regularly a component of heterosexual mating; however, may be scored regardless of the reproductive potential of the participants.
7. *Self-directed and maintenance:* any behavior directed toward self.

Behaviors included in each of these seven categories were mutually exclusive: i.e., they could be scored in one, and only one, functional class.

For purposes of this study, only data from the four adult males were analyzed; furthermore, only the interactions they initiated to other group members were used for most analyses. Although this approach does not allow for the complete analysis of the male-immature interactional system, it provides insight into some of its key features.

RESULTS

Demographic Characteristics

Before examining adult male-immature interactions in detail, the distribution of male-initiated acts to all age-sex classes must be considered to determine if frequency of interaction might be a function of the number of available partners. Put differently, can the observed distribution of behaviors initiated by adult males simply be accounted for by the composition of the group? The distribution of male-initiated interactions to all group members during the 8 time blocks comprising the entire study period is shown in Table 4-3. Could the high frequency of initiation of interactions to adult females (41% of the total) be because there were more adult females available in the study group? The question was further complicated when we considered the demographic changes of the study group, which obviously affected the number of available interactants in any age-sex class. Therefore, the hypothesis that the distribution of male-initiated acts was simply a function of the number of available partners must be considered.

When these data were analyzed using the technique presented by Altmann and Altmann (1977, pp. 365-66, case two example), theoretically expected values for the frequency of interactions between adult males and all other age-sex classes were derived. Given the number of individuals in any age-sex class and the duration of that sample period as a proportion of the total available animal hours during the study, theoretically expected values for each behavior class for each age-sex class were calculated.

Data were analyzed in a similar manner for each of the separate behavior classes and the theoretically expected frequencies were derived (Table 4-4). Table 4-4 also shows the results of a Chi-square analysis for each behavior class, as well as total social interactions. Significant differences ($p \leq .05$) between observed and expected frequencies were found for total social interactions, as well as for each of the six constituent behavior classes. These data indicated that factors other than simple group demography must be implicated in the distribution of adult male interactions.

Table 4-3. Distribution of Adult Male-Initiated Social Interactions to All Group Members.

Time Block		Adult Male*	Adult Female	Subadult	Juvenile	Immature	Total
1	No. of individuals	4	18	4	5	5	36
	Observed frequencies	921	940	189	156	154	2360
	Sample time, 9 hrs						
2	No. of individuals	4	17	4	5	5	35
	Observed frequencies	1925	3909	753	736	948	8271
	Sample time, 31 hrs						
3	No. of individuals	4	17	4	6	4	35
	Observed frequencies	302	676	137	157	248	1520
	Sample time, 6 hrs						
4	No. of individuals	4	16	4	6	4	34
	Observed frequencies	436	620	148	179	182	1565
	Sample time, 6 hrs						
5	No. of individuals	4	16	4	7	3	34
	Observed frequencies	3837	7523	2065	2988	2194	18607
	Sample time, 74 hrs						
6	No. of individuals	4	16	4	7	4	35
	Observed frequencies	767	1582	499	761	456	4065
	Sample time, 15.5 hrs						
7	No. of individuals	4	17	3	7	4	35
	Observed frequencies	108	459	134	183	69	953
	Sample time, 4.5 hrs						
8	No. of individuals	4	16	3	7	5	35
	Observed frequencies	1463	3349	777	1808	1315	8712
	Sample time, 27.5 hrs						
	Total Observed Frequencies	9759	19058	4702	6968	5566	46053
	(% Total)	21.19	41.38	10.21	15.13	12.09	

*All four males pooled as subjects.

Table 4-4. Observed and Expected Frequencies for Male-Initiated Interactions by Behavior Class and Age-Sex Class.

Behavior Class	Frequency	Adult Males	Adult Females	Subadults	Juveniles	Immatures	χ^2
Total social	Observed	9759	19058	4702	6968	5566	4385.73*
Interactions	Expected	5325	21758	5080	8614	5276	
Aggressive	Observed	186	601	185	361	171	42.80*
	Expected	174	711	166	281	172	
Submissive	Observed	198	88	24	66	87	503.54*
	Expected	54	219	51	87	53	
Affiliative	Observed	2347	3702	867	876	1118	2143.51*
	Expected	1030	4210	983	1667	1021	
General social	Observed	7021	14341	3581	5636	4188	2647.92*
	Expected	4020	16426	3835	6503	3983	
Play	Observed	0	0	0	14	2	46.16*
	Expected	2	8	2	3	2	
Sex	Observed	7	326	45	15	0	227.55*
	Expected	45	187	43	74	45	

*Significant $p \leq .05$.

The distribution of theoretically observed and expected acts was then examined for each age-sex class separately to see if there were significant departures. In order to evaluate each cell in our Chi-square table, a procedure outlined in Bishop et al. (1975, pp.136-37) was employed that allowed us to compute standardized cell residuals and compare derived values to the appropriate Chi-square distribution.[1] Specifically, it was found that adult males initiated significantly more ($p \leq .05$) interactions with immatures than expected. These interactions included significantly more ($p \leq .05$) submissive behavior, general social behavior, and affiliative behavior, although they initiated significantly less ($p \leq .05$) sexual behavior than expected (an expected result due to the maturational status of the immatures). No significant departures from theoretically expected values were found for aggressive behavior or play.

Figure 4-3 shows a comparison of the types of behaviors that adult males exhibit toward all age-sex classes. From a qualitative point of view, the adult males were consistent in exhibiting different types of behaviors across age-sex classes, initiating general social and affiliative behaviors predominantly to all age-sex classes. Specific behaviors that adult males were observed to initiate to immatures included a variety of types of threat faces, displace, hit, push, grab, pin, bite, chase, jerk, avoid, approach, proximity, look, watch, inspect, behind lift (called bridging in Estrada and Sandoval, 1977), groom, touch, hold, grasp, huddle, pout face, pull in, lip-smack, gutteral, dorsal and ventral hold, dorsal and ventral carry, play face, play mouth, and play grapple. When the observed and expected percentiles of different classes of interactions were graphically depicted across age-sex classes, significant departures in all age-sex classes were found for at least one behavior class (Figure 4-4).

Intermale Variability

As has been shown, males directed a disproportionately large number of interactions toward immatures. The next question was to look at each of the four adult males to see if there were individual differences in rates of

---

[1]Standardized cell residuals (Z_i) were calculated as follows:

$$Z_i = \frac{o_i - e_i}{e_i}$$

where o_i = observed value and e_i = expected value. Then the absolute value of the square of the standarized cell residuals was compared to the square root of 5% significance level of the appropriate Chi-square distribution divided by the number of cells. See Fagen and Mankovich (1980) for a review of problems associated with this type of analysis.

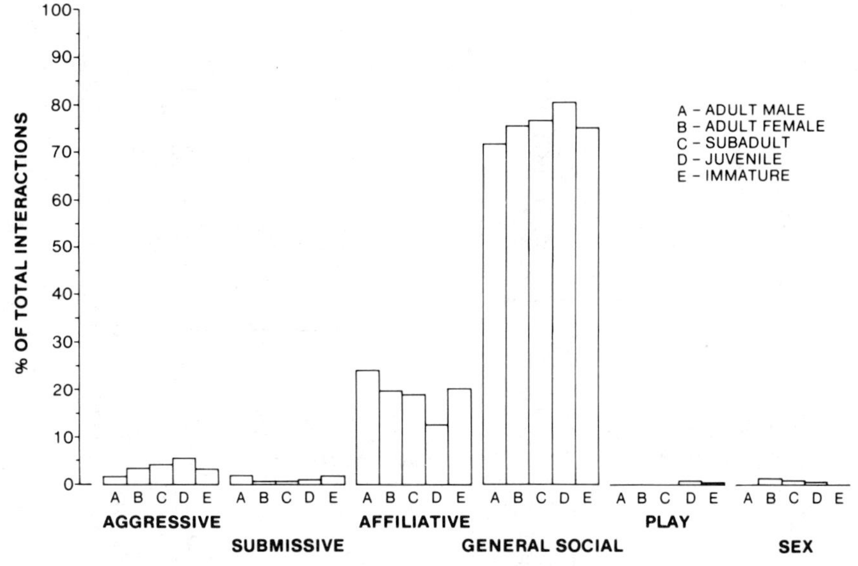

Fig. 4-3. Adult male-initiated behaviors to all age-sex classes.

interaction with immatures, as compared to their frequency of interaction. To examine specific rates of interaction, individual rates of interaction for each adult male-immature dyad were compared. Since maturation and birth affected the composition of the immature age class during the study, we based our calculation of rates on the number of hours each immature was available for interaction with a given male. Therefore, rates were expressed as a frequency of acts per immature hour for a specific dyad. This then allowed the utilization of all the data from immatures who, during the course of the study, graduated from the immature age class, and the data from immatures born during the study. Table 4-5 shows the average rate of interaction per immature hour during the study for total social interactions and four behavior classes (play and sexual behavior could not be analyzed because of their low frequency of occurrence). Significant differences ($p \leq .05$) among the four males were found for affiliative behavior, general social behavior, and total social interactions. As can be seen qualitatively, male 10 was the most predominantly involved of all four males in positive interactions with immatures.

Dominance Rank

Given that significant differences existed among the adult males in the rate of interaction with immatures, data were analyzed to determine if

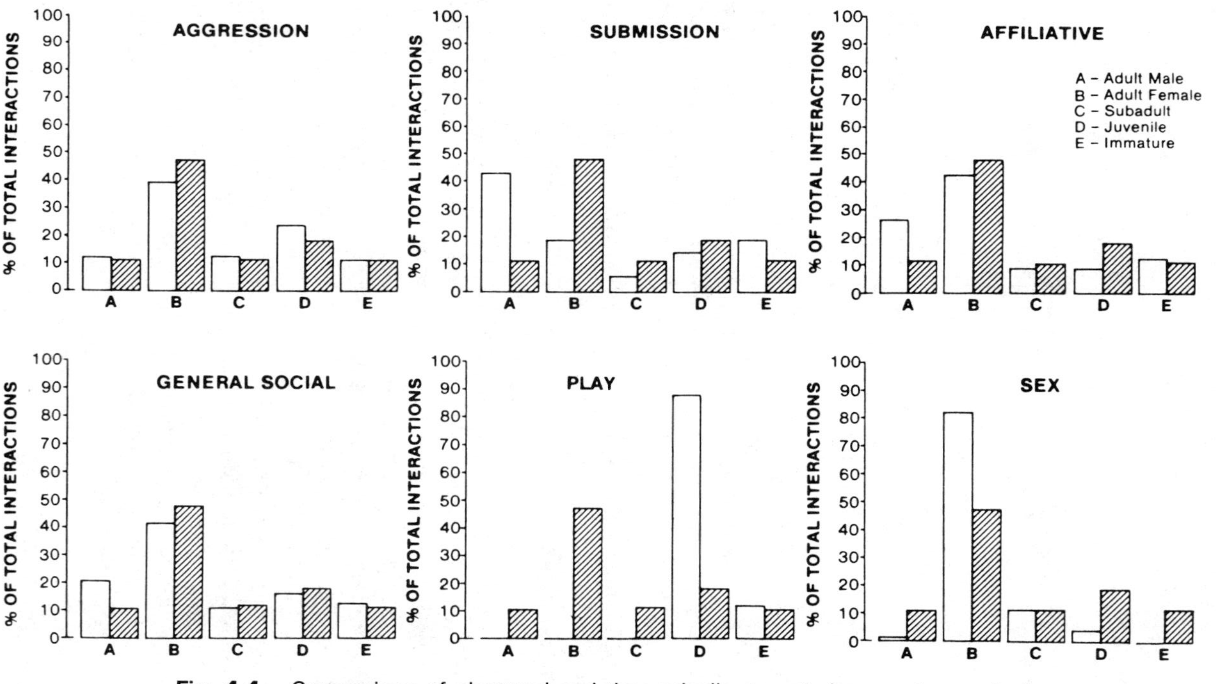

Fig. 4-4. Comparison of observed and theoretically expected percentages of all behavior types by age-sex class. Crosshatched bars indicate theoretically expected values.

Table 4-5. Adult Male Rates of Interaction Initiated to Immatures (Frequency/Immature Hour) for Four Behavior Classes and Total Social Interactions.

	Adult Male				
Behavior Class	**06**	**10**	**13**	**18**	χ^2
Aggressive	0.22	0.46	0.13	0.18	0.257
Submissive	0.03	0.17	0.09	0.22	0.164
Affiliative	0.38	5.49	0.23	0.43	12.185*
General social	3.52	11.87	5.79	3.22	7.925*
Total social interactions	4.15	17.99	6.24	4.05	16.454*

*Significant $p \leq .05$.

dominance rank might account for this variability. The adult males were ranked based on the outcome of aggressive and submissive interactions for each of the eight separate blocks of the study, and Kendall's Coefficient of Concordance for these ranks was calculated ($W = 0.9125$, sig. $p \leq .001$) (Siegel, 1956, p. 231) to determine the consistency of rank across the study. Given the high degree of similarity across the blocks, the males were then ranked according to their frequency of inititation of each behavior class that could be analyzed, as well as total social interactions. Spearman rank order correlation coefficients for the rate of total social interactions and rate of all behavior classes were not significant with dominance rank ($p \leq .05$). Therefore, it was concluded that dominance rank of the adult male was not an important factor in accounting for the intermale variation.[2]

Individual Male-Immature Dyads

In order to precisely delineate the nature of the adult male-immature interactions, the rates of interaction between each of the adult males and each immature were calculated. Aggressive and submissive interactions, as well as play and sexual behavior, were not analyzed in this manner because of their infrequent occurrence (see Table 4-6). For total social interactions, males 06 and 10 showed significant deviations ($p \leq .05$) from expected rates for total social interactions when the observed individual rate for each male was compared to the average rate for each male with all immatures combined. Further examination of the data presented in Table 4-6 revealed that male 06 initiated over 2.5 times as many social interactions to immature 39 than

[2]In a previous study (Peffer-Smith, 1978), however, high-ranking males were found to interact significantly more ($p \leq .05$) with immatures than low-ranking males. This could possibly be explained by the association between kin and rank in the study group.

Table 4-6. Adult Male Rates of Interaction to Immatures for Total Social Interactions, Affiliative, and General Social Behavior.

	Immature								
Adult Male	**33**	**34**	**35**	**36**	**37**	**38**	**39**	$\overline{X}$	χ^2
				Total Social Interactions					
06	2.00	1.46	5.36	2.99	4.51	2.92	11.86	4.44	18.25*
10	9.00	8.69	10.34	36.23	17.16	8.77	2.81	13.29	48.62*
13	6.80	6.31	5.40	6.76	5.31	9.33	7.71	6.80	2.22
18	0.60	1.77	2.84	4.25	5.83	5.87	5.19	3.76	6.52
				Affiliative					
06	0.30	0.23	0.51	0.39	0.25	0.00	1.29	0.42	2.72
10	2.40	1.62	1.23	16.32	3.10	0.17	0.00	3.55	40.85*
13	0.50	0.00	0.16	0.39	0.14	0.42	0.00	0.23	1.10
18	0.00	0.15	0.12	0.53	0.79	0.60	0.30	0.36	1.27
				General Social					
06	1.30	1.08	4.57	2.39	40.20	2.57	9.86	3.68	15.57*
10	6.40	6.69	8.81	18.87	13.62	6.47	2.81	9.10	19.13*
13	6.30	5.92	5.08	6.18	5.06	8.00	7.43	1.11	1.58
18	0.60	1.15	2.36	3.47	4.76	3.40	4.74	1.64	5.18

*Significant $p \leq .05$.

would be theoretically expected (sig. $p \leq .05$).[3] Also, male 10 directed over twice the expected rate of total social interaction to infant 36. In both cases, theoretically expected rates were the average for each male across all immatures.

When total social interactions were partitioned into the major behavior classes (affiliative and general social behavior), no significant differences were found in the distribution of affiliative acts among the immatures by male 06, although immature 39 received over 2.5 times more affiliative behavior than the next highest immature. Male 10, however, showed significant variation among infants, with infant 36 receiving over six times the rate of affiliative behavior for any other immature (sig. $p \leq .05$). In fact, immature 36 received almost twice the rate of interactions of all other infants combined. For general social behavior, the pattern persisted; namely, the unusually high rate of interaction between animals 06 and 39, and 10 and 36 (all sig. $p \leq .05$). These results suggested some sort of special relationship between these particular male-infant dyads.

To further clarify the nature of these relationships and to assess the possibility that these males were initiating behaviors to these immatures in order to associate with the immatures' mothers, we examined the interactions between these two males and the mothers of the respective infants. Figure 4-5 shows the rate of interaction between male 06 and all females with infants. Of special interest, of course, is the rate of interaction between male 06 and female 22, mother of immature 39. Clearly, male 06 had, prior to the birth of immature 39, a consistently high interaction rate with female 22, except for the time block immediately preceding the birth of immature 39. When male 06's rate of interactions was compared across all females (both with and without infants), it was found that his rate of interaction with female 22 was the fourth highest overall, suggesting that the relationship formed with immature 39 was a byproduct of an existing strong, positive relationship to female 22.[4]

The interactions between adult male 10 and immature 36 presented an entirely different picture. Figure 4-6 shows the hourly rate of interaction between male 10 and immature 36 across time blocks. The significant point was the marked increase in the rate of interaction in block 2 and the consistently high rate thereafter. It should be noted that the demographic change demarcating block 2 was the permanent removal of adult female 08, mother of immature 36. From the data presented in Figure 4-6, it was clear that the permanent removal of immature 36's mother significantly altered the pattern of interactions initiated by adult male 10 to immature 36.

---

[3]Individual male-immature dyads were analyzed in the manner noted on pages 95 through 98.
[4]Similar results were noted in a previous study (Peffer-Smith, 1978).

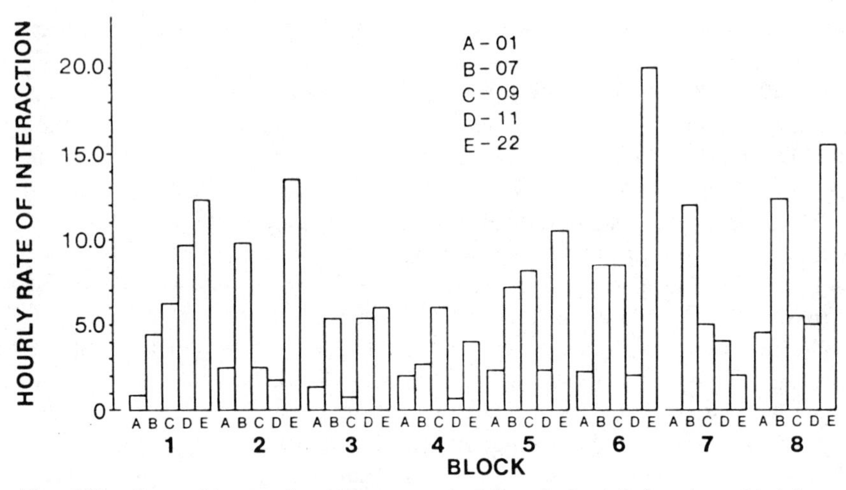

Fig. 4-5. Rate of interaction between male 06 and all adult females with infants across each of the eight time blocks of the study.

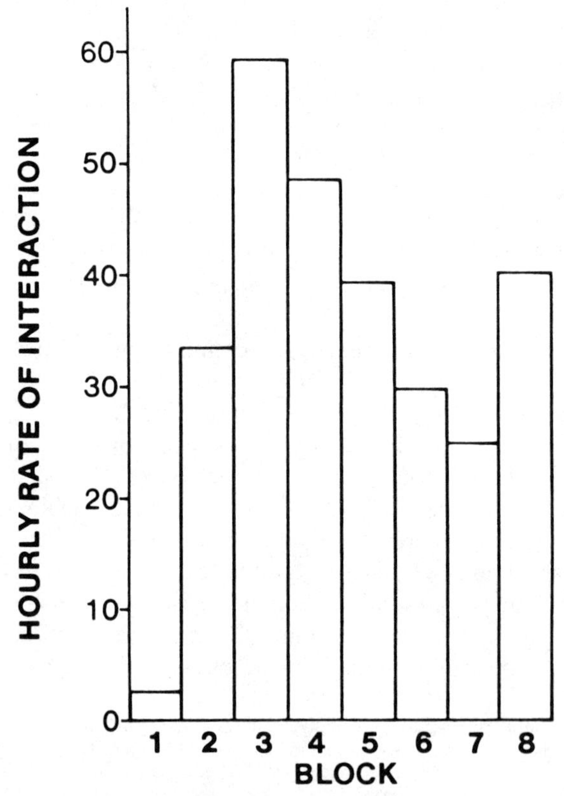

Fig. 4-6. Rate of interaction between adult male 10 and immature 36 across the eight time blocks of the study.

Given these data, the patterns of interactions initiated by immatures 36 and 39 toward the adult males were examined in order to determine if the special relationships could be characterized as symmetrical, i.e., immatures initiating interactions to the adult males as well as vice versa. Table 4-7 shows the rate immature 36 initiated interactions to each of the adult males. Given the null hypothesis of no significant differences among the adult males, a Chi-square analysis revealed significant differences ($p \leq .05$) for total social interactions, as well as affiliative and general social interactions. Adult male 10 received over six times as many social interactions from immature 36 as the next most popular male. Therefore, it was concluded that the relationship between male 10 and immature 36 was symmetrical and reflected clear preferences on the part of both members of the dyad.

To further examine the relationship between male 06 and immature 39, the rates of interaction initiated by immature 39 to each adult male were examined and no significant differences ($p \leq .05$) were found, given the same null hypothesis. These results tended to support the interpretation of the interactions between male 06 and female 22.

Finally, given the nature of the relationships between particular adult males and immatures, the total male rate of interaction for each immature was calculated. The possibility existed that even though a special relationship may have been noted for one male-immature dyad, other immatures may have received interactions from several males in such a manner that their cumulative rate of interaction would equal or exceed that of those immatures with special relationships. The cumulative distribution of rates of interaction with adult males for each individual immature is shown in Table 4-8. A

Table 4-7. Immature 36 Initiated Interactions to Adult Males.

	Adult Male				
Behavior Class	**06**	**10**	**13**	**18**	χ^2
Total social interactions	5.47	91.86	14.41	9.43	168.18*
Affiliative	1.45	42.96	0.83	1.60	111.22*
General social	3.84	45.94	12.51	7.70	63.79*

*Significant $p \leq .05$.

Table 4-8. Cumulative Rates of Interaction for Immatures with All Adult Males.

	Immature						
Adult Male	**33**	**34**	**35**	**36**	**37**	**38**	**39**
06	2.00	1.46	5.36	2.99	4.51	2.92	11.86
10	9.00	8.69	10.34	36.23	17.16	8.77	2.81
13	6.80	6.31	5.40	6.76	5.31	9.33	7.71
18	0.60	1.77	2.84	4.25	5.83	5.87	5.19
Total	18.40	18.23	23.94	50.23	32.81	26.89	27.57

Chi-square test revealed significant differences ($p \leq .05$) among the immatures, and indicated that immatures did not receive equivalent amounts of male interaction no matter how many males interacted with them. In fact, immature 36 received more interactions from male 10 than each of the other immatures received from all males combined.

DISCUSSION

Interactions between adult male and immature nonhuman primates present a fascinating area of inquiry in the study of social organization because of the implications of these interactions for both adult males and the immatures. To put these interactions in a broader perspective, we initially asked an important question concerning the overall pattern of male interactions in our captive stumptail group: Do adult males interact with immatures more or less frequently than would theoretically be expected based on the demography of the group? Our results indicated that adult male stumptails do, indeed, initiate more interactions with immatures than would theoretically be expected, and, excluding the adult males themselves, immatures are the only age-sex class that received more interactions than would theoretically be expected based on the demography of the group. In fact, when total social interactions were partitioned into the constituent behavior classes, males exhibited significantly more submissive, affiliative, and general social behavior to immatures than would be expected. This suggested that infants were a powerful feature of the social life of adult male stumptails, and that adult males may have important roles in the social development of the young.

Adult male-immature interactions can be generally characterized as positive and were composed primarily of general social interactions and affiliative behavior (95.3% of total). Brandt et al. (1970) found that 527 of 534 (98.6%) male-immature interactions in their stumptail group were positive social behavior; these could be accounted for predominantly by huddling, passive contact, and proximity. Comparatively little aggressive, submissive, or play interactions were initiated by the adult males to immatures. No sexual behavior was initiated by the adult males to the young. These results generally concur with other studies (Estrada, 1977; Estrada and Sandoval, 1977; Gouzoules, 1975; Hendy-Neely and Rhine, 1977).

Intermale Variability

It is clear, however, that individual adult males vary considerably in their rates of interaction with immatures. Hendy-Neely and Rhine (1977) also found significant differences among adult males in the amount of parental care exhibited. This was not unexpected, given intermale variability in other aspects of behavior. The individual variability was of interest, however, since one male in this study interacted with immatures more than all the other males combined did.

Since individual adult males can be differentiated in their overall rate of interaction with immatures, a careful examination of the relationship of each adult male with each immature was important. Our results indicated two interesting patterns emerging. Specifically, a significantly higher rate of male-initiated interactions between adult male 06 and immature 39, as well as male 10 and immature 36, was found. Bertrand (1969) also noted that some males developed special relationships with specific immatures. The "special relationship" was found for affiliative behavior and general social interactions for both male-immature dyads. This was important, since it demonstrated the potential for adult males to form distinct bonds with immatures who, as an age class, received considerable male attention.

Adult Male-Adult Female Relationships

An adult male who associates closely with a female may have extensive interactions with her offspring. Nash and Ransom (1971) and Southwick et al. (1965) found this to be true with animals in consort. Altmann (1978; 1980, p. 117), Ransom and Ransom (1971) and Seyfarth (1978) also noted enduring social bonds between adult males and the immatures of favored females. Similarly, elevated levels of interaction between one adult male and female, and the subsequent involvement of the female's infant, were noted. It was suggested that the interactions between male 06 and immature 39 were a by-product of the relationship of male 06 and female 22 and were not the result of active development of social bonds to the immature. This view was further reinforced when the interactions initiated by infant 39 to all the adult males were examined. These results showed that infant 39 was not directing a significant proportion of his interactions toward male 06 nor any other particular adult male.

Male 10, on the other hand, presented a different picture in his relationship with immature female 36. Prior to the permanent removal of female 08 (mother of immature 36), male 10 initiated relatively little behavior toward immature 36 or any other immatures in the group. Removal of female 08 (the event which demarcated the second time block) brought a significant change in the relationship. From block 2 onward through the remainder of the study, male 10 and immature 36 maintained consistently high rates of interaction (see Fig. 4-6). (Alexander, 1970; Hasegawa and Hiraiwa, 1980; and Taylor et al., 1978 also observed intense male-young interactions with orphaned infants.)

The nature of this relationship was further revealed when the distribution of immature 36's interactions was considered. A statistically significant difference among the males in their rate of reception of interactions from immature 36 was observed. Clearly, immature 36 was directing her interactions preferentially toward adult male 10 and, unlike immature 39, demonstrated a symmetrical relationship with adult male 10.

To put the relationship between male 10 and immature 36 into a broader perspective, the total male interaction with each immature was examined. Interestingly, there were significant differences among the immatures, with immature 36 exceeding other immatures in rate of interaction with all males. It should be emphasized that although the relationship in one dyad stands out, all immatures received a considerable number of interactions from all of the males.

Adaptive Significance

The existence of a system of interactions of the magnitude outlined in this paper suggests certain implications for both the adult male and the immature. As described in this paper, adult male interactions with immatures may take several different forms. As such, adult males may have multiple roles in the socialization and development of young (Rowell, 1975). Two distinctly different types of interactions have been observed in the study group. On one hand, an adult male formed a close and intense association with an immature after the death of the immature's mother, while, on the other hand, an adult male associated with an immature as a byproduct of an ongoing relationship with the immature's mother. In either case, it may be argued that the net result was similar; the interactions with the adult may promote the welfare of the infant.

As Rosenblum and Coe (1977) point out, the socializing influence of a particular agent may not be a direct corollary of the frequency of interaction with that agent. A relatively infrequent experience with some class of individuals, such as adult males, may have a more significant impact on socialization than more frequent experiences with other classes of individuals. Infants may have received a different type of experience from the combined interactions with four adult males than the intense, prolonged interactions with a single adult female—mother.

The proximity of an adult male to an immature may have important consequences for predator protection under some circumstances and may also attenuate roughness from other group members (Hendy-Neely and Rhine, 1977). Bernstein (1976) noted that although male care of infants varies widely across species, in most cases, males are responsive to situations requiring the protection of infants in distress. Itani (1959) observed that among Japanese macaques, other group members treat an infant just as they do the adult male protector beside it. For example, an adult female does not take a choice piece of food from an infant when the infant is near an adult male. When situations like this have occurred several times, "the infant grows very strong willed . . ."; infants protected by adult males for a long time have "become more domineering than other infants of the same age" (Itani, 1959, p. 83). This male protector relationship was observed in the present study, and its ramifications in the socialization of one immature in particular will be a focus of continued investigation.

Adult males may also accrue advantages from the close proximity of an immature. Males may associate closely with an immature to stop attacks or to decrease the likelihood of aggression exhibited toward them (termed agonistic buffering; Deag and Crook, 1971; Gilmore, 1977; Gouzoules, 1975; Kummer, 1967; Nash and Ransom, 1971; Rowell, 1967; Stoltz, 1972; Stoltz and Saayman, 1970), to increase the likelihood that they will acquire or maintain high rank in the group (Deag and Crook, 1971; Nash, 1973; Russell and Russell, 1971), or to increase their own integration into the group (Itani, 1959; Poirier, 1969).Through their exploratory behavior, young develop new ways of interacting with the environment which can be transmitted to adult males. Itani (1958) described the probable routes of acquisition of a new food habit in *Macaca fuscata* and found that adult males could learn a new, and possibly beneficial, food habit from infants or from younger siblings.

It can also be argued, however, that male-immature interactions are merely the byproduct of the social nature of primates and are not, by themselves, adaptive. Rowell (1979) suggested that the variance in the expression of male-immature interactions can be explained as an example of cultural drift, with no adaptive significance under present conditions. Perhaps these interactions represent selection for tolerable social conditions and not the optimal, when intertwined with other features of the social organization. Alternatively, the interactions observed in this captive group may reflect the potential or capacity for parental care (Redican, 1978) not typically expressed by individuals in their undisturbed, free-ranging state.

CONCLUSION

We have demonstrated in this study that adult male stumptail macaques do not distribute their interactions within the group in a random manner that could be predicted based simply on the demography of the group. Adult males initiate more interactions to immatures than would be theoretically expected. Differences exist among the adult males in the amount of interactions they have with immatures, and certain males may form "special relationships" with immatures. Benefit to both the immatures and the adult males may accrue from these interactions and, as such, promote and foster the continuation of the bonds.

ACKNOWLEDGMENTS

This research was supported by U.S. Public Health Service Grants DA-02128, RR-00165, and RR-00167 (Division of Research Resources, National Institutes of Health). The authors wish to thank Dr. D. M. Taub for helpful discussions of the data presented in this chapter. We wish to thank Drs. P. J. Brown and J. F. Dahl for comments on an earlier version of this manuscript. Also, we thank F. Kiernan, G. Mason, and P. Plant for assistance with the manuscript and illustrations.

REFERENCES

Alexander, B.K. 1970. Parental behavior of adult male Japanese macaques. *Behaviour*, 36: 270-85.

Altmann, J. 1974. Observational study of behavior: Sampling methods. *Behaviour* 49: 227-65.

———. 1978. Infant independence in yellow baboons. In G. M. Burghardt, and M. Bekoff (eds.), *The Development of Behavior: Comparative and Evolutionary Aspects*, pp. 253-277. New York: Garland Publishing Company.

———. 1980. *Baboon Mothers and Infants*. Cambridge: Harvard University Press.

Altmann, S. A., and Altmann, J. 1977. On the analysis of rates of behaviour. *Animal Behaviour* 25: 364-72.

Bernstein, I. S. 1970. "Paternal" behavior in nonhuman primates. *American Zoologist* 10: 480.

———. 1976. Dominance, aggression and reproduction in primate societies. *Journal of Theoretical Biology* 60: 459-72.

Bertrand, M. 1969. The behavioral repertoire of the stumptailed macaque. A descriptive and comparative study. *Bibliotheca Primatologica* 11: 1-273.

Bishop, Y. M. M.; Fienberg, S. E.; and Holland, P. W. 1975. *Discrete Multivariate Analysis: Theory and Practice*. Cambridge: MIT Press.

Brandt, E. M.; Irons, R.; and Mitchell, G. 1970. Paternalistic behavior in four species of macaque. *Brain Behavior and Evolution* 3: 415-20.

Burton, F. D. 1972. The integration of biology and behavior in the socialization of *Macaca sylvana* of Gibraltar. In F. E. Poirier (ed.), *Primate Socialization*, pp. 29-62. New York: Random House.

Chevalier-Skolnikoff, S. 1974. The ontogeny of communication in the stumptail macaque (*Macaca arctoides*). *Contemporary Primatology* 2: 1-174.

Deag, J. M., and Crook, J. H. 1971. Social behavior and 'agonistic buffering' in the wild Barbary macaque *Macaca sylvanus* L. *Folia Primatologica* 15: 183-200.

Estrada, A. 1977. A study of the social relationships in a free-ranging troop of stumptail macaques (*Macaca arctoides*). *Boletin de Estudios Medicos y Biologicos*, 29: 313-94.

Estrada, A.; Estrada, R.; and Ervin, F. 1977. Establishment of a free-ranging colony of stumptail macaques (*Macaca arctoides*): Social relations I. *Primates* 18: 647-76.

Estrada, A., and Sandoval, J. M. 1977. Social relations in a free-ranging troop of stumptail macaques (*Macaca arctoides*): Male care behaviour I. *Primates* 18: 793-813.

Fagen, R. M., and Mankovich, N. J. 1980. Two-act transitions, partitioned contingency tables, and the 'significant cells' problem. *Animal Behaviour* 28: 1017-23.

Gilmore, H. B. 1977. The evolution of agonistic buffering in baboons and macaques. Paper presented at the Forty-sixth Annual Meeting of the *American Association of Physical Anthropologists*, Seattle, Washington.

Gouzoules, H. 1975. Maternal rank and early social interactions of infant stumptail macaques, *Macaca arctoides. Primates* 16: 405-18.

Hasegawa, T., and Hiraiwa, M. 1980. Social interactions of orphans observed in a free-ranging troop of Japanese monkeys. *Folia Primatologica* 33: 129-58.

Hendy-Neely, H., and Rhine, R. J. 1977. Social development of stumptail macaques *(Macaca arctoides):* Momentary touching and other interactions with adult males during the infant's first 60 days of life. *Primates* 19: 589-600.

Hrdy, S. B. 1976. Care and exploitation of non-human primate infants by conspecifics other than the mother. In J. S. Rosenblatt; R. A. Hinde; E. Shaw; and C. Beer (eds.), *Advances in the Study of Behavior,* vol. 6, pp. 101-58. New York: Academic Press.

Itani, J. 1958. On the acquisition and propagation of a new food habit in the natural group of the Japanese monkey at Takasaki-Yama. *Primates* 1: 84-98.

———. 1959. Paternal care in the wild Japanese monkey, *Macaca fuscata fuscata. Primates* 2: 61-93.

Jones, N. G. B., and Trollope, J. 1968. Social behavior of stump-tailed macaques in captivity. *Primates* 9: 365-94.

Kummer, H. 1967. Tripartite relations in hamadryas baboons. In S. A. Altmann (ed.), *Social Communication among Primates,* pp. 63-71. Chicago: University of Chicago Press.

Lindburg, D. G. 1971. The rhesus monkey in North India: an ecological and behavioral study. In L. A. Rosenblum (ed.), *Primate Behavior: Developments in Field and Laboratory Research,* vol. 2, pp. 1-106. New York: Academic Press.

Makwana, S. C. 1978. Paternal behaviour of the rhesus macaque, *Macaca mulatta,* in nature. *Journal of the Bombay Natural History Society* 75: 475-76.

Mitchell, G. D. 1969. Paternalistic behavior in primates. *Psychological Bulletin* 71: 399-417.

———. 1977. Parental behavior in nonhuman primates. In J. Money, and H. Musaph (eds.), *Handbook of Sexology,* pp. 749-59. Amsterdam: Elsevier.

Nash, L. T. 1973. Social behavior and social development in baboons *(Papio anubis)* at the Gombe Stream National Park, Tanzania. Ph.D. dissertation, University of California, Berkeley.

Nash, L. T., and Ransom, T. W. 1971. Socialization in baboons at the Gombe Stream National Park, Tanzania. Paper presented at the *17th Annual Meeting of the American Anthropological Association,* New York.

Peffer-Smith, P. G. 1978. Adult male-immature interactions in a captive stumptail macaque *(Macaca arctoides)* group. M.A. thesis, Emory University, Atlanta.

Poirier, F. E. 1969. The Nilgiri langur troop: Its composition, structure, function, and change. *Folia Primatologica* 10: 20-47.

Ransom, T. W., and Ransom, B. S. 1971. Adult male-infant relations among baboons *(Papio anubis). Folia Primatologica* 16: 179-95.

Redican, W. K. 1978. Adult male-infant relations in captive rhesus monkeys. In D. J. Chivers, and J. Herbert (eds.), *Recent Advances in Primatology,* Vol. 1: *Behaviour,* pp. 165-67. New York: Academic Press.

Redican, W. K., and Taub, D. M. 1981. Adult male-infant interactions in nonhuman primates. In M. E. Lamb (ed.), *The Role of the Father in Child Development,* 2nd ed., pp. 203-58. New York: John Wiley and Sons.

Rhine, R. J., and Hendy-Neely, H., 1978. Social development of stumptail macaques *(Macaca arctoides):* Momentary touching, play, and other interactions with aunts and immatures during the infant's first 60 days of life. *Primates* 19: 115-23.

Rosenblum, L. A. 1978. The creation of a behavioral taxonomy. In G. P. Sackett (ed.), *Observing Behavior,* vol. II: *Data Collection and Analysis Methods,* pp. 15-24. Baltimore: University Park Press.

Rosenblum, L. A., and Coe, C. L. 1977. The influence of social structure on squirrel monkey socialization. In S. Chevalier-Skolnikoff, and F. E. Poirier (eds.), *Primate Bio-Social Develop-*

ment: Biological, Social, and Ecological Determinants, pp. 479-99. New York: Garland Publishing Company.

Rowell, T. E. 1967. Quantitative comparison of the behaviour of a wild and a caged baboon group. *Animal Behaviour* 15: 499-509.

———. 1975. Growing up in a monkey group. *Ethos* 3: 113-28.

———. 1979. How would we know if social organization were not adaptive? In I. S. Bernstein, and E. O. Smith (eds.), *Primate Ecology and Human Origins: Ecological Influences on Social Organization*, pp. 1-22. New York: Garland Publishing Company.

Russell, C., and Russell, W. M. S. 1971. Primate male behaviour and its human analogues. *Impact of Science on Society* 21: 63-74.

Seyfarth, R. M. 1978. Social relationships among adult male and female baboons. II. Behaviour throughout the female reproductive cycle. *Behaviour* 64: 227-47.

Siegel, S. 1956. *Nonparametric Statistics for the Behavioral Sciences*. New York: McGraw-Hill.

Smith, E. O., and Begeman, M. L. 1980. BORES: behavior observation recording and editing system. *Behavior Research Methods and Instrumentation* 12: 1-7.

Smith, E. O., and Peffer-Smith, P. G. Unpublished manuscript. Behavioral inventory of the stumptail macaque (*Macaca arctoides*, I. Geoffroy, 1831).

Southwick, C. H.; Beg, M. A.; and Siddiqi, M. R. 1965. Rhesus monkeys in North India. In I. DeVore (ed.), *Primate Behavior: Field Studies of Monkeys and Apes*, pp. 111-59. New York: Holt, Rinehart and Winston.

Spencer-Booth, Y. 1970. The relationship between mammalian young and conspecifics other than mother and peers: A review. In D. S. Lehrman; R. A. Hinde; and E. Shaw (eds.), *Advances in the Study of Behavior* vol. 3, pp. 119-94. New York: Academic Press.

Stoltz, L. P. 1972. The size, composition and fissioning of baboon troops (*Papio ursinus* Kerr, 1792). *Zoologica Africana* 7: 367-78.

Stoltz, L. P., and Saayman, G. S. 1970. Ecology and behavior of baboons in the Northern Transvaal. *Annals of the Transvaal Museum* 26: 99-143.

Taub, D. M. 1980. Testing the 'agonistic buffering' hypothesis. I. The dynamics of participation in the triadic interaction. *Behavioral Ecology and Sociobiology* 6: 187-97.

Taylor, H.; Teas, J.; Richie, T.; Southwick, C.; and Shrestha, R. 1978. Social interactions between adult male and infant rhesus monkeys in Nepal. *Primates* 19: 343-51.

Whiten, A., and Rumsey, T. J. 1973. "Agonistic buffering" in the wild Barbary macaque, *Macaca sylvana L. Primates* 14: 421-25.

Wilson, E. O. 1975. *Sociobiology: The New Synthesis*. Cambridge: Harvard University Press.

5

Free-Living Rhesus Monkeys: Adult Male Interactions with Infants and Juveniles

Stephen H. Vessey
Department of Biological Sciences
Bowling Green State University
Bowling Green, Ohio

Douglas B. Meikle
School of Interdisciplinary Studies
Miami University
Oxford, Ohio

INTRODUCTION

Rhesus monkeys *(Macaca mulatta)* might be considered "typical" mammals in that males invest little in offspring compared to females and that male-infant interactions are infrequent (Taylor et al., 1978). On the other hand, Suomi (1977) and Redican (1975) have shown in laboratory studies that males have the potential to play an important role in the social development of infants. In this chapter we shall present quantitative data on the type and frequency of male-infant interactions among free-living rhesus monkeys and consider several hypotheses concerning the possible adaptive significance of such interactions.

Trivers (1972) suggested that the optimum number of offspring for males is usually higher than for females. Since females by definition produce larger gametes (eggs), they have a greater initial investment in each offspring and therefore can produce fewer of them. Males, on the other hand, invest little in each mating effort and are likely to seek more than one mate to attain their optimum number of offspring.Thus polygyny, where males compete to monopolize access to females, should be the general trend (Emlen and Oring, 1977).

Mammals are further predisposed toward polygyny because the female is solely responsible for the developing embryo through gestation and for its nourishment after birth (Orians, 1969). Many of the Cercopithecine primates, including macaques, are semiterrestrial, generalist feeders, utilizing resources that are of relatively low quality but seasonally abundant. This resource distribution, coupled with the risk of predation, favors living in larger groups and a polygynous mating system (reviewed in Wittenberger, 1981).

Rhesus macaques live in multi-male groups that show a pronounced dominance hierarchy, both among and within groups. Females remain within the group for life, whereas most males join a new group, at, or shortly after, puberty (Drickamer and Vessey, 1973). Females have several estrous cycles during a discrete mating season; each may last a week or more, during which time the female consorts with several males in succession, giving the males little confidence of paternity. Because of the absence of extended pair bonding, rhesus monkeys might be considered promiscuous, but males compete for access to females and show other characteristics associated with polygyny: sexual dimorphism is substantial, with males weighing about 50% more than females (Napier and Napier, 1967) and having much larger canines (Harvey et al., 1978). Males engage in fights during the mating season, suffer wounds, and as adults have higher mortality rates than females (Drickamer, 1974). Most importantly, the variance in reproductive success is at least five times greater for males than for females (Meikle, 1980).

The above data suggest that males should show little paternalistic behavior or other types of interactions with infants (Trivers, 1972; Alexander et al., 1979). However, at least one aspect of rhesus macaque social organization suggests that there may be some parental investment by males: although the mating season lasts only about five months, most adult males stay in the group year round. High-ranking males are active in group defense and intervene in fights, playing a "control role" (Bernstein and Sharpe, 1966; Vessey, 1971). We also know that these latter males have highest seniority, i.e, years in group (Drickamer and Vessey, 1973), and this observation, coupled with the positive correlation between rank and reproductive success (Smith, 1980), suggests that males may be investing more than sperm in their offspring.

The purposes of this study were (1) to quantify the amount and types of male-infant interactions; (2) to place these in perspective with the other social interactions of infants; and (3) to test the following alternative hypotheses about the possible adaptive significance of such male-infant interactions.

I. Male-infant interactions are phenotypically selfish for the male. Males may become proximate to or carry infants in order to reduce the frequency and consequences of agonistic interactions with others ("agonistic buffering" Deag and Crook, 1971). Also, males may approach and groom infants as a means of promoting social integration with adult females. For both of the above cases we would predict that the least senior and lowest-ranking males should most frequently interact with infants.

II. Male-infant interactions are parental on the part of the male. Such behavior would be phenotypically altruistic but genotypically selfish. Here we would predict that the most senior and highest-ranking males would interact most frequently with infants and that those interactions should be protective or otherwise positive.

III. Male-infant interactions are based on inclusive fitness other than parent-offspring. Such behavior would also be phenotypically altruistic but genotypically selfish. Males still in their natal group should have the highest rates of interaction since many of the infants are their relatives. Among nonnatal males, those with brothers or other male relatives in the same group should interact more frequently with infants than those males without relatives, since the former are more likely to have infant relatives in the group.

IV. Interactions with adult males are important to the development of infants. Here we predict that infants should more frequently initiate interactions with males and should groom them more than the reverse. Female infants may gain experience useful later in life when they choose mates. Infant males may develop skills in interacting with males within the relative safety of their natal group where they receive agonistic aids from relatives (Kaplan, 1978). Male mortality increases markedly after sexual maturity, when males leave their natal groups and engage in escalated fights with other males (Drickamer, 1974).

SUBJECTS AND METHODS

Observations were conducted at the La Parguera facility of the Caribbean Primate Research Center in Puerto Rico. Monkeys were introduced onto two 80-acre islands in 1962. Food hoppers provided with commercial monkey pellets and water were available at several locations. Each monkey had its birth date and mother's identity recorded and was individually marked at about one year of age; its group affiliations were tracked throughout its life. For a history of the early stages of the colony see Vandenbergh (1967).

The data reported here were collected in two phases. Focal animal observations were conducted on 15 infants born in 1967 and 1968 among three social groups. The group of infants was divided as equally as possible with respect to sex, to mother's rank, and to maternal parity. Each infant was observed for one 30-minute period per week for the first year of life (26 hours each infant). During point samples taken each minute, the infant's distance from or type of contact with the mother, mother's activity, and behavior of monkeys less than one meter from the infant were recorded. Social interactions of the infant during the one-minute interval between point samples were also recorded.

The second set of data is from focal animal observations of 27 adult males older than 32 months of age that had transferred from their natal groups to 1 of 4 other social groups. From February 1978 through February 1979 each subject was observed for 51 periods, 20 minutes each (17 hours per subject). During each observation period, the duration of the subject's grooming and proximity (less than 2m.) to other animals, direction of his approachers, the nature of the interaction (e.g., agonistic encounter), and the identity of the interactor were recorded. In the analysis of these data we consider interaction between the subjects and immature animals that were up to 32 months old (i.e., those born in 1976, 1977, and 1978).

RESULTS AND DISCUSSION

Focal Observations of Infants

These data provide unbiased perspective on the sex and age of monkeys with which infants interact in the first year of life. The time spent in ventral contact with the mother declined rapidly from almost 100% in the first week of life to less than 50% by 10 weeks, and then more slowly to nearly zero by the end of the first year (Figure 5-1). Sex of the infant, rank, or parity of the mother did not significantly affect the results, which are similar to those reported by Berman (1980).

Ventral carrying of infants by monkeys other than the mother was relatively rare, reaching 1.9% of one-minute intervals during the first 10 weeks, then declining to near zero by 27 weeks (Table 5-1). Most of these cases involved carrying of the female infant of the alpha female by the 3-year-old nulliparous daughter of a high-ranking female. No instances of carrying by adult males were recorded and only three cases by juvenile (two- to three-year-old) males. Besides the mothers, it was generally nulliparous females, two to three years old, that carried infants.

Handling or otherwise manipulating infants also was done mostly by young females, primarily during the first 10 weeks of the infants' lives (Table 5-2). Interestingly it was nonrelatives who carried and handled infants more often than relatives (Tables 5-1 and 5-2). No adult males were seen handling infants during the focal infant samples. Only two instances of grooming of infants by adult males were recorded (1% of the total grooming by nonmothers).

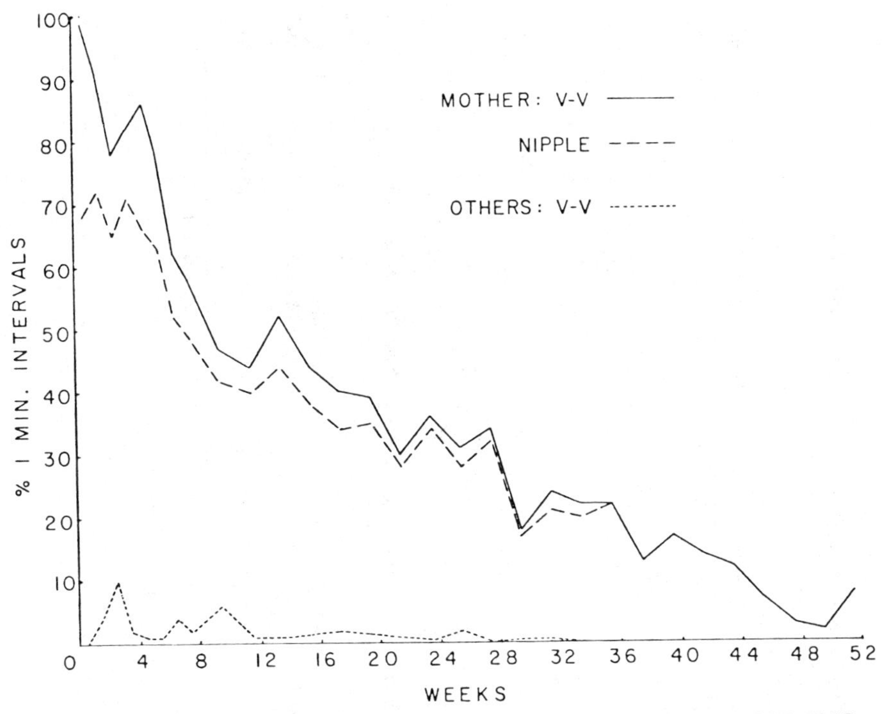

Fig. 5-1. Mean ventral clinging and nipple contact scores for 15 infants, 1967-1969. "Others" refers to females that were not the mothers but that carried subjects.

Aggression by males toward infants was also very rare; of only six instances, four occurred while the aggressor male was in consort with the infant's mother. No instances of aiding infants during agonistic encounters were noted during focal observations.

Infants played primarily with other infants of the same sex, less often with juveniles, and not at all with adults of either sex (Table 5-3). Juvenile-infant play was highest between older male infants and male juveniles, but was also relatively high between older female infants and female juveniles.

In 2.6% of the point samples, an adult male was within 1 m. of a subject infant. For more than half (59%) of these instances, the proximity was due to the male's interaction with the infant's mother during sexual consorts when the male followed, groomed, and copulated with the mother.

In sum, the results of focal infant observations suggest that direct interactions between infants and adult males make up only a fraction of 1% of the total. The decrease in proximity between infants and their mothers during the first year (Fig. 5-1) was followed by a concomitant increase in the time that 12- to 32-month-old animals spent within 2 m. of the focal male

Table 5-1. Carrying of Infant by Monkeys Other Than the Mother, 1967-69.

	Relatives				Nonrelatives			
	Juvenile		Adult		Juvenile		Adult	
	Male	Female	Male	Female	Male	Female	Male	Female
Total Carries	1	25	N.A.	0	2	87	0	15
Percent of 1-Minute Intervals	1	19	N.A.	0	2	67	0	11

Weeks	Total Carries	% of 1-Minute Intervals
1-10	84	1.9
11-26	41	0.7
1-26	125	1.2
27-52	5	0.1

N.A. means none available.
Weeks denote age of infant.

Table 5-2. Handling of Infant by Monkeys Other Than the Mother, 1967-69.

	Relatives				Nonrelatives			
	Juvenile		Adult		Juvenile		Adult	
	Male	Female	Male	Female	Male	Female	Male	Female
Total Handles	3	66	N.A.	0	9	84	0	87
Percent of 1-minute Intervals	1	26	N.A.	0	4	34	0	35

Weeks	Total Handles	% of 1-minute Intervals
1-10	131	3.0
11-26	94	1.5
1-26	225	2.1
27-52	24	0.3

Table 5-3. Wrestling between Subject Infants and Nonrelatives Per Hour of Observation, 1967-69, Corrected for Monkeys Available.

	Play Partners			
Subject	Infants		Juveniles	
Age in Weeks	Male	Female	Male	Female
Females				
9-30	0.66	0.94	0.02	0.07
31-52	0.32	1.02	0.02	0.21
Males				
9-30	0.70	0.38	0.04	0.06
31-52	2.39	0.32	0.26	0.10

subjects in the second year (Table 5-4). In the following analysis, we utilize focal observations of both infants and adult males, as well as ad libitum field observations where appropriate, to test alternative hypotheses regarding the possible adaptive significance of interaction between adult males and immatures.

Hypotheses

Hypothesis I, that male-infant interactions are phenotypically selfish on the part of the male, is rejected on the following grounds. During the 459 hours of focal observations of adult males there were no carries of infants associated with any of the more than 650 agonistic encounters involving those males. Only one carry was recorded (while the male was grooming the infant's mother).

There was no indication that low-ranking males associated with infants more than did high-ranking males so as to "buffer" their involvement in agonistic interactions. In fact, there was a nonsignificant trend toward dominant males' associating with infants more than did subordinate males. Taylor et al. (1978) and Lindburg (1971) also reported that the greatest interaction rates were with dominant males.

There was no evidence that males of relatively short tenure and low rank initiated interactions with immatures in order to establish proximity to their mothers and facilitate their own integration into the group. On the contrary, as stated above, males of relatively high rank were more frequently near immatures. Furthermore, 25 of the 27 subjects ($P < 0.01$, sign test) were more frequently approached by immatures ($\bar{x} = 32.7$ approaches per subject) than they approached immatures ($\bar{x} = 7.2$, Table 5-4). Also, immatures groomed the subjects about four times more frequently than subjects groomed immatures (1.5% versus 0.4% of total time, respectively). However, it may be

Table 5-4. Average Number of Approaches and Duration within 2 m. Between Infants and Focal Males During 1978-79 Observations.

Infant Birth Year	Infant Approach	Subject Approach	Minutes within 2 m.
1978	0.88	0	2.75
1977	17.14	3.15	63.12
1976	14.66	4.07	50.36
Total	32.74	7.22	

that very peripheral males who have just begun to associate with the group initiate interactions with immatures more frequently than vice versa. If so, then this sample of 27 males would not be appropriate for testing this alternative hypothesis since all of the subjects had tenure of at least eight months and were not the most peripheral males in their groups.

If adult male interactions with infants are parental (Hypothesis II), then males with more offspring should accrue more time in proximity to infants than do those with few offspring. If we assume that males of relatively long tenure and high rank have high reproductive success (Smith, 1980), then a positive relationship should exist between the tenure and rank of males and their interaction rates with immatures. Dominance rank was positively but not significantly correlated with either number of infants within 2 m. or the total minutes infants spent within 2 m. (Table 5-5). However, seniority (time in group) of adult males *was* positively correlated with number of infants and time within 2 m. of infants (Spearman rho = 0.486 and 0. 527, respectively, $P < 0.005$) (Table 5-5). Seniority was also highly correlated with rank (Spearman rho = 0.81, $P < 0.001$), a relationship also found by Drickamer and Vessey (1973). Males that have spent many years in one group could be expected to have fathered more offspring than those who have recently joined, and, therefore, this finding is consistent with the notion that male-infant interactions are parental.

Several types of interactions that could be construed as parental were recorded in field notes. High-ranking, central males occasionally intervened in fights on behalf of infants. Most often, an adult female bit or chased an infant of a lower-ranking female; the male would then chase the higher-ranking attacking female. In one instance, an adult male (161) "adopted" a 6-month-old female infant (L6) whose mother had died one week earlier. This middle-ranking male groomed and huddled with the infant until it was more than one year old. During a four-month period, male 161 was with 1 m. of infant L6 for 7% of the point samples (32/459). Only the score between L6 and her older sister was higher. On several occasions one of the alpha males

Table 5-5. Focal Male Subjects for 1978-79 Listed by Group in Decreasing Order of Dominance Rank, with Years in Group, Minutes Spent within 2 m. of Immatures, and Number of Immatures Observed within 2 m.

Group	Subject*	Years in Group	Minutes within 2 m.	No. Infants within 2 m.
A	N2	5.5	170.5	24
	4G	3.5	271.8	31
	5C[a]	3.0	221.0	21
	Y2	2.5	76.3	15
	V7	2.5	206.0	18
	9I[a]	2.0	362.0	18
C	E4[b]	9.5	150.0	19
	I2	7.5	117.0	20
	08[b]	3.5	165.0	21
	1D[b]	2.5	88.5	18
	7P[b]	1.5	168.8	15
	2T[b]	1.5	278.0	25
	5K	1.5	81.0	7
	5A	1.0	59.0	16
	8Y	0.7	91.0	19
E	8F	3.5	158.3	24
	5P	2.0	67.0	18
	7Z[c]	1.5	44.0	10
	CV7[c]	1.0	0.3	1
	AC9	1.0	9.3	5
I	V5	3.5	70.8	13
	X0[d]	3.0	60.5	6
	2L[d]	2.0	12.3	7
	1L[e]	1.5	47.5	9
	8G	2.5	41.3	8
	7S[e]	1.5	29.3	6
	W8	1.0	32.5	8

*Males with same superscript are maternally related half-siblings.

(140) carried the female infant of B7, a primiparous female. Female B7 had a habit of leaving her infant behind while moving with the group when the infant was only three months of age. On four occasions, male 140 retrieved the crying infant and carried it off in the direction in which the rest of the group was headed. Taylor et al. (1978) describe two instances of an alpha male's adopting a neonate after the mother had died, but the infant was too young to feed itself and died also. It may be significant that in all these cases males aided, or attempted to aid, female infants. Berenstain et al. (1981) demonstrated in a captive group of rhesus monkeys that offspring associated more with their fathers than with other males.

There seems to be a tendency for male care to occur in the nonbreeding season, which is also when the males' testes are in a regressed state. Wilson and Vessey (1968) noted that some castrated males showed greater interest in infants than did intact males. Protection of infants was evident on several occasions as unrestrained dominant males attacked and bit humans. In all but one case, the humans were handling infants of that male's group. The remaining attack occurred when one of us (S.H.V.) was carrying an anesthetized adult female out of the male's group.

In a recent report Busse and Hamilton (1981) conclude that male baboons *(Papio ursinus)* carry their own offspring to protect them against intruding males. However, parental investment can be much more subtle than direct aid and nourishment of infants. By remaining in the group year round and defending the group against outside threats, high-ranking males may be investing in their offspring. At this point it is impossible to reject Hypothesis II.

We briefly examine a less parsimonious (Williams, 1966) explanation, Hypothesis III, that males interact with immatures because they are apt to be nondescendent kin. Fifteen of the 27 subjects listed in Table 5-5 had at least one maternally (and perhaps paternally) related brother in their social group. Hence, many of the immature animals that those males interacted with may have been their nephews and nieces.

There was no evidence that nonnatal males in the same group as a brother were in proximity to more infants or for more time than those without relatives in the group (Table 5-6). In group C (Table 5-5) half the adult nonnatal males were brothers and it is likely that many of the infants were sired by these five, yet there was no evidence that they showed more interest in infants than the other males in the group did.

Studies of the care of young usually emphasize the costs and benefits to the adults, but it seems likely from the above data that many interactions are essentially neutral for the male but are of benefit to the immature animal (Hypothesis IV). As we have already shown (Table 5-4), immatures approached males much more frequently than the reverse and groomed them more frequently as well. Furthermore, male immatures did so more often than did female immatures. Seventeen of the male subjects had more male than female immatures approach them, while only six had more females than males approach (17++, 6−−, 4 ties, $P < 0.05$, sign test). The frequency of approaches to within 2 m. of males was almost twice as high for male immatures as for female immatures (Table 5-7). In addition to the sex difference, it is evident that adult males were not close to immature males as a consequence of being close to their mothers; most such approaches took place in the absence of the mother (Table 5-8).

Table 5-6. Average Number of Infants Observed within 2 m. and Average Years Spent in Group (in Parentheses) for Males with or without One or More Siblings in Their Group in 1978-79. The Longer the Males Had Been in a Group, the More Often Were They Near Infants.

Group	A	C	E	I
Males with Sibs	19.5 (2.5)	19.6 (3.7)	5.5 (1.3)	7.2 (2.0)
No Sibs	22.7 (4.3)	15.5 (2.7)	21.0 (2.2)	10.5 (2.3)

Table 5-7. Average Number of Approaches within 2 m. of Adult Male Subjects by Male and Female Immatures in 1978-79.

	Approaches by Immatures	
Infant's Birth Year	Males	Females
1978	0.50	0.38
1977	11.44	5.70
1976	8.96	5.70
Total	20.90	11.88

Table 5-8. Average Number of Observation Periods in 1978-79 When an Immature Was within 2 m. of a Subject Male and Mother Was or Was Not within 10 m. of Immature. In the Absence of the Mother, Male Immatures More Often Associated with Males.

	With Mother		Without Mother	
Immature's Year of Birth	Male	Female	Male	Female
1977	2.52	2.55	5.37	3.04
1976	1.48	1.52	7.00	2.07
Total	4.00	4.07	12.37	5.11

The fact that immature males approach adult, nonnatal males more frequently than do immature females suggests that the young males benefit more from such interactions. Just before or during sexual maturity, most males leave their natal group (Drickamer and Vessey, 1973) and tend to experience higher mortality (Drickamer, 1974; Meikle, 1980). After leaving his group, a male interacts with unfamiliar males. As he begins to associate with a new social group, he receives aggressive, xenophobic responses from resident adults of both sexes. Previous experience with adult males during the first few years of life may prepare these young males for the challenge of becoming members of a new social group.

Finally, we should consider the possibility that some of the interactions between males and immatures require *no* adaptive explanation. Gould and Lewontin (1979) have argued the dangers of always making up stories to explain the adaptiveness of traits. We know that adult male rhesus monkeys have the potential to invest substantially in their offspring (Suomi, 1977; Redican, 1975). Such potential could exist because male parental investment was adaptive in environments in the past. Alternatively, male care of infants may be merely the result of unusual circumstances, such as the death of the infant's mother or hormonal abnormalities of the males. Sex differences in behavior are known to be influenced by prenatal hormones. Testosterone in the embryo produces species-typical masculine behavior and its absence, feminine behavior (Phoenix, 1974). Presumably, subtle effects on the fetal environment could produce equally subtle effects on later sex-specific behavior patterns.

CONCLUSIONS

When placed in the context of the multitude of individuals that can interact with a rhesus monkey infant, we find that adult male-infant interactions are rare. Nevertheless, given the potential for paternalistic behavior as demonstrated in the laboratory, we seek to understand the significance of those few interactions that do occur.

The notion that males use infants to reduce the likelihood that they will be attacked by others was rejected because we never saw males carry or get close to infants in connection with agonistic interactions. Nor was there indication that males use infants to gain access to females.

Several lines of evidence suggest that males may interact with infants as their fathers. Males with the greatest seniority in the group were most often close to infants. These males could be expected to have sired more of the group's infants than those males that had recently joined. The fact that most males stay in the group year round could be a form of parental investment, as could be cases of males' "adopting" infants, carrying lost infants, and attacking humans who handle infants.

Because most males leave the natal group at or shortly after sexual maturity, there is little opportunity for them, as adults, to care for their siblings. There was no evidence that brothers that had joined the same nonnatal group cared for each other's offspring; thus kin selection seems an unlikely explanation for male care. Some instances of male care for infants may have no adaptive significance, resulting from unusual circumstances or from hormonal abnormalities during development, whereby feminized adult males are produced.

Most of the interactions were initiated by the immatures, suggesting that these young animals may profit from alliances with adult males and develop skills necessary for later reproductive success. Young males, who initiate most of the interactions with adult males, may gain the most since they must enter a new social group to breed.

ACKNOWLEDGMENTS

This study was supported in part by the National Institute of Mental Health (MH 20339-01), Sigma Xi, and Bowling Green State University. The cooperation of the Caribbean Primate Research Center is appreciated. K. Vessey assisted with data analysis and the manuscript.

REFERENCES

Alexander, R. D.; Hoogland, J. L.; Howard, R. D.; Noonan, K. M.; and Sherman, P. W. 1979. Sexual dimorphisms and breeding systems in pinnipeds, ungulates, primates, and humans. In N. A. Chagnon and W. Irons (eds.), *Evolutionary Biology and Human Social Behavior: An Anthropological Perspective*, pp. 402-35. North Scituate, MA: Duxbury Press.

Berman, C. M., 1980. Mother-infant relationships among free-ranging rhesus monkeys on Cayo Santiago: A comparison with captive pairs. *Animal Behaviour* 28: 860-73.

Berenstain, L.; Rodman, P. S.; and Smith, D. G. 1981. Social relations between fathers and offspring in a captive group of rhesus monkeys *(Macaca mulatta). Animal Behaviour* 29: 1057-63.

Bernstein, I. S., and Sharpe, L. G. 1966. Social roles in a rhesus monkey group. *Behaviour* 26: 91-104.

Busse, C., and Hamilton, W. J. 1981. Infant carrying by male chacma baboons. *Science* 212: 1281-83.

Deag, J. M., and Crook, J. H. 1971. Social behaviour and "agonistic buffering" in the wild Barbary macaque, *Macaca sylvana* L. *Folia Primatologica* 15: 183-200.

Drickamer, L. C., 1974. A 10-year summary of reproductive data for free-ranging *Macaca mulatta. Folia Primatologica* 21: 61-80.

Drickamer, L. C., and Vessey, S. H. 1973. Group changing in male free-ranging rhesus monkeys. *Primates* 14: 359-68.

Emlen, S. T., and Oring, L. W. 1977. Ecology, sexual selection and the evolution of mating systems. *Science* 198: 215-23.

Gould, S. J., and Lewontin, R. C. 1979. The spandrels of San Marco and the Panglossian paradigm: a critique of the adaptationist programme. *Proceedings of the Royal Society of London* B205: 581-98.

Harvey, P. H.; Kavanaugh, M.; and Clutton-Brock, T. H. 1978. Sexual dimorphism in primate teeth. *Journal of Zoology*, London, 186: 475-85.

Kaplan, J. R. 1978. Fight interference and altruism in rhesus mokeys. *Am. J. Phys. Anthrop.* 49: 241-50.

Lindburg, D. C. 1971. The rhesus monkey in North India: An ecological and behavioral study. In L. A. Rosenblum (ed.), *Primate Behavior: Developments in Field and Laboratory Research*, vol. 2. New York: Academic Press.

Meikle, D. B. 1980. Sex ratio and kin selection in rhesus monkeys. Ph.D. dissertation, Bowling Green State University, Ohio.

Napier, J. R., and Napier, P. H. 1967. *A Handbook of Living Primates*. New York: Academic Press.

Orians, G. H. 1969. On the evolution of mating systems in birds and mammals. *American Naturalist* 103: 589-603.

Phoenix, C. H. 1974. Prenatal testosterone in the nonhuman primate and its consequences for behavior. In R. C. Friedman; R. N. Richart; and R. L. Vande Wiele (eds.), *Sex Differences in Behavior.* New York: Wiley.

Redican W. K. 1975. A longitudinal study of behavioral interactions between adult male and infant rhesus monkeys *(Macaca mulatta).* Unpublished doctoral dissertation, University of California, Davis.

Smith, D. G. 1980. Paternity exclusion with six captive groups of rhesus monkeys *(Macaca mulatta). American Journal of Physical Anthropology* 53: 243-49.

Suomi, S. J. 1977. Adult male-infant interactions among monkeys living in nuclear families. *Child Development* 48: 1255-70.

Taylor, H.; Teas, J.; Richie, T.; Southwick, C.; and Shrestha, R. 1978. Social interactions between adult males and infant rhesus monkeys in Nepal. *Primates* 19: 343-51.

Trivers, R. L. 1972. Parental investment and sexual selection. In B. Campbell (ed.), *Sexual Selection and the Descent of Man, 1871-1971,* pp. 136-79. Chicago: Aldine.

Vandenbergh, J. G., 1967. The development of social structure in free-ranging rhesus monkeys. *Behaviour* 29: 179-94.

Vessey, S. H. 1971. Free-ranging rhesus monkeys: Behavioural effects of removal, separation and reintroduction of group members. *Behaviour* 40: 216-27.

Williams, G. C. 1966. *Adaptation and Natural Selection: A Critique of Some Current Evolutionary Thought.* 1968. Princeton, NJ: Princeton University Press.

Wilson, A. P., and Vessey, S. H. 1968. Behavior of free-ranging castrated rhesus monkeys. *Folia Primatologica* 9: 1-14.

Wittenberger, J. F. 1981. *Animal Social Behavior.* Boston: Duxbury Press.

6

Social Relations of Males and Infants in a Troop of Japanese Monkeys: A Consideration of Causal Mechanisms

Harold Gouzoules
The Rockefeller University, New York

INTRODUCTION

Junichiro Itani contributed the first in-depth analysis of the social relations between adult males and infants in a nonhuman primate species. He observed the phenomenon of "paternal care" in the Takasakiyama troop of Japanese monkeys *(Macaca fuscata)* in May 1953; that his classic paper (Itani, 1959) continues to be so universally cited in the ethological literature is testimony to the significance of his findings and of the phenomenon itself.

Itani noted that during the birth season, which at Takasakiyama ran from May through August, some of the adult males of the troop engaged in special relationships with one- and two-year-old juveniles: males carried, groomed, and protected these young monkeys, and this constituted a striking seasonal change in their behavior. Adult males could be differentiated on the basis of age and dominance. Males estimated to be around 20 years old were designated "leaders." Other males, 4 to 10 years old, were considered "peripherals." Each class differed somewhat from the others with reference to the dominance and spatial relationships of its members. Only males 10 years or older were observed to have these special relationships with juveniles, and leader and subleader males accounted for almost all of the incidents recorded.

Less was known about the juveniles that received this attention because individual recognition of these monkeys was not always possible. However, it was clear that males did not select among one-year-olds on the basis of sex;

in the case of two-year-olds, however, females were more frequent recipients of male attention, as were individuals that were particularly "undergrown and weak."

Of special interest was the fact that the pattern of male care described was not common to all troops of Japanese monkeys, and Itani suggested that it represented a "protocultural" difference. Only 3 of 18 troops surveyed manifested this behavior pattern consistently, while it was considered rare in 7 and absent in 8 others.

Alexander (1970) reported adult male behavior in the Oregon troop of Japanese monkeys that was very similar to that described by Itani although dominance did not influence the males' behavior patterns. It was suggested that male parental care was partly "a ramification of a more general change in behavior" of the males that is seasonal and corresponds with the "effects of androgen withdrawal" following the mating season.

The Problems of Causality

Why should male Japanese monkeys display the type of care detailed by Itani and Alexander? Alexander's suggestion that seasonal hormonal changes might be involved provides a possible proximate stimulus for the timing of the behavior. Several attempts to integrate this behavior pattern into a sociobiological (in the sense of Wilson, 1975) framework, i.e., to consider behavior in terms of the postulates and tenets of population genetics, have been offered. Does care of infants by Japanese monkey males increase individual or inclusive fitness?

Wilson (1975, p. 351) noted that "in species characterized by a single male in the group or at least one or a very few dominant males likely to be fathers, the males tend to show an almost maternal solicitude toward infants." Itani's (1959) observations on the care given yearlings during the birth season were cited as evidence. Perhaps disqualifying Japanese monkeys from consideration, however, is the fact that, to date, no studies have indicated that only dominant males father offspring in Japanese monkey troops; in fact, data suggesting that this is not the case have been presented (Eaton, 1974; Fedigan and Gouzoules, 1978).

Wilson (1975, p. 351) also considered "situations where males are less likely to be the fathers, [where] the forms of interactions are different and appear to serve other ends." Itani's observations were cited as illustrating this situation as well, but this time it was the male care performed toward two-year-old females that was called to attention. Hrdy (1976, p. 114) more openly asked the logically seductive question as to whether these young females were more likely to breed with their former caretakers; if this were the case, it would provide for directly enhanced reproductive success and explain, in an ultimate sense, the males' behavior. It is unlikely this scenario is an accurate one. At least in this species, persistent non-mating season affiliative behavior between males and both immature and sexually adult females lessens the probability of mating between the individuals (Alexander, 1970; Enomoto, 1975, 1978; Fedigan and Gouzoules, 1978; Baxter and Fedigan, 1979).

Further sociobiological speculation was offered by Crook (1977; p. 32). Agonistic buffering, the use of infants by adults to inhibit aggression by other adults, might be one explanation for adult male interest in infants. Itani (1959) was cited by Crook as providing evidence in support of this contention. In actual fact, however, Itani (1959) stressed the docile, nonaggressive behavior of males caring for infants (p. 83). Only two males ever attacked other individuals while carrying infants, and in the two described incidents, the "individuals" attacked were human observers that had been perceived as a threat to the infant (p. 85). Although one subleader did improve his status by showing interest in infants, there was no suggestion that aggression was deferred as a result of care for infants; instead, Itani (p. 81) interpreted that male's "gathering of infants around him" as an active effort to identify himself with the central portion of the troop. Citations of Itani concerning "agonistic buffering" in Japanese monkeys would seem to be inappropriate and probably stem from misinterpretations of Deag and Crook's (1971) use of the term "passport" in reference to the behavior of this one individual male (e.g., see Wilson, 1975; Hrdy, 1976).

Pursuing the topic of "exploitation" of infants by Japanese monkey males, Hrdy (1976, p. 117) asserted, again citing Itani (1959), that "closely ranked subleader males sometimes *vie* [emphasis mine] with one another for the first infants born each season. Incidents in which males drop the infant that they are carrying or else pull them about by force were reported." Again, I believe Itani has been misread. The source for the alleged male competition was Itani's statement (1959, p. 72), "We felt there might exist a sort of sense of rivalry"; thus, no overt competition was implied. Also, the care Itani referred to involved the yearlings and two-year-olds of the troop (not neonates), presumably stimulated by the first births of the season. The inferred "nonchalance and self-absorption" claimed by Hrdy (1976, p. 117), based on a single observation described by Itani (1959, p. 72), does not seem merited; in fact, the behavior described could equally well be interpreted as falling well within the bounds of typical mother-infant "conflict."

A still more extreme speculation regarding exploitation of infants by males was offered by Kurland (1977, p. 83) who interpreted his observations of "extreme forms of aggression" directed toward infants by a few males (most of whom were not three years of age and were thus sexually immature) as "nascent infanticide." Kurland argued that such behavior might occur "infants represent competition with other animals in that their very existence entails dilution of the competitor's genes in that generation." That virtually all young males leave their natal troop and thus do not have a reproductive investment in that troop was not related to this "strategy."

The above commentary is offered, not as polemic criticism of previous attempts to interpret the relations between Japanese monkey males and infants, but, instead, to underline the need for more data from additional populations. The present study considers the relations between males and infants of the Arashiyama West troop. Unlike previous studies on this topic, the focus here is on infants under one year of age, the age-class most generally defined as "infant" in the primate literature.

METHODS

The Study Troop

The monkeys of Arashiyama, near Kyoto, Japan, have been studied since 1954, when provisioning of the free-ranging troop began. The population grew with provisioning, from 47 monkeys in 1954, to 125 in 1964, and 163 in 1966, when a fission of the troop into two groups of about equal size occurred in June (Koyama, 1967, 1970). By the spring of 1971, one of the two troops, Arishiyama A, numbering 150, had outgrown the Iwatayama Monkey Sanctuary at Arashiyama and was encroaching on neighboring temples and gardens. A decision was made to move the troop and the monkeys were relocated to a 108-acre (42.9 ha.) enclosure on a ranch about 30 miles (48 km.) north of Laredo, Texas, on 22 February 1972.

The vegetation and climate of the Texas site represent a considerable contrast to the monkeys' native habitat in Japan (see Clark and Mano, 1975; Gouzoules et al., in press). The vegetation of the enclosure can be characterized as south-Texas brushland (sometimes referred to as mesquite-chapparal or mesquite-savannah). Dominant plants include *Opuntia* and *Acacia* species and mesquite trees *(Prosopis glandulosa)*.

A number of native mammalian species have periodically gained entry into the enclosure, including coyotes *(Canis latrans)* and bobcats *(Lynx rufus)*. Bobcat predation upon the monkeys has occurred (Gouzoules et al., 1975). The monkeys respond to snakes (including rattlesnakes, *Crotalus atrox*), but no losses have been attributed to these reptiles.

There were 127 members of the Arashiyama West troop at the start of the 1974 birth season. Table 6-1 provides a breakdown of the age-sex classes present in the troop at that time. The 1974 birth season resulted in 31 infants that served as subjects, 29 of which survived for the duration of the study (one infant that died within the first three days after birth is not shown on Table

Table 6-1. Arashiyama West Age-Sex Composition (Post-Birthseason 1974).

	Adult	Subadult	Juvenile	Infant
	7 Yrs.	4-6 Yrs.	1-3 Yrs.	1 Yr.
Male	9	18	13	12
	5 Yrs.	3-4 Yrs.	1-2 Yrs.	1 Yr.
Female	59	15	12	18

6-1). Although no live births were witnessed, infants were observed from the day of birth through at least 10 months of life. Background information for each infant included knowledge of matrilineal genealogical relations dating back to 1954; the 31 infants were born to females of varying rank and age (Table 6-2), as well as of diverse matrilineal kin networks.

Data Collection

The Arashiyama West troop was studied between May 1973 and May 1975 and from July 1977 to August 1978. Data for statistical analysis in this paper were collected using focal animal sampling techniques (Altmann, 1974). Each of the 1974 infants was observed for 10 ten-minute sessions (5 morning and 5 afternoon) per month for each of the first 10 months of life. During each test-period, the focal individual was observed and its behavior patterns, in terms of frequencies and durations of the behavioral categories employed (see Gouzoules, 1980), were recorded by hand onto a printed data sheet. A standard stopwatch marked the time.

In addition, less systematic observation increased the average time spent observing the troop to about six hours per day, usually six days a week. Statistical analysis concerned only the focal animal data, while descriptions for behavioral patterns and case histories made use of all observations.

RESULTS

A detailed descriptive account of social development during the first 10 months of life for infants of the Arashiyama West troop has been given elsewhere (Gouzoules, 1980), and only those aspects of early social interaction involving males are considered here.

Some individual mothers with infants only several days old spent more time than usual in the vicinity of a high-ranking male of the troop, particularly during foraging. This was noticeable because subgroups within the troop are discernible and quite consistent. The increased spatial proximity was not accompanied by a concomitant increase in other forms of social interaction

Table 6-2. Description of Subjects.

Infant Name	Intra-Mother Rank Months		Intra-Troop* Female Rank Months		Gender***	Birth Order	Mother's Age	Spatial** Status Months		Recurrent Male Care
	1-3	3-10	1-3	3-10				1-3	3-10	
1. R6974	2	3	4	12	1	27	5	1	1	No
2. R606774	28	28	55	56	2	13.5	7	3	3	Alpha male (Dai)
3. R586974	1	2	2	10	2	16	5	1	1	Alpha male (Dai)
4. Ro6974	3	4	5	13	2	28	5	1	1	Alpha male (Dai)
5. Ro6774	4	5	6	14	1	4	7	1	1	No
6. Ro6674	5	6	7	15	1	3	8	1	1	No
7. N6374	9	10	19	27	2	25	11	1	1	No
8. N6274	10	11	20	28	2	6	12	1	1	4th ranking male (Bus62); brother (N6272)
9. N6174	8	9	16	24	2	1	13	1	1	No
10. Ku74	13	14	26	34	1	22.5	19	2	2	No
11. Ku6774	14	15	28	36	1	17	7	2	2	No
12. Me74	15	16	29	37	2	25	20	3	3	No
13. Me6774	16	17	30	38	2	13.5	7	2	2	No
14. Pe6474	11	12	23	31	1	6	10	2	2	Brother (Pe6470)

15. Pe646974	12	13	24	32	1	29	5	2	2	No
16. B16774	23	24	43	46	2	15	7	2	1	No
17. Ran74	30	29	63	60	2	21	18	3	3	2 brothers (Ran68, Ran72)
18. B636874	27	7	52	22	1	8	6	3	2	No
19. B596674+	7	—	15	—	1	19	8	1	1	No
20. B586574	26	1	50	3	2	2	9	3	1	No
21. Deko6574	6	8	14	23	2	6	9	1	1	No
22. M6474	17	18	32	40	2	10	10	2	2	Alpha male (Dai)
23. M646974	18	19	33	41	2	30	5	2	2	No
24. M616774	19	20	36	44	1	25	7	3	3	No
25. M616674	20	21	37	45	2	22.5	8	3	3	No
26. M606874	21	22	40	47	1	20	6	3	3	No
27. M5874	24	26	44	51	2	9	16	3	3	Brother (M5870)
28. M586374	25	27	46	53	2	18	11	3	3	2nd ranking male (Ko59)
29. P6074	22	23	42	49	1	11	14	1	1	No
30. Wa6574	29	25	60	50	2	12	9	2	2	Uncle (Wa70)

+Died at three months of age.

*Intra-troop female rank is the rank of the mother considered among all adult females, while intra-mother rank considers only that subset of females with 1974 infants.

**Key to spatial status: 1 = central; 2 = ordinary; 3 = peripheral.

***Key to gender: 1 = male; 2 = female.

with the males and, after a few days, these females resumed their more typical spatial deployment patterns with reference to the males and the rest of the troop. These temporary postpartum associations did not seem to influence later interactions between males and infants of those females involved.

As early as the second week of life, some infants touched and climbed on their mothers' grooming partners or on monkeys sitting nearby. This pattern continued through the third week of life, during which the less restrictive mothers allowed their infants to venture considerable distances away from them, often more than 30 feet. It was during these excursions that some infants received their first aggression from monkeys other than their mothers. Infants this young were clumsy and frequently wandered into or attempted to climb on monkeys in their vicinities. Although most monkeys, including adult males, at first showed striking fear responses toward three-week-old infants that approached in this fashion, this reaction diminished over the course of several weeks. Infants that continued to approach adult males too closely were swatted or threatened (Fig. 6-1), usually eliciting retrieval by their mothers.

Agonistic and affiliative interaction rates between males and infants varied over the first 10 months of life (Fig. 6-2). Most aggression was directed toward infants during their second and third months of life. This period of peak aggression from adult males seemed to stem from the inability of young, marginally coordinated infants to discriminate among troop members (Gouzoules, 1980). Another lesser peak corresponded to the start of the mating season. Certain males tended to be intolerant of the proximity of an infant (or other kin) of the female with whom they were in consort. Despite this early mating season tendency, there was a general decrease in aggressive interactions as the infants aged ($r_s = -.643$, $n = 10$, $p < .05$).

Affiliative interaction between males and infants peaked at three months of age, but decreased prior to the commencement of the mating season. A second peak, at seven months, was also evident. Differences among males participating at these two times are considered below under determinants in the selection of infants for male care.

Males of the Arashiyama West troop could be classified in terms of their relations with infants into: (1) those ($n = 10$) that directed no affiliative behavior toward infants (although they would, on occasion, tolerate infant proximity); (2) those ($n = 10$) that had transitory affiliative interactions (i.e., were observed to groom or protect an individual infant only infrequently); and (3) those ($n = 9$) that established relatively long-duration relationships with particular infants. Included among the latter males were two nonnatal (immigrated into the troop) adult males and seven natal males (born into the troop). These nine males were involved in 12 different male-infant dyads.

Fig. 6-1. Alpha male punishes a two-month-old infant that approached him. Note that the male holds the infant at arm's length and that the facial expression, which was noticeably prolonged, is exaggerated with an exceptional open-mouth component, suggesting perhaps compensation by the male for the social naiveté of the infant.

Case Accounts of Persistent Relations

Active affiliative infant-directed behavior by an adult male was first observed in the alpha male who groomed a 10-week-old female infant (M6474) that had approached him. During the following two weeks, they were recorded sitting in contact during two focal animal samples and at other times during ad libitum observations. However, it was not until the infant was 33 weeks old that further affinitive behavior between them was again seen. After this

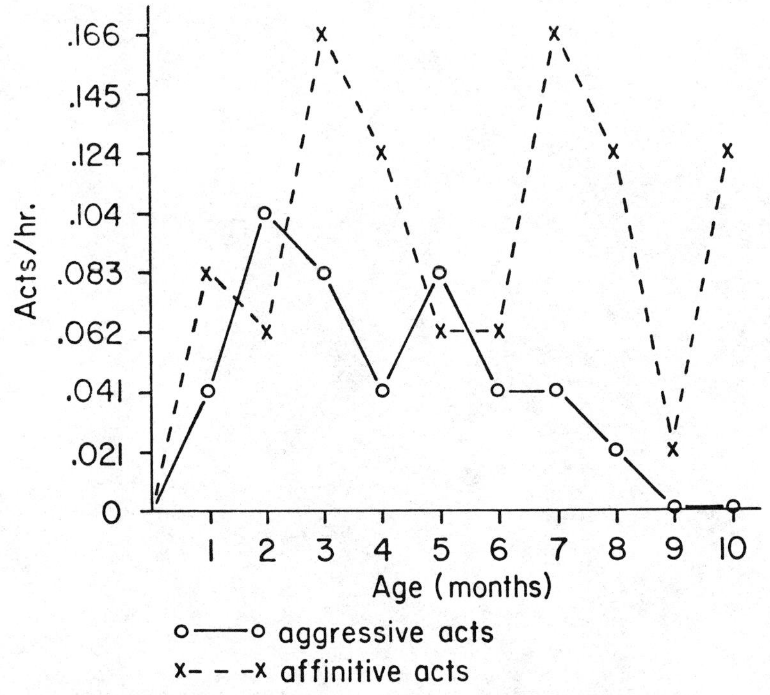

Fig. 6-2. Affinitive and aggressive interaction rates: rate of affiliative and aggressive acts by males toward infants over the first 10 months of life. (Both males and infants are considered as classes.) Plots are derived from infant focal data, which included 48 affiliative and 22 agonistic encounters with males.

recurrence, over the following eight weeks, four incidents were observed in which the alpha male groomed, carried in the ventral position, and aided this infant when it was aggressed (Figs. 6-3 and 6-4). Each case of ventral carriage was preceded by the infant's having given "lost-call" vocalizations while away from its mother.

The alpha male was by far the most active of the troop males regarding care of infants, and he was involved similarly in three other infant relationships. He initiated care of another female infant (Ro6974), nine weeks old; this persisted for a week and included carrying and grooming. He did not interact with her again until she was 32 weeks old, and then his behavior consisted merely of aid during an agonistic encounter (observed during ad libitum observations).

The most intense relationship, in terms of the frequency of affiliative interactions, between the alpha male and an infant began when this infant (R586974) (also a female) was 10 weeks old. This relationship consisted at first of grooming bouts of up to 10 minutes, but two days later, she was observed being carried by this alpha male. This relationship was maintained over the

Fig. 6-3. Alpha male caring for a 10-week-old infant.

next five weeks. The infant was permitted to feed in close proximity to the alpha male during provisioning, was defended by him, groomed by him frequently, and carried on five occasions, including one instance that occurred in response to screams given by the infant after it had fallen from a tree. By the end of September, this relationship had waned, and it was not resumed following the mating season.

The second-ranking male developed a close relationship with another female infant (M586374), 15 weeks of age. He tolerated her proximity during provisioning and aided her so persistently that her ability to dominate other infants during this period eventually exceeded the rank that her mother's position should have permitted. The infant behaved quite aggressively in the presence of the male and even initiated agonistic episodes with older monkeys, including adult females; the male interceded on behalf of the infant whenever she directed enlisting screams at him. While this relationship disappeared

Fig. 6-4. Alpha male with the same infant at age 35 weeks. (Hair length varies seasonally).

during the mating season, it was renewed the following spring and persisted for three weeks. This male was never observed to carry the infant, nor did he ever comfort or retrieve her when she gave "lost-call" vocalizations.

The above accounts provide examples of the longer-duration relationships between adult nonnatal males and specific infants. While other adult males exhibited affiliative behavior toward infants, none established the longer-term associations described for the two highest-ranking males.

No male was ever observed to carry an infant during an agonistic encounter (neither during the focal animal samples for this study, nor at any time during the three years during which I carried out research on the troop). Males did not use infants as agonistic buffers. One male's movement from peripheral to central spatial status was temporally correlated with an increased interest in infants. His change in spatial status did not result in an increase in dominance (agonistic) rank, however. More generally, interaction with infants was not associated with upward mobility of rank status in the Arashiyama West troop.

Other instances of consistent affiliative interaction between particular male-infant pairs involved siblings and are discussed below.

Consideration of Determinants

Paternity. Paternity could not be ascertained with confidence in this study. However, data collected during the 1973-74 mating season allowed some assessment of the likelihood that males selected infants on the basis of paternity. These data were the result of a project investigating factors that influence mating season partner choice and were obtained through systematic observation by two researchers (see Fedigan and Gouzoules, 1978). Thus data concerning the identity of consort partners for males were used to determine whether selection of particular infants for care was related to recent consort relationships. This was of particular interest because our long-term consort data had revealed that on average, a statistically significant one-third of a male's partner selections were similar to those of previous years, i.e., that partner selection showed similarities from year to year, suggesting long-term partner preferences (Fedigan and Gouzoules, 1978). Thus the proximate basis for male discrimination among infants was demonstrable at least in terms of the infants' mothers as sexual partners during the mating season.

The four central males of the Arashiyama West troop are considered in Table 6-3. There was no tendency for these high-ranking males to bias their care in favor of the infants of females with whom they had consorted (Fisher exact test, all p's › .05).

Table 6-3. Relationships between Male Care and Previous Sexual Associations.

	Consorts*		Nonconsorts*		
Male's Rank	Care	No Care	Care	No Care	
1	2	6	4	17	$p = .696$
2	0	5	1	23	$p = .828$
3	0	6	1	22	$p = .793$
4	1	4	0	24	$p = .172$

*Only females that had infants in 1974.
There was no tendency for high-ranking males to bias their care in favor of infants of females with whom they had consorted (Fisher exact test).

Matrilineally Related Males. Itani (1959, p. 65) reported that two- to three-year-old males usually had no interest in infants and "seldom protect them." Among Japanese macaques this age-class is comprised of natal males who consequently have matrilineal kin in the troop. During the present study, 7 of 29 infants had brothers two years of age or older; these 11 brothers ranged in age from two to nine years.

Seven of these 11 males demonstrated care and affiliative interactions with infant maternal half-siblings, including holding, ventral and dorsal carriage, grooming, agonistic aiding, retrieval, and play (Figure 6-5). Of particular interest was the apparent influence of infant age on the occurrence of these interactions. For nonmatrilineal males, 52% of care incidents took

Fig. 6-5. A 4½-year-old male comforts his 7-month-old brother during the mating season.

place during the first 3 months of life, while 48% fell during months 4 through 10. As noted above, there was a decrease in infant interaction among these males as the mating season approached. Natal males, however, performed only 10% of their interactions with infant maternal half-siblings during the first 3 months of life and 90% during months 4 through 10. They accounted for much of the care that resulted in the second peak in affiliative interaction (Fig. 6-2) observed during the seventh month of life. Although infants were observed to interact with related males other than brothers, they did so much less frequently and no such care was recorded during the focal animal samples.

Condition of Infants Selected (Weight). Itani (1959, p. 75) observed that the paternal care given one- and two-year-olds at Takasakiyama was directed preferentially toward "undergrown and weak" individuals.

On 8 February 1975, 14 infants were captured and weighed at the annual capture in Arashiyama West, during which tattooing and routine health checks were performed. The mean age of these infants was 7.8 months and the mean weight was 1537.1 g. (standard deviation = 247.8 g.). There was no correlation between weight and age ($r_s = -.016, p > .05, n = 14$). Neither was there a significant correlation between weight and the frequency of male care received ($r_s = .039, p > .05, n = 14$).

Overall Popularity. At Arashiyama West, incidents of alloparental care were far more commonly performed by individuals other than the sexually mature males of the troop. Other individuals performed a total of 330 incidents, while these males performed only 48 incidents (but demographically composed a smaller fraction of the troop). To determine whether males were selecting infants on the basis of criteria similar to those used by nonmale alloparents, a correlation coefficient was calculated between nonmale incidents of alloparental care received and male care received. This proved nonsignificant, suggesting that infants receiving male care were not additionally those most popular with other alloparents ($r_s = .083, p > .05, n = 29$). (Factors influencing alloparental care are considered in detail elsewhere [Gouzoules, 1980]).

Other Variables. Several other variables were examined as possible determinants of male choice. There was no correlation between maternal dominance rank and the incidents of male care received ($r_s = .154, p > .05, n = 29$). Sex of the infant was not a significant determinant of aggression received from males ($X^2 = 2.06, p > .05, df = 1$, with expected values corrected for the number of male and female infants). Female infants did receive more affiliative behavior from males than did male infants ($X^2 = 7.77, p < .05, df = 1$).

Infants as Initiators. As early as the second month of life, infants began to perform a behavior pattern that was characteristic only of young monkeys. Labelled the "male warble-approach," it consisted of a complex of behaviors given by infants as they approached, often in pairs or groups, an adult male of the troop. Included was a pout face (lips pursed), carriage of the tail in an upright position, and a warblelike vocalization.

Warblelike vocalizations were performed by adults as well and appeared to serve as an affinitive signal, given when a monkey approached another and seemed unsure of what response the approach would elicit. For example, it was often given as a low-ranking female approached a higher-ranking mother and neonate. The combination of tail carriage and facial expression with a warble vocalization was, however, characteristic of young monkeys and was given almost exclusively by them to the males of the troop.

This interaction between infants and males was clearly infant initiated and it was reasoned that perhaps the basis for the observed preference of males for female infants lay in a tendency for female infants to perform this pattern more frequently. However, no sex difference in the performance of "male warble-approach" was detected (months 1-3: $X^2 = 2.99$, $p > .05$, $df = 1$; months 4-6: $X^2 = .385$, $p > .05$, $df = 1$; months 7-10 $X^2 = 1.21$, $p > .05$, $df = 1$), and, therefore, at least on the basis of this behavior pattern, an infant-initiated sex difference in male care could not be substantiated.

DISCUSSION

The picture of male relations with infants that emerged from the Arashiyama West troop was substantially different from that described by Itani (1959) and Alexander (1970) for the Takasakiyama and Oregon troops, respectively. Itani himself took note of the fact that intertroop variability with respect to male relations with infants was characteristic of Japanese monkeys. While Arashiyama West males did not perform the kind of "paternal care" described by Itani, i.e., care of yearlings and two-year-olds during the birth season, distinct care patterns and relatively long-term relationships between some individual males and infants were revealed in the study of infants born during 1974.

It is worth noting that Itani (1959, p. 88) reported that K. Nakajima, who was among the first to study the Arashiyama monkeys, did not observe "distinct paternal behavior." It would be incorrect to conclude on the basis of the results reported in the present paper, however, that the Arashiyama monkeys had acquired patterns of male care over the years since 1957: the two nonnatal males that demonstrated persistent care toward specific individuals in this study did not consistently show this pattern on a year-to-year basis. Although the alpha male had on one previous occasion in 1973 been observed to carry an infant during a troop disturbance caused by visitors to the site, and had once assumed a protective relationship with an orphaned yearling, he had never before shown interest of the kind described in this paper. Nor did he continue this interest in later years. Thus, during his tenure as alpha male (February 1972 to April 1978), he displayed interest in and developed these care relationships with infants during only one year. Our data on his performance during five different mating seasons, including the one preceding his interest in infants, did not indicate a source for this variation (e.g., no great variation in mating activities). The same was true for the second-ranking male, the other nonnatal male to demonstrate persistent care toward a specific 1974 infant.

Considered separately, the data from 1974 are enigmatic. The hypothesis that probable paternity should influence which infants were selected for care by nonnatal males was not supported by the data, although kin-selection theory could more satisfactorily explain the care given by males to their infant maternal half-siblings. However, the problem of trying to fit behavioral phenomena observed during a relatively short time frame (patterns that may or may not prove consistent over long periods) to hypothetical, ultimately derived behavioral strategies should be recognized. It could be argued that the Arashiyama West males are merely showing facultative paternal behavior—male care that is capable of being adaptively varied—similar to that which apparently occurs in the hoary marmot *(Marmota caligata),* studied by Barash (1975). Behavorial scaling (Wilson, 1975, p. 19), i.e., behavioral variation that covaries with, for example, population density, stage of life cycle, or certain parameters of the environment, might thus be proposed to account for the observations. This explanation seems unlikely in this case, however, as the factors most likely to lead to adaptive facultative paternal behavior, namely variation concerning confidence in paternity or variation in mating success on the part of the males involved, could be discounted.

It is not, however, necessary to relegate the described nonnatal male care for infants to idiosyncrasy; I would like to suggest that it may be an epiphenomenon stemming from a more general tendency for "senior" high-ranking Japanese monkey males to adopt infants that have lost their mothers. Recently Hasegawa and Hiraiwa (1980) reported on the social interactions of "orphans" in the Takagoyama-I troop, many members of which had been shot or captured because of troop raids on cultivated fields. Consequently, more than 40 orphans were observed over a four-year period and patterns of care directed toward these orphans were noted. Infants under one year of age that lost their mothers were said to desperately search for a substitute for the mother, and that adult males were likely to respond. Immature close relatives of the orphan, including siblings, also attempted to care for these orphans, but the infants preferred the adult males to their immature kin. (Adult females were generally indifferent to the orphans regardless of whether they had infants of their own or not.) The older a young monkey was when it lost its mother, the less likely it was that a male would care for it.

Why should a young orphan prefer an adult male? There should be no reason for an infant to discriminate *against* males, since it is apparently unlikely that a lactating female would adopt it. Important immediate benefits in terms of increased rank and access to food can be accrued by orphans cared for by adult males (Hasegawa and Hiraiwa, 1980) and even infants under a year old quickly acquire the social skills to make use of male protectors during agonistic squabbles (as described for the second-ranking male and his protected infant in the Arashiyama West troop). As for a high-ranking adult male, there is little cost to the kind of care given: no energetic demands resulting from lactation; little risk in supporting the infant during agonistic encounters that more often than not are against other juveniles or adult females; no interference with mating, since care diminishes during the mating season. Thus, from an ultimate perspective, it is likely that the cost of infant care is sufficiently low for the males to enable them to benefit even when the probability of genes shared is quite small. Proximate benefits to males might include the learning of new behavior patterns from the less conservative infants (e.g. Itani, 1959, p. 83). Perhaps herein lies the basis for the occasional tendency to care for infants with living mothers as well: the small cost to the male weighed against the potentially significant benefits, albiet short-term ones in most cases, for the infants.

Infants can derive significant benefits from associations with males, especially should their mothers die, but also as a result of even short-term relationships. In effect, in the appropriate circumstances, individual infants should exploit the low cost of care given by the males. Given the social structure of the troop, with dominance and spatial hierarchies, not all infants enjoy equal opportunities to establish relations with senior males: access and opportunism are precursors, and with these, even infants under a year old seem to have the necessary social skills to take advantage of the social umbrella potentially offered by a senior male.

ACKNOWLEDGMENTS

All of us who have studied the Arashiyama West troop are very grateful to the Japanese primatologists whose years of work with the monkeys prior to translocation provided us with unparalleled research opportunities. Partial support for my work was provided by NIH Grant RR00167 to the Wisconsin Regional Primate Research Center. S. Gouzoules provided constructive criticism of the manuscript.

POSTSCRIPT

This chapter was completed in November, 1980.

REFERENCES

Alexander, B. K. 1970. Paternal behavior of adult male Japanese monkeys. *Behaviour* 36: 270-85.

Altmann, J. 1974. Observational study of behavior: Sampling methods. *Behaviour* 49: 227-67.

Barash, D. 1975. Ecology of paternal behavior in the hoary marmot *(Marmota caligata):* An evolutionary interpretation. *Journal of Mammalogy* 56: 612-15.

Baxter, J., and Fedigan, L. M. 1979. Grooming and consort partner selection in a troop of Japanese monkeys *(Macaca fuscata). Archives of Sexual Behavior* 8: 445-58.

Clark, T. W., and Mano, T. 1975. Transplantation and adaptation of a troop of Japanese macaques to a Texas brushland habitat. In S. Kondo; M. Kawai, and A. Ehara (eds.), *Contemporary Primatology,* pp. 358-61. Basel: S. Karger.

Crook, J. H. 1977. On the integration of gender strategies in mammalian social systems. In J. S. Rosenblatt, and B. R. Komisaruk (eds.), *Reproductive Behavior and Evolution,* pp.17-38. New York: Plenum.

Deag, J. M., and Crook, J. H. 1971. Social behaviour and 'agonistic buffering' in the wild Barbary macaque, *Macaca sylvana* L. *Folia Primatologica,* 15: 183-200.

Eaton, G. G. 1974. Male dominance and aggression in Japanese macaque reproduction. In W. Montagna, and W. A. Sadler (eds.), *Reproductive Behavior,* pp. 287-98.

Enomoto, T. 1975. The sexual behavior of wild Japanese monkeys: The sexual interaction pattern and the social preference. In S. Kondo; M. Kawai; and A. Ehara (eds.), *Contemporary Primatology,* pp. 275-79. Basel: Karger.

———. 1978. On social preference in sexual behavior of Japanese monkeys *(Macaca fuscata). Journal of Human Evolution* 7: 283-93.

Fedigan, L. M., and Gouzoules, H. 1978. The consort relationship in a troop of Japanese monkeys. In D. J. Chivers, and J. Herbert (eds.), *Recent Advances in Primatology* vol. 1, pp. 493-95. London: Academic Press.

Gouzoules, H. 1980. Biosocial determinants of behavioral variability in infant Japanese monkeys *(Macaca fuscata).* Ph. D. dissertation, University of Wisconsin-Madison.

Gouzoules, H. Fedigan, L. M.; and Fedigan, L. 1975. Responses of a transplanted troop of Japanese macaques *(Macaca fuscata)* to bobcat *(Lynx rufus)* predation. *Primates* 16: 335-49.

Gouzoules, H.; Gouzoules, S.; and Fedigan, L. M. In press. Japanese monkey translocation: Effects on seasonal breeding. *International Journal of Primatology.*

Hanby, J. P., and Brown, C. E. 1974. The development of sociosexual behaviors in Japanese macaques, *(Macaca fuscata). Behaviour,* 49: 152-96.

Hasegawa, T., and Hiraiwa, M. 1980. Social interactions of orphans observed in a free-ranging troop of Japanese monkeys. *Folia Primatologica* 33: 129-58.

Hrdy, S. B. 1976. Care and exploitation of nonhuman primate infants by conspecifics other than the mother. In J. S. Rosenblatt; R. A. Hinde; E. Shaw; and C. Beer (eds.), *Advances in the Study of Behaviour* pp. 101-58. New York: Academic Press.

Itani, J. 1959. Paternal care in the wild Japanese monkey, *Macaca fuscata fuscata. Primates* 2: 61-93.

Koyama, N. 1967. On dominance rank and kinship of a wild Japanese monkey troop in Arashiyama. *Primates* 8: 189-216.

———. 1970. Changes in dominance rank and division of a wild Japanese monkey troop in Arashiyama. *Primates* 11: 335-90.

Kurland, J. 1977. Kin selection in the Japanese monkey. *Contributions to Primatology* vol. 12. Basel: S. Karger.

Wilson, E. O. 1975. *Sociobiology: The New Synthesis.* Cambridge: Belknap Press.

7

Why Males Use Infants

Shirley C. Strum
Department of Anthropology
University of California, San Diego

INTRODUCTION

In primate groups, males and infants engage in two major kinds of interactions: a male can "care" for an infant (paternal care, Itani, 1959) similar to the way its mother cares for it; or a male can "use" an infant to his own benefit during social interactions. Male care has been observed in varying forms and degrees in many different species (see chapters 14, and 15). By contrast, male use of infants has been observed in only a few species and may be restricted to one family and two genera of Old World monkeys, baboons, and macaques (e.g., Itani, 1959; Deag and Crook, 1971; Taub, 1980; Deag, 1980; Ransom and Ransom, 1971; Gilmore, 1977; Popp, 1978; Packer, 1980; Altmann, 1980; Busse and Hamilton, 1981; but see Hrdy, 1976, for other examples).

At present, uncertainties about male use of infants extend to every interesting aspect of the phenomenon: what it is, how it works, when it happens, why it works, its costs and benefits, and the evolutionary basis of the behavior. Moreover, these uncertainties have hindered the analysis of male care and its possible relationship to male use (Hrdy, 1976).

Itani (1959) was the first to report males using infants as "passports" to gain entry into the central region of Japanese macaque troops. Subsequently, Deag and Crook (1971) suggested that Barbary macaque males might use infants as "agonistic buffers" in interactions with one another; the authors applied the term to "the deliberate use of a baby as a buffer in a situation where an approach without the buffer would lead to the increased likelihood of an aggressive response by a dominant male (p. 196)."

In more recent usage, however, "agonistic buffering" has come invariably to refer to cases where the use of an infant turns off the *actual* aggression of an adversary (Ransom and Ransom, 1971; Gilmore, 1977; Popp, 1978; Packer, 1980; Altmann, 1980; Taub, 1980). The recent usage serves to highlight the relative absence of aggression in the cases cited by Deag and Crook (1971), and more fully by Deag (1980), and suggests that what these authors call "agonistic buffering" may actually be instances of what Itani referred to as passport use of infants. Furthermore, Taub's studies of Barbary macaques (1978, 1980) also reveal little actual aggression in male interactions involving infants' use. He suggests that the majority of these interactions in this species can be interpreted as part of a pattern of "a shared, common and special care-taking relationship with the same infant" (1980, p. 196).

Among baboons, cases of agonistic buffering involving real antagonism do occur, and so do cases of males using infants as passports (Ransom and Ransom, 1971; Gilmore, 1977; Popp, 1978; see below). Baboon males also care for infants, however, and often the very infant for whom a male cares is the one he uses as a passport or a buffer. This conjunction of caring for infants and using them in aggressive or potentially aggressive contexts raises questions about the possible relationship between the two kinds of behavior.

In order to clarify and resolve a number of questions, this paper focuses on males' use of infants in one troop of olive baboons which the author has studied over a period of eleven years. It will be necessary to consider which males use infants in what contexts and against whom; which infants are chosen; the success of strategies involving infants; and the determinants of success. Clarification of these issues will lead to an interpretation of why males use infants, the relative costs and benefits for a male and for an infant, and the evolutionary basis of the behavior. The paper will also consider data and interpretations from other baboon populations, with the aim of developing a broadly applicable theory of infant use. In the general discussion and conclusions, the central interpretive concept of social strategies will serve to link the phenomenon of male use of infants with paternalistic behaviors.

METHODS

Study Site and Group

Observation of one troop of baboons, the Pumphouse troop, covered 3,850 hours during four study periods from 1972 to 1979 (Table 7-1). The study site near Gilgil, Kenya, described by Blankenship and Qvortrup (1974) and

Harding (1976), is a savannah habitat at high altitude with low rainfall. At any one time, the troop contained an average of 7 to 10 adult males and 19 to 21 adult females (Table 7-1).

Monitoring of polyadic agonistic interactions was part of daily research on several topics. Since episodes of male use of infants occurred relatively infrequently, the data for this analysis were gathered in an ad libitum fashion. Only data on adult males are included. Males under the age of 8 to 9 years used infants less frequently than did adults, and those under the age of four seldom used infants. Strum (ms. a) has described the ontogeny of this behavior in males.

In total, 21 different males resided in the troop during the study periods. Each male was observed for at least three months, but some were observed in all study periods. Although males can be classified according to a number of basic variables such as age, size, dominance rank, and length of residency with the troop, the latter proved the most useful parameter. Assessment of the other variables was either too unreliable or, when possible, revealed no significant relationship to the male behavior under consideration. Determining the ages of males, for example, was extremely difficult since all but one of the males in the sample were immigrants of unknown histories. Assigning putative ages to each male failed to produce any significant correlation with male behavior. Similarly, weighing experiments showed that it was not possible to determine male size accurately from observation alone, but even the use of actual weights yielded no significant relationship between differences in male size and differences in male behavior.

Dominance in this baboon troop does not fit the traditionally reported pattern. Agonistically high-ranking males did not get limited resources while agonistically low-ranking males did. Furthermore, dominance rank and length of residency were negatively correlated (Strum, 1982). This inversion of the expected pattern has made it preferable to rely primarily upon length of residency in comparing males, although reference to dominance ranks will be included where it is appropriate.

Accordingly, males have been assigned to residency categories that reflect a male's social integration into the troop relative to other males. The time intervals for the categories are somewhat arbitrary, but the purpose of the categories is to provide a relative time scale for residency. Long-term residents (LTR) are males who have been with the troop for more than 3 years; short-term residents (STR) have been in the troop between 1½ and 3 years; newcomers (N) are those of residency shorter than 1½ years. Males have been classified on the basis of their status at the beginning of each study period and have not been moved between categories during a study period. Depending on how long they were with the troop, some males contributed data to only one or two categories, while others contributed to all three. Data

Table 7-1. Distribution of Infant Use During Study Periods.

Study Period SCS	Hours of Observation	No. Episodes of Infant Use	Mean Troop Size	Mean Number						Research Topics
				m	s.am	f	s.a.f	j	i	
Dec. 1972 to Jan. 1974	1200	98	53	7	—	19	—	14	13	Social networks, roles & dynamics
June 1975 to Sept. 1975	250	6	82	10	2	20	1	29	20	Consistency & stability of social networks
August 1976 to Sept. 1977	1200	107	87	7	10	21	6	21	28	Transition to adulthood in adolescent males & behavioral effectiveness of social networks
Sept. 1978 to July 1979	1200	82	97	7	7	21	6	35	19	Kinship & dominance: cost-benefit analysis of differing resource contexts
Total	3850	293								

from several individuals have been pooled in comparisons of residency categories. When all males in a category were not present for the same amount of time during a study period, it has been necessary to make appropriate adjustments in some comparisons. Where tests of unpooled and pooled data produced different results, both sets of results have been presented. Furthermore, the analysis has made use of two different statistical tests: one that assumes that the behaviors followed a Poisson distribution because they were relatively infrequent (J. Kozial, pers. comm.); and another, a Chi-square test, that makes no assumption about the underlying distribution. The results of the Chi-square tests provide the conservative statistic.

Infants are born with a black natal coat, and a gradual transition to the brown adult color usually is complete around seven months of age. The transition is variable and in 36 recorded cases occurred from 5 to 12 months of age. Individuals were classed as juveniles at 24 months, by which time they usually had a younger sibling. Juveniles were not used by adult males as agonistic buffers or passports, although immature males occasionally attempted to use juveniles in these ways.

CLASSIFICATION AND TERMINOLOGY

Males "use" infants as agonistic buffers and as passports. In either case, the "user" picks up an infant and places it on his ventrum. Since baboon males do not normally carry infants, this behavior is a good first criterion by which to identify the relevant interactions. However, males sometimes do have infants in this position during male "care," and to distinguish between this and male "use" it has therefore been necessary to add two other identifying criteria: *before* picking up an infant, the user must be either interacting with or close to another male; and the situation must demonstrate tension on the part of one or both males.

An interaction can count as agonistic buffering when there is actual aggression between the participants or when aggression is only threatened. The user focuses on the infant and, by hunching over it, avoids the glance of the other male. The user may sit with the infant as the other male approaches, avoid the other male's approach while carrying the infant, or directly approach the other male. Sometimes the second male walks off as soon as the user obtains the infant.

An interaction can count as passport use of infants when the user approaches another male without aggression (passport tension) or when the user's approaching behavior changes from submissive to aggressive after picking up the infant (passport aggression). A passport user invariably approaches another male and then sits, threatens, or quickly moves away. Whichever the behavior, the infant clings quietly while the user focuses his attention on the other male, rather than on the infant as during agonistic buffering.

A third set of interactions, termed "reassurance" interactions, does not fit neatly into this scheme. Here a male picks up an infant after his interaction with another male is effectively over. Again, however, the presence or absence of aggression in the interaction can serve to delineate the behavior as either post aggression reassurance or tension reassurance.

When an infant follows a male much as it would follow its mother, grooms him, and is groomed by him, and the two are frequently found in proximity to one another, they may be considered to have a special relationship and be called "affiliated." Similarly, males have such affiliated relationships with certain adult females of the group (Strum, in pr. a). Thus frequent proximity, grooming, and following was a combination unique to affiliated infant-male pairs and such a classification was used as a relevant analytical category (Strum, ms. a).

QUALITATIVE DESCRIPTION

Although male use of infants involves a variety of behaviors, the sequence of behaviors is consistent (Figure 7-1). The sequence begins in the midst of an interaction between two or more males *(Phase I)*. Either there is aggression between them or the proximity or approach of one male disrupts the normal activity of another. One of the males then gets an infant *(Phase II)*. The infant may be distant or near, alone or with another baboon, sometimes its mother. Infrequently, the infant may initiate contact by moving toward the male, but the initiator is most often the male. Upon contact, the infant may or may not cooperate, and the mother may or may not restrain it. Sometimes, a male's approach prompts an eager and willing infant to run to him. The infant clings ventrally, although a male will occasionally carry an infant on his back. Throughout his approach, the male grunts at the infant, and once it is on his ventrum, he grunts, lip-smacks, and holds it gently.

What happens next *(Phase III)* depends on whether the infant is used as a passport or as an agonistic buffer. These interactions rarely last less than 2 minutes and may continue for up to 20 minutes. The outcome *(Phase IV)* is marked by the disengagement of the user and the infant. This phase is often punctuated by periods when the two males sit in relative proximity, staring at the baby or glancing at each other. Total disengagement occurs after the males cease interacting. Grooming and grunting between user and infant then stops, but the infant may remain around the user instead of running off. In no instance do the males groom each other or exchange the infant, nor do they remain in proximity after the interaction (See Deag, 1980, and chapter 2 for comparison with Barbary macaques).

Fig. 7-1 *(above and at right).* Sequence of agonistic buffering. *Phase I:* aggression between two males. *Phase II:* A male tries to use an infant but the mother restrains it. *Phase II:* A male obtains an affiliated infant. *Phase III:* A male with an infant as the other male approaches. *Phase IV:* A male grooms the infant after supplanting the other male.

MALE USE OF INFANTS

The Pattern

Two hundred and ninety-three cases of males using infants occurred in four study periods (Table 7-1). These cases can be distinguished according to the presence or absence of limited resources, the nature of the resources, and the degree of aggression. In the majority of cases, infant use was unrelated to any resource (Table 7-2). Relevant resources included estrous females and prey carcasses. Infant use did not occur in other food contexts. Of the 233

Phase II

Phase III

Phase IV

Table 7-2. Context of Infant Use.

Study Period		Contexts			
		Resource-related		Unrelated	
		Meat	Estrous Females	Male Proximity, Approach	Sum
1973	1	6	9	83	98
1975	2	1	2	3	6
1976	3	9	8	84	101*
1978	4	2	16	63	81*
		18 (6%)	35 (12%)	233 (82%)	286

*n = 6 and n = 1 ambiguous cases not included, respectively.

cases of infant use in a context unrelated to a resource, a male's proximity accounted for 44% while the approach of another male accounted for 29% and actual aggression for 27%.

Infants were used primarily as buffers (87%) and less frequently as passports (12%). The use of an infant for reassurance was rare (1%).

In general, males of longer residency used infants more, and these LTR males used them against males who were newer to the troop (Table 7-3), both STR and N males.

Consistent with this residency pattern, short-term residents used infants less than long-term residents but more than newcomers. Similarly, short-term residents were more likely to have infants used against them than long-term residents but less likely than newcomers.

Males used both black and brown infants. Infants were first used at two to seven months of age (n = 50). Despite being less numerous, black infants were used more often than brown infants for all years except 1978-79 (Table 7-4). Individual males, however, had preferences at variance with this general trend. Some males never used black infants, even when they were available, and some males switched from using mainly black infants to using mainly brown ones, even when new black infants were available. In 1978-79, the majority of males used brown infants, although black ones were no less available than in previous periods (Table 7-5).

There was no discernible relationship between the relative availability of infants of different colors and the preferences each male exhibited. Nor was there any apparent pattern when males were considered as members of specific residency classes. Rather, each male relied upon a limited and nearly unique set of infants out of the total available pool (Table 7-6).

Table 7-3. Male Interactants in Infant Use.

Study Periods	1973	1975	1976	1978
Males Who Use Infants Are	Long-term residents (χ^2, $p < .001$, $n = 98$)	n.s. (long-term residents) (Kolmorogov Smirnov $.05 < p < .10$, $n = 6$)	Long-term residents (χ^2, $p < .05$, $n = 107$)	Long-term residents (χ^2, $p < .01$, $n = 82$)
Infants Used Against	Newcomers (χ^2, $p < .005$, $n = 98$)	n.s. (newcomers) (Kolmorogov Smirnov $.05 < p < .10$, $n = 6$)	Short-term residents (χ^2, $p < .001$, $n = 107$)	Newcomers (χ^2, $p < .005$, $n = 82$)
Demography:				
No. Newcomers	4	3	4	4
No. Short-term Residents	—	2	3	2
No. Long-term Residents	4	5	2	5
Total Events	98	6	107	82

Table 7-4. Infant Availability.

Study Periods	1973	1975	1976	1978
Age At First Use				
Mean	4 months	—	3.3	4.1
Range	2-7	—	2-7	2-7
Black Infants				
Mean Per Month	5.0	4.2	7.3	4.1
Range	4-9	4-5	2-13	2-6
Brown Infants				
Mean Per Month	10.6	15.5	20.3	15.6
Range	7-12	12-18	19-22	9-19
Mean No. Black Infants Per Male Per Month $\bar{X}$	.83	.47	1.04	.51
Proportion of All Infants Used That Were Black	85%	80%	61%	35%

Table 7-5. Infant Coat Color and Frequency of Use.

Study Periods	Most Frequently Used Infants	Difference in Effectiveness between Black and Brown Infants
1973	Black (χ^2, $p < .001$, $n = 98$)	n.s. (χ^2, $p > .30$, $n = 98$)
1975	Black (Kolmorogov Smirnov, $p < .05$, $n = 6$)	n.s. (Kolmorogov Smirnov, $p > .05$, $n = 6$)
1976	Black (χ^2, $p < .005$, $n = 107$)	n.s. (χ^2, $p > .30$, $n = 107$)
1978	Brown (χ^2, $p < .005$, $n = 82$)	n.s. (χ^2, $p > .30$, $n = 82$)

A male's choice of an infant appears to have been influenced mainly by whether or not they were affiliated. Outside of the context of agonistic buffering, males did not groom unaffiliated infants, and infants almost never groomed unaffiliated males. A male who was affiliated with a black infant would continue to use that infant when it turned brown. Furthermore, a male who was affiliated only with brown infants would almost always use them regardless of the availability of unaffiliated black infants. The rare exceptions will be discussed below.

It would be useful to know the biological relationship between a male and

Table 7-6. Distribution of Infants Used by Individual Males.*

	1973								1975		
Males	**SU**	**CA**	**BS**	**RD**	**RA**	**AR**	**ST**	**Males**	**BS**	**RD**	**ST**
Infants								Infants			
AD		SR[1]			Blk i[2]		Blk i	AL	SR		
CL		Brn i[3]						NI	Blk i	SR	
CE			Brn i					TE			SR
CR	SR	SR	SR								
DA				SR	SR						
FO	SR	Blk i	SR	SR			Blk i				
LY				SR							
ME	SR			SR							
NA		Blk i			SR						
VA		SR		Blk i		SR					
WI			Brn i								
Mean Infants Available	15.6	15.6	15.6	15.6	15.6	14.3	16.3		20.25	20.25	20.25
Number Infants Used	3	6	4	5	3	1	2		2	1	1
% Affiliated Infants Used	100	50	50	80	67	100	0		50	100	100

*Males who did not use infants are not listed.
[1]SR = special affiliative relationship between male and infant.
[2]Black infant.
[3]Brown infant.

(continued)

Table 7.6 *(continued)*

				1976				
Males	**BS**	**RD**	**ST**	**BR**	**DR**	**MQ**	**GA**	**RE**
Infants								
AX	SR				SR			
CO	SR				SR			
DI			SR			Blk i		
DM			SR				SR	
FI		SR		SR				
JH			Brn i			SR	SR	
KI	SR				SR			
MH		SR	Blk i			SR		
NE				Blk i		Blk i		
PR			Brn i		SR			
QT							SR	
SB	Blk i	Blk i						
SP	Brn i							
TD	SR		SR		SR		SR	SR
TE								SR
TL							SR	
TT	Blk i							
VT				Blk i	SR			SR
WL							Blk i	
MG	SR							
Mean Infants Available	27.7	27.7	27.7	27.7	27.7	27.7	27.7	27.7
No. Infants Used	8	3	6	2	7	3	7	3
% Affiliated Infants Used	63	67	50	50	86	67	71	100

					1978						
Males	**BO**	**DU**	**GA**	**BS**	**DV**	**MQ**	**AG**	**HI**	**RD**	**CH**	**AT**
Infants											
AU	SR			SR							
BU		Blk i			SR						
CY				SR							
EW	Brn i										
JM	SR		SR								
KT	SR										
MR	SR	Blk i		SR							
NE							Brn i	Brn i	SR		
PT		Brn i	Brn i								
RI		SR			SR						
SS	Blk i		Blk i								
VE					SR						
WR	SR	Brn i	Brn i	Brn i		SR					
WW			SR	Blk i		Blk i					
ZA			SR								
ZH			SR								
Mean Infants Available	20.2	21.1	20.2	20.2	20.2	20.2	21.5	21.1	20.2	21.1	16.0
Number Infants Used	7	5	7	5	3	2	1	1	1	0	0
% Affiliated Infants Used	71	20	57	60	100	50	0	0	100	—	—

*Males who did not use infants are not listed.
[1]SR = special affiliative relationship between male and infant.
[2]Black infant.
[3]Brown infant.

the infants he uses. Paternity is difficult to assess from behavioral data, particularly in this troop, where several males consorted with each female prior to conception and on the most likely days of fertilization. Although it is possible that affiliation between males and infants sometimes reflected biological relatedness, the consort data are not helpful in resolving the point. It is certain, however, that affiliations occurred that could not have been based on biological relatedness because the infants in these relationships were conceived before the males entered the troop. Furthermore, these males showed the same preference for using these obviously unrelated infants that other males showed in regard to infants with whom they might have been related. Thus relatedness does not appear to be a primary or even necessary component of affiliation.

In summary, males of longer residency used infants against males who were newer to the troop. In the majority of cases, resources were not involved and the infant was used primarily as a buffer. While there was a general preference for using black infants (see below), the primary factor affecting a male's choice of an infant appears to have been the nature of the social relationship between them and only secondarily the infant's coat color.

Determinants of Successful Infant Use

What counts as success when a male uses an infant in interactions with another male? In cases of agonistic buffering, criteria of success are relatively easy to identify in terms of shifts in the interactions that favor the user. Either aggression that had been directed at the user stops when he obtains an infant, or the user, previously supplanted, now becomes the supplanter. In passport cases, the relative weakness of male interaction prior to infant use presents a difficulty. Conceivably, one could score an episode of passport use as successful if, after picking up an infant, the user merely gains greater proximity to another male. A more conservative measure would be to score as a success only those episodes in which the user actually becomes more aggressive in approaching another male or the latter noticeably moves out of the way to avoid the user. The latter is the scoring method chosen here. Reassurance cases present another difficulty since here interactions between the males seem already "decided" before one of them obtains an infant. In these cases, success is defined as a reversal in the "decided" outcome. Thus

Fig. 7-2 *(at right).* Examples of male care. *(a)* A male caring for two affiliated infants who are sitting nearby; *(b)* a male with two infants ventral on him as mothers feed nearby; and *(c)* a male grooming an affiliated infant.

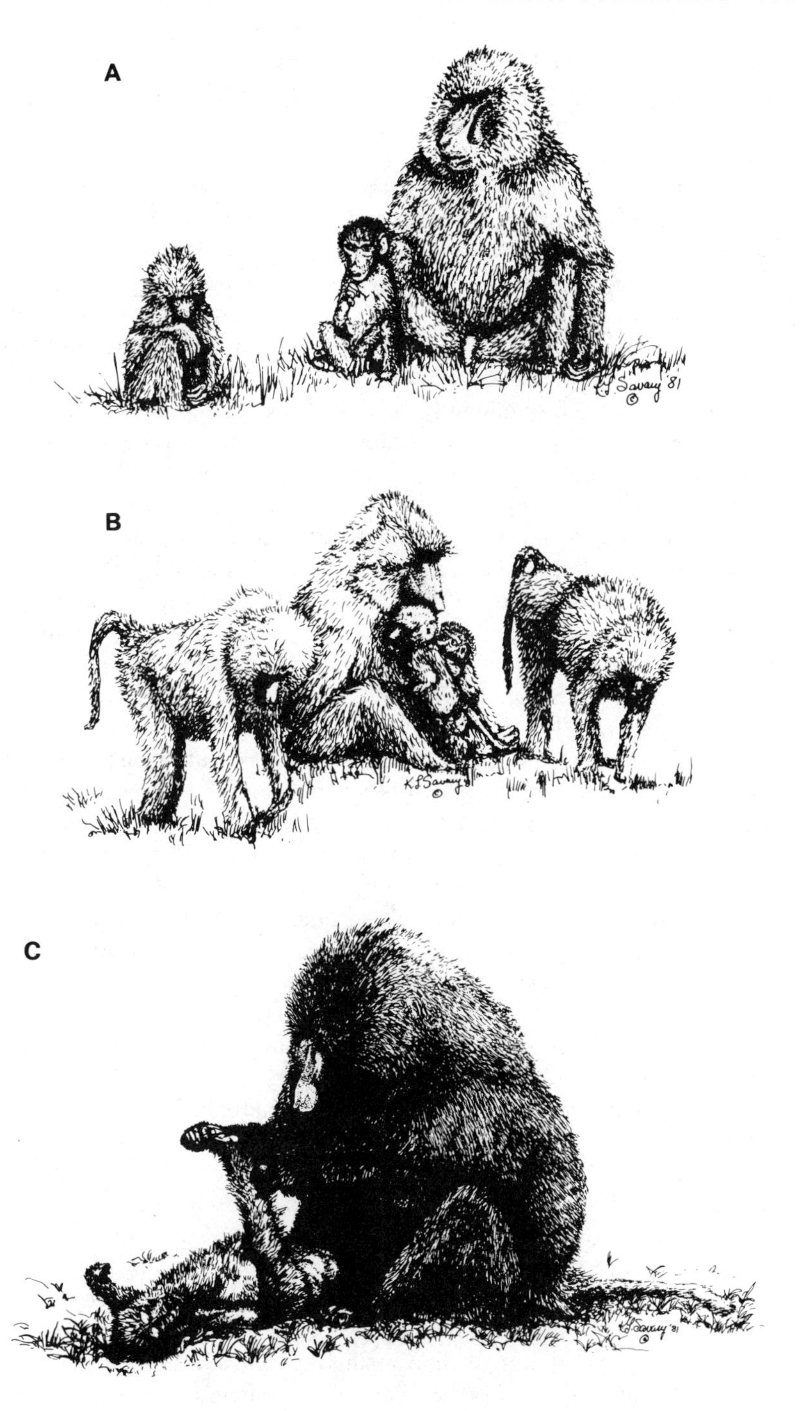
A
B
C

if the supplanting male suddenly runs off or the winner in an aggressive encounter begins to give submissive gestures when the supplanted or defeated male obtains an infant, then these cases are counted as successful instances of infant use.

Given these criteria, the success rate associated with a strategy of infant use was relatively constant between study periods (Figure 7-3). And although the overall success rate was low, about one-third of all cases, the rate of success was consistently higher in aggressive than in nonaggressive contexts (Binomial test, $n = 13$ males, $p = .046$). Use of brown infants was equally as effective as use of black infants (Table 7-5), and some males used brown infants with greater success than they used black infants.

Although it was predominantly males of longer residency who used infants against newer males, a male was somewhat more likely to be successful if he used an infant against a male who was of longer residency than himself (Binomial test, $n = 17$ males, $p = .025$). There seems to be no correlation, however, between a male's successful use of an infant and the possibility that the opponent might have been the infant's father or a likely relative (sibling). In only 4% of all the cases of success ($n = 107$) was the opponent a likely relative of the infant. If these same cases are examined under the conservative assumption that any male who resided in the troop when the infant was conceived could be a relative, then the possible role of a kinship factor increases to 31% of the cases but still represents a minority influence on success.

By far the greatest influence on a male's success in using an infant seems to have been the relationship of affiliation between the male and the infant. Affiliated infants cooperated with a male significantly more than did unaffiliated ones (Figure 7-4). A cooperative infant did not try to run away from its user. Either it remained voluntarily in proximity to the user during the male-male interaction, or it clung quietly to the user's belly without squirming or giving distress vocalizations. This cooperation correlated strongly with success (Figure 7-5; Binomial test, $n = 11$ males, $p = .033$). Apparently, males select infants based on affiliation, affiliation insures greater cooperation, and cooperation increases the success of the strategy of infant use in male-male interactions.

The preference in most years for black infants can also be interpreted within this framework. Discussions of infant use (Ransom and Ransom, 1971; Gilmore, 1977) often cite the lesser mobility of black infants as a critical factor in their selection. It seems from the Pumphouse data, however, that mobility is only secondarily important. Of greater importance is that black infants are less socially experienced and, as a result, seem to have a naive trust that a male can play upon in a way not possible with older infants. Consequently, a black infant is usually more cooperative than a brown infant, but less cooperative than an infant that is affiliated with a male. Because an infant's cooperation is crucial to a male's success, a male ought

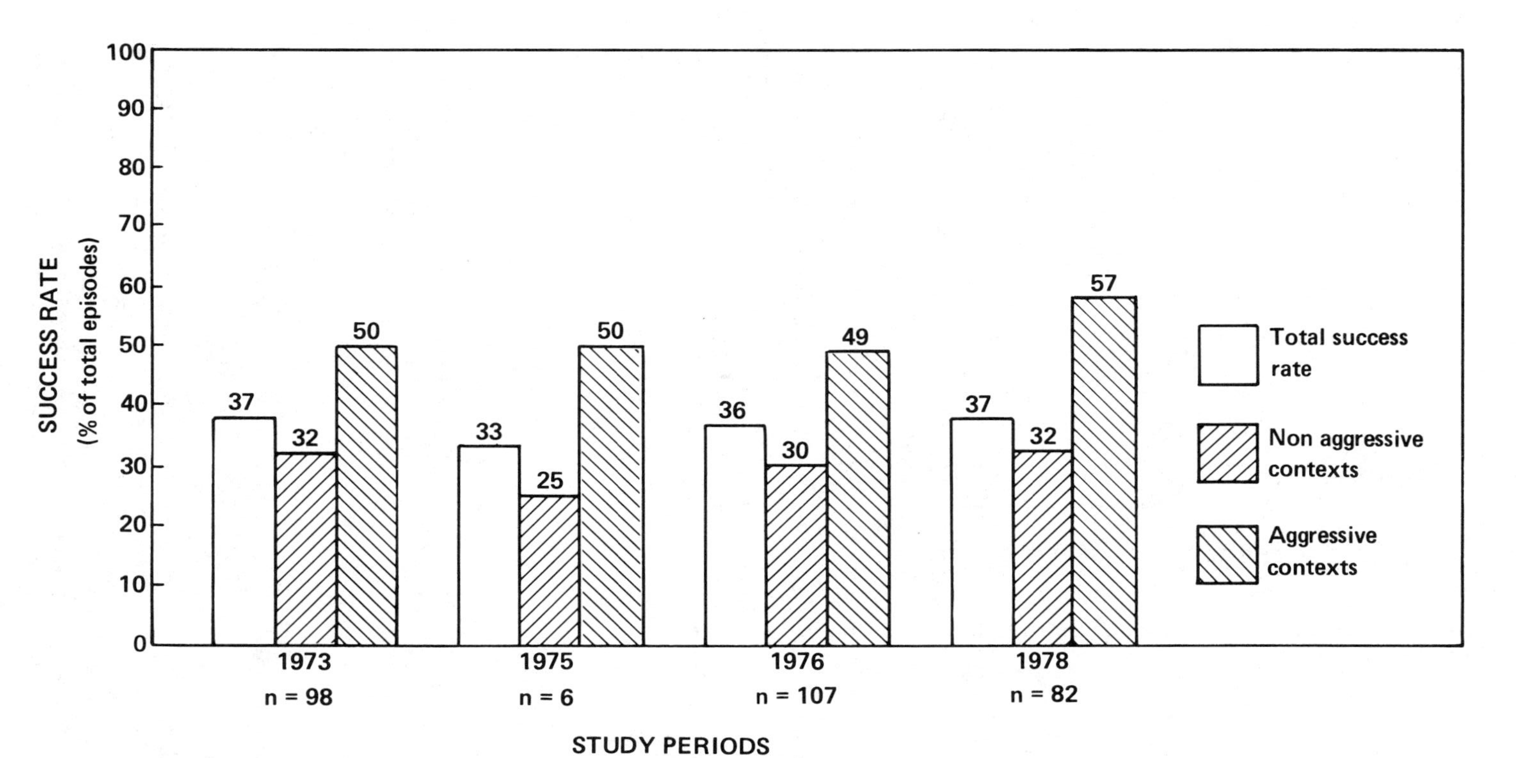

Fig. 7-3. Success of using infants. Significant differences in success rates are shown between aggressive and nonaggressive contexts when each male is considered over all years, Binomial test, $n = 13$ males, $p = .046$.

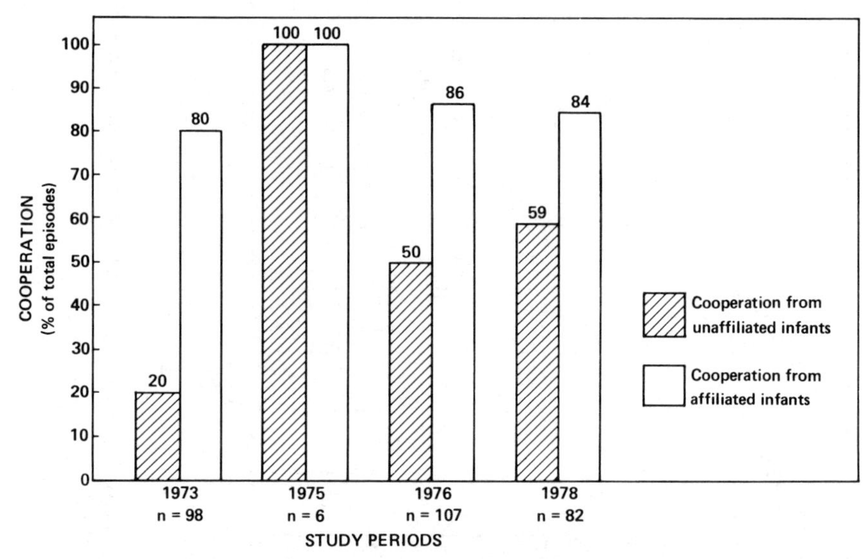

Fig. 7-4. Relationship between affiliation and cooperation in infant use. The significant difference between affiliated and unaffiliated infants when pooled across males is: *1973*, χ^2, $p < .001$, $n = 98$; *1975*, Kolmorogov Smirnov, $p > .05$, $n = 6$; *1976*, χ^2, $p < .01$, $n = 107$; *1978*, χ^2, $p < .02$, $n = 82$. The significant difference between affiliated and unaffiliated infants when each male is considered individually over all years is: Binomial test, $n = 8$ males, $p = .035$.

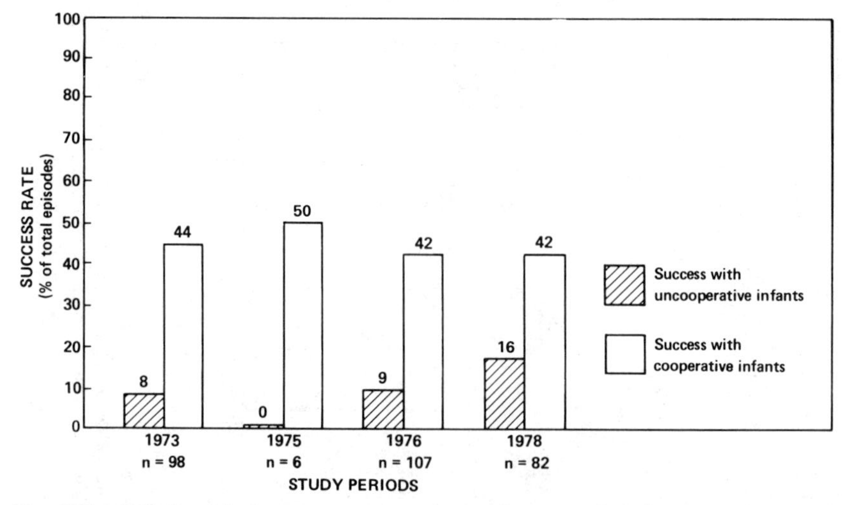

Fig. 7-5. Relationship between cooperation and success in infant use. The significant difference between the rate of success with cooperative and uncooperative infants when pooled across males is: *1973*, χ^2, $p < .005$, $n = 98$; *1975*, Fisher's exact, $p = .40$, $n = 6$; *1976*, χ^2, $p < .01$, $n = 107$; and *1978*, χ^2, $p < .05$, $n = 82$. The significant difference in rate of success depending on cooperation when each male is considered individually over all years is: Binomial test, $n = 11$ males, $p = .033$.

to prefer, and to be most successful in using, black affiliated infants and then, in decreasing order of preference and success, brown affiliated infants, black unaffiliated infants, and brown unaffiliated infants. This is just what the Pumphouse data show (Figure 7-6). The data also suggest that if no affiliated infant is available, then a black unaffiliated infant should be used in preference to a brown unaffiliated one. In general, the frequency with which particular infants were used correlates with the degree of cooperativeness of differing categories of infants ($r_s = 1.0$, $n = 4$, $p = .05$).

In light of the above, the preferred and most successful combination of elements should involve a male using a black affiliated infant against an opponent of longer residency than himself. Yet, this combination represents only a small proportion of the cases (17% for all years together). This suggests that, from the point of view of the user, infant use involves a number of parameters and benefits which have not yet been considered.

Why Males Use Infants

Data from the Pumphouse troop at Gilgil lend little support to any of the previous hypotheses about why males use infants or the reasons for the success of the strategy (e.g., Itani, 1959; Deag and Crook, 1971; Gilmore, 1977; Popp, 1978; Packer, 1980; Taub, 1980). A brief review of these hypotheses in light of the present data may be helpful before offering an alternative hypothesis.

The earliest and still the most pervasive hypothesis about agonistic buffering sees the infant as providing a special stimulus, primarily through its natal coat, which inhibits aggression in the male to whom it is presented (Deag and Crook, 1971; Ransom and Ransom, 1971; Hrdy, 1976; Alley, 1980). It follows that, for baboons, black infants should be the only ones used by males, they should be used indiscriminately, and they should be highly effective since the concept of a sign stimulus (Manning, 1967) implies an "innate releaser."

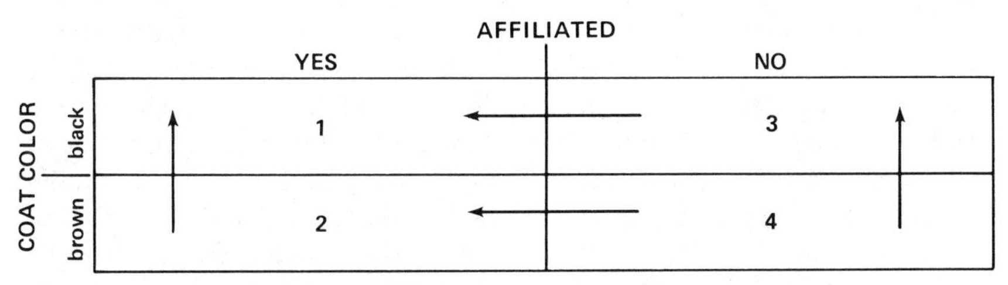

Fig. 7-6. Affiliation and coat color as factors in successful use of infants. The arrow indicates the direction of more successful use, comparing adjacent categories; numbers *(1, 2, 3, 4)* indicate order of success rate. The significant difference between success rate of different cells is: for *1* and *3*, χ^2, $p < .001$, $n = 176$; *2* and *4*, χ^2, $p < .025$, $n = 117$; *1* and *2*, χ^2, $p < .005$, $n = 215$; *3* and *4*, χ^2, $p < .05$, $n = 78$.

Pumphouse males do seem to have a preference for black infants, but they also used brown infants in all years and even used them predominantly in one study period. As a "releaser" in male-male interactions, the presentation of a black infant had a remarkably low success, indeed far too low to suggest an evolutionarily derived innate response. Finally, although males were selective in the infants they used, selection seemed to occur discriminately on the basis of affiliation rather than indiscriminately on the basis of coat color.

As an alternative to the special stimulus hypothesis, several authors have tried to explain success in using an infant in terms of biological relatedness between the infant and the male the user is opposing. The question then arises as to whether or not, and by what means, males can assess their relatedness to infants in their troop. In one version of this approach, Popp (1978) has argued for direct assessment of paternity. He suggests that coat color serves as the proximate cue that helps males identify their offspring since, as he interprets the data from his baboon population, only dominant males sire offspring. On this reasoning, when a subordinant male uses an infant against a dominant, the latter's aggression is inhibited because he will not risk injuring his own offspring. In a somewhat different version, Gilmore (1977) has proposed that infant use is successful because the mating system of baboons makes it likely that any adult male may be the father of an infant and therefore each male must act accordingly.

Pumphouse baboons differ markedly from Popp's population. Since infants are used primarily by residents against newcomers, males use infants who are more likely to be their own against males who have the least, or no, likelihood of being related to the infant. Furthermore, the likelihood that any male can assess paternity is negligible, given the number of males who consort with a female during the cycle in which she conceives and the high male turnover rate in the population. Finally, it is the relationship of the *user* to the infant, not that of the opponent to the infant, that most heavily influences the choice of the infant to be used in this population.

By contrast, the Pumphouse data do lend some minimal support to Gilmore's (1977) hypothesis. On the one hand, possible fathers often use infants against unlikely fathers, and this is not what one would predict on the basis of Gilmore's hypothesis. On the other hand, in those instances where an infant is used against a long-term resident, and thus where the likelihood of *his* relatedness to the infant is greater, the probability that the outcome of the interaction will favor the user is also greater. Still, the overwhelming pattern in the longitudinal data from the Pumphouse troop (the same troop from which Gilmore's short-term data derive) is that infants are used against males who could not possibly have mated with the mother because they were not resident in the troop at that time.

Some authors see males using infants as a means of gaining access to desired resources. In his analysis, Packer (1980) excluded infant use that occurred around consorts, but he emphasized the ability of a male with an infant to displace or supplant an opponent at food resources. Popp (1978) has argued for both increased access to food and increased access to estrous females. In the latter cases, he thinks the presence of the infant permits the user a closer approach to the consort male, thereby establishing a better position from which to steal the consort male's female.

Pumphouse males do use infants around consorts (12% of all cases of infant use) and around consumption of prey carcasses (6% of all cases of infant use). In the majority of cases (82%), however, use of infants is the result of the proximity, approach, or aggression of another male in a context that is unrelated to a resource (Table 7-2). When females or meat are at stake, it is more often than not the possessor of the resource that uses an infant. Thus the strategy in these cases appears to enable a male to retain possession of an item and not, or at least not so much, to appropriate an item from another male.

Itani (1959) originally proposed that males use infants to raise their status in the group. Popp (1978) and Packer (1980) have since made related arguments. The instability of dominance rank and the marginal significance of high dominance as compared to long residency among Pumphouse males have been treated elsewhere (Strum, 1982). Nevertheless, if each interaction is assessed independently, then it can be admitted that infants were used most often by a male who could be classified as subordinant in *that* interaction. In a number of cases, however, it was a dominant individual who used an infant against a subordinant. More significantly, in no instance did the use of an infant result in a permanent shift in dominance rank. The status function of infant use as proposed by Itani and others seems little supported by these data.

Finally, in a recent and novel interpretation, Busse and Hamilton (1981) suggest that males use infants not to "exploit" them for the males' own benefit but to protect them against immigrant males with infanticidal tendencies (see also chapter 8). The Pumphouse data are in general agreement with these authors' data concerning the major category of infant users (i.e., long-term residents) and the major category of males whom infants are used against (i.e., immigrant or STR males). Nevertheless, newly immigrated Pumphouse males do use infants against resident males and they do so as soon as they can (see Strum, in pr. a). Moreover, Pumphouse brown infants, past the age of weaning, are frequently selected for use and carried. If carrying provides protection from infanticide, these brown infants are being protected at a noncritical time when the mother has already resumed cycling or is actually pregnant with her next baby. Obviously, immigrant males would gain little reproductive benefit from killing such infants. All this makes the protection hypothesis a much less compelling interpretation of male use of infants at Gilgil than it may be for the Botswana population (see below for further discussion).

On the basis of this brief review, the need for an interpretation of male use of infants that is more consistent with the Pumphouse data should be apparent. The following interpretation not only seems the most consistent with these data but also helps to explain anomalies in the data from other baboon populations.

A male may derive several benefits from using an infant, and it is the sum of these benefits that must be considered. In seeking the main benefit or benefits, previous interpretations have looked primarily at the effect that the use of an infant ought to have on the male who is being opposed. The present interpretation reverses this focus by suggesting that a primary consideration, from the point of view of the user, is the effect of the infant's presence on the user himself. *Physical contact with an infant is a way for a male to change his own emotional state.* This is the core benefit of infant use. All other benefits, however important, can be seen as deriving from or at least additional to this core benefit.

"Contact comfort" is important to infant monkeys (Harlow and Harlow, 1965). As a youngster matures, the physical presence of others is sought at times of distress or tension (see discussions of grooming in primates). An adult male, however, faces severe constraints on the individuals available to him for this purpose. His size and status limit his access to lower-ranking individuals, and with other adult males, at least among *anubis* baboons, there is seldom contact or affiliative proximity. "Contact comfort" for these males is most possible with infants, particularly affiliated infants (see below), and with certain adult females (Strum, in pr. a).

Since the nature or even the outcome of an interaction may change if the emotional state of one of the interactants changes (Marler and Hamilton, 1967), getting an infant would seem a good strategy for an adult male. This is so, in the first instance, not because the strategy changes his adversary's emotional state but because it changes his own. Secondarily, this change in the user's state may cause his adversary to reevaluate the costs and asymmetry of the interaction and modify his expectations with respect to his own and the user's future course of action.

Additionally, bringing an infant into an interaction can serve to redirect attention by providing an alternative focus for the interactants. (I am grateful to D. Manzolillo for discussions on this point.) In the presence of an infant, one or both males frequently look, grunt, and generally exchange attention-to-the-infant behavior. At a minimum, such shifts in attention create a temporary hiatus in the male-male interactions. Sometimes, they offer one or both males an easy exit from what might otherwise become an escalating encounter.

Finally, an infant can serve as a male's ultimate protection since he can use it to mobilize the troop in its defense. When an infant's distress is serious enough, the entire troop or a large segment of it will mob the source of the offense, whether or not the responding individuals are related to the infant. This defense response of the troop is a discriminating one. Which male will be attacked depends on the residency status of the two males, not on who actually has the infant ($n = 8$). If a male of established residency uses an infant against a newer male, particularly a newcomer, the troop will mob the newer male. If, on the other hand, a newcomer uses an infant against a resident, the troop will likely mob the newcomer when the infant screams. Thus it pays a socially established male to use an infant as insurance against an all-out attack, particularly in interactions with relative strangers whose behavior is less known and predictable. This insurance factor helps us understand why resident males so frequently resort to infant use against newcomers even though the strategy does not always, or even nearly always, reverse the outcome of the interactions. Merely to lessen the possible severity of an outcome may be a sufficient measure of success for the user.

The defense response of the troop can also help us to understand why the use of infants works to turn off or inhibit aggression in male-male interactions. An opponent is not innately inhibited because of the infant's coat color or because of any possible relatedness to it. Rather, he has learned to be inhibited because he knows that the entire troop may be mobilized in defense of the infant. Newcomer males often elicit a fear reaction in females and immatures simply through proximity. When the distress is great enough, the troop will mob the male. Thus a male learns the power of this response from the time of his entry into the troop.

In summary, males use infants for at least three reasons. A primary consideration for the user is that of equilibrating his own emotional state through contact with an infant. This change of state may favorably influence the interaction for him in two ways: (1) he may now be able to behave in a manner previously thwarted by his tension, and (2) there may be inhibiting consequences for his adversary's course of action. To make a simple human analogy, we would respond to someone who was overcome by fear differently than we would if he was relaxed and confident, and that individual would himself behave differently if anxious or relaxed. Beyond this, using an infant can change the nature of the interaction between males by (1) providing the infant as an alternate point of focus and (2) providing through the infant a resource that allows a male of established residency to enlist the support of the rest of the troop against a newcomer should the situation become really serious.

Resolving Some Paradoxes

The above interpretation of male use of infants makes sense of some otherwise paradoxical findings that emerge from the data on Pumphouse baboons, including:

1. *Why males use affiliated infants regardless of their coat color.* If the purpose of the infant is to provide "contact comfort" for the male, the infant must be cooperative. An uncooperative infant is useless, or even worse than no infant at all, since he creates yet another point of conflict and tension, forcing the male to divide his time and energies further under circumstances that, we may assume, are already stressful. Affiliated infants consistently cooperate with a male during use (Figure 7-4). Therefore, a male will gain a greater advantage from using a brown affiliated infant rather than a black unaffiliated one when both are available. Following the same principle, however, it would be preferable for a male to use a black rather than a brown unaffiliated infant because younger infants are more naively cooperative and less able to disengage themselves (for a similar suggestion on infant naiveté, see Gilmore, 1977). The key to success when using an infant is the infant's cooperation and not its biological relationship to the male who is being opposed or the color of its coat, except in so far as black infants are intrinsically more cooperative than brown ones.

2. *Why newcomer males infrequently use infants on occasions when they might benefit from doing so:* New males have difficulty forming affiliations with infants. For a newcomer to use an unaffiliated infant would be a high-risk/low-benefit behavior since the infant will not be cooperative and the troop might mob the male. The few infants actually used by newcomers had a characteristic in common: they were spatially available because they were weaned early or neglected for some reason by their mothers. These infants seemed more willing to allow themselves to be used by any male, including newcomers. With time, newcomers formed affiliations with infants (as they became short-term residents) and then used infants more.

3. *Why is it generally the possessor of a resource who uses an infant, although rarely, and not the contesting male?* When used in a resource-competitive context, infants are only used in contests over estrous females or prey carcasses (Table 7-2). In these situations the male in possession of the resource is often of longer residency, and lower agonistic dominance rank, than the challenger. It can be assumed that the possessor of a resource is the one under the greatest degree of stress and also the least mobile. He

will therefore be the one most in need of the equilibration that an infant can provide. On the other hand, his relative immobility might prevent him from obtaining any infant who was not immediately available. The challenging male might occasionally benefit from using an infant, but if he is a newcomer his opportunities to do so are severely restricted. Thus several factors appear to contribute to the relative rarity of infant use in connection with resources. The immobility of the challenged male and the residency status of the challenging male clearly play a role. But the evidence suggests that behind this, most agonistic interactions between males, which are the most appropriate context for infant use, do not occur during competition over resources (Strum, 1982).

4. *Why the success rate of infant use (i.e., changing the outcome of an interaction in favor of the user) is lower than might be expected from previous interpretations* (but not from the data, see below, Table 7-7). If the success of using an infant does not derive from evolutionarily fixed reactions but depends instead on complex interactions involving learned skills and judgment, it is to be expected that the strategy, although a good one and certainly better than no strategy at all, produces less than perfect results. It may also be that the present criteria of success are too crude to measure the subtle benefits of "contact comfort" and "insurance." If using an infant merely reduces the intensity of aggression directed at a user, or simply allows him, by changing his emotional state, to enter into an interaction in the first place or to remain in one in progress, these effects may count as benefits for the user, but they will be very difficult for observers to measure under field conditions.

5. *Why males also use adult females as buffers in situations of tension and aggression* (Strum, in pr. a). This behavior obviously cannot be explained in terms of any inhibition of male aggression due to the female's small size or color or to her possible biological relatedness to the opponent. Yet the pattern of use parallels that of infant use in many important ways, and the present interpretation is consistent with both patterns. Clutching a female and attempting to groom or be groomed by her provides contact comfort similar to that provided by an infant. The troop can also be mobilized in defense of a female in much the same way as for infants (see Strum, in pr. b for a fuller comparison of the two types of buffering).

How Males Choose Infants—Some Examples

It would be instructive to know how males actually go about selecting infants for use and, in particular, whether or not they are aware of the parameters of success as outlined in the present interpretation. The following examples of sequences of infant use would seem to indicate that male baboons actually do understand these parameters, although they must often make compromises between the optimal strategy and what is available (Strum, ms. b).

Example 1. During an agonistic interaction with a newcomer, a resident male tried to use a nearby infant to whom he was not affiliated. The infant did not cooperate, and the male finally gave up trying to hold onto it. The infant ran away and sat near an affiliated male. The resident then quickly scanned the troop and walked directly to an affiliated infant while the newcomer continued his harassment. Reaching the infant, the resident took it ventrally, grunted, and groomed it. The newcomer threatened for a few more seconds and then walked off. This was a frequent type of sequence, which sometimes included several attempts to use nearby infants until the proper one was found.

Example 2. Two males were engaged in an escalating agonistic interaction. The aggressor was a newcomer, the object of aggression a resident. The newcomer tried to use an infant that was affiliated to the resident, but the infant refused to allow this. He ran to the resident male, who took the infant ventrally. This act terminated the newcomer's aggression. The importance of infant cooperation is epitomized by this sequence.

Example 3. In complicated male interactions, males who were affiliated to the same infant could be forced to choose other infants because their primary affiliation was already being used. In one episode, a newcomer successively harassed several resident males. The first two males sequentially used the same infant, who was affiliated to both of them, although other infants were potentially available for use. Both residents succeeded, in their turn, in directing the newcomer's aggression toward another male. When another newcomer male entered the fray, however, one of the initial resident participants was forced to use a less affiliated infant since the desired one was still being groomed by the other resident. In this case, the presence of the (less affiliated) infant did not stop the newcomer's harassment and the resident ran off with the infant.

Example 4. In one and the same incident, a male will sometimes make successive choices between using an infant and using a female. These choices seem to follow the above pattern: that is, a male first tries to use the individual nearest at hand and, if this turns out to be an inappropriate choice in terms of affiliation and cooperation (or differential effectiveness, see Strum, in pr. a), he then selects a more appropriate individual. In these cases, it appears as if the act of choosing an infant or a female, even if it is the wrong choice, buys the male some time to locate and obtain a more suitable individual.

Costs of Infant Use

Now that we know some of the benefits of infant use, we may consider its costs. What are the costs to the user and to the infant? Why should an infant allow itself to be used by a male?

The costs to the user depend on the investment he makes in establishing and maintaining an affiliative relationship with an infant. Pumphouse males actively initiate these relationships, which are often the outgrowth of affiliative bonds between the male and the infant's mother (see Ransom and Ransom, 1971; Ransom and Rowell, 1972; Packer, 1980). Aside from grooming the infant, the male also defends it from conspecifics and may defend it from external threats. Injury for the male is unlikely when he intervenes for the infant against females and immatures. The risk is higher if the infant's aggressor is an adult male, but no male was observed to be injured in this way during any of the study periods. The chances of injury are probably greatest when defending an infant against external threats such as humans and other predators, but there was no injury in the two observed instances.

Many reports allude to the high cost to the infant, in terms of injury or death, of being used by a male (e.g., Gilmore, 1977; Popp, 1978; Packer, 1980; Hrdy, 1976). Nevertheless, there is little direct evidence from other populations to support this contention (see discussion below). The Pumphouse data also suggest that the infant's cost in terms of injury is quite minimal. In 3,850 hours of observation, during which time 293 cases of infant use were recorded, no infant was injured; yet 29% of these cases involved aggression between two males. Injury to the male who used the infant occurred twice during these episodes, although the wound was superficial in both cases. By comparison, the injury rate in baboon play is high enough to make it likely that at some time during its "play" life, each young Pumphouse baboon will sustain an injury. The cost of play would seem to be higher for Pumphouse baboons than the cost of being used by a male in interactions with other males.

Infant costs could be expanded to include being frightened, losing feeding time, or occasionally being deprived of a preferred resource. The relatively infrequent use of infants makes it seem unlikely that these costs would ever accumulate to a significant level.

Infants do not benefit directly or immediately from the interactions in which they are used by males, except perhaps for the grooming they receive. Yet, the affiliative relationship between the male and the infant does provide the infant with deferred benefits, both direct and indirect. Paternalistic behaviors such as grooming and defense may be counted as direct benefits; since these involve the proximity of a friendly adult male, they may in turn yield indirect benefits for the infant, such as increased access to food and increased influence over conspecifics, as Hrdy (1976) has suggested.

Infant cooperation with male users can thus be explained in terms of a two-level scheme. At a proximate level, cooperation is the result of trust in an affiliated male. At the level of ultimate evolutionary processes, cooperation is the consequence of the improved fitness an infant acquires through the paternalistic benefits the male provides. It is therefore not difficult to understand why an infant would cooperate in being used by an affiliated male, especially when the costs are so minimal. In contrast, unaffiliated infants rarely cooperate, and though the costs to them may be minimal, the benefits may not outweigh the costs.

DISCUSSION

Comparison of Findings across Baboon Populations

The present interpretation fits the Pumphouse data, but is it also compatible with data on infant use from other baboon populations? That is, can the interpretation offered here serve as a general theory of infant use among baboons? It will be helpful to approach this question in two steps. First, to what extent does the general pattern of findings from Pumphouse correspond to patterns found elsewhere with respect to: (1) which males use infants against which other males; (2) what infants are used; (3) the context and success of the strategy; and (4) the costs and benefits that have been assigned to the behaviors. These comparisons are summarized in Table 7-7. Second, if the general patterns of findings agree, does the present interpretation of these findings offer advantages over previous interpretations?

In most of the populations studied, the males who use infant's are primarily residents of long enough duration to have been in the troop at the time the infant was conceived or at least at the time it was born. Recent immigrants don't use infants at all (Packer, 1980; Busse and Hamilton, 1981) or rarely use them (Ransom and Ransom, 1971). In all but one study (Popp, 1978), infants are used primarily against newcomer males.

The infants selected for use range in age from 2 to 10 months, the period coinciding with a black natal coat and preceding weaning. It is difficult to determine how frequently older infants or individuals of other age-sex classes are also used because either these categories are excluded from analysis or no quantitative comparisons are presented. Except for one

Table 7-7. Comparison of Baboon Populations.

Populations	Ransom & Ransom, 1971	Gilmore, 1977	Popp, 1978	Packer, 1980	Busse and Hamilton, 1981	This Study
Location habitat	Gombe Tanzania Forest	Gilgil Kenya Savannah	Masai Mara Kenya Savannah	Gombe Tanzania Forest	Moremi Botswana Savannah/Swamp	Gilgil Kenya Savannah
Number of episodes of infant "use"	Unstated	53	39	Unstated	112 in two troops	293
Hours of observation	2,500	1,372	348	1,400	10 months	3,850
Males who use infants	Residents	Residents or immigrants	Lower ranking (Immigrants)	Residents	Residents	Residents
Most frequent males infants are used against	Immigrants	Nontenured	Higher ranking (Residents)	High ranking or young (Dangerous)	Immigrants (Dangerous)	Immigrants (Newer males)
Infant's age Color	Young Black or brown	3-64 weeks Mostly black	< 10 months Mostly black	2-8 months (By definition) Black	2-7 months Black	2-24 months Black or brown
Relationship between infant and user	Same subgroup Affiliated or unaffiliated	Offspring of lower-ranking females	Unrelated to user (Less likely related to user than opponent	Infants cared for	Related (Offspring or sibling)	Affiliated with user

(continued)

Table 7-7 *(continued)*

Populations	Ransom & Ransom, 1971	Gilmore, 1977	Popp, 1978	Packer, 1980	Busse and Hamilton, 1981	This Study
Context of infant use	Proximity of other male	Encounters with other males (Not necessarily related to resources)	Resource context (primarily estrous females)	Encounters with other males (Resource = food by definition)	Proximity of male (Unrelated to resources)	Proximity of male (Can be used to retain possession of resource)
Success rate of infant use	No data presented Black infants more effective than brown	37% ignore male with infant 21% of males with infants avoided 16% aggression is redirected	In 1 of 39 cases aggression did escalate	20 of 22 dyads threatened less In 17 of 20 dyads supplanter had infant	No deaths of infants in 112 episodes	Overall success is 33–37% of episode Success in aggression is 49–57%
Cost to infant		0		1		
Observed:	No data	(2 injuries inferred)	No data	(3 injuries inferred)	0	0
Assumed:	High cost	High cost	High cost	High cost	No cost to infant	Low cost
Mechanism of success	Black coat color protects infant	Paternity uncertain in promiscuous species	Paternity ascertained Father inhibited	Infant coat color?	Willingness of kin caretaker to fight for infant	See discussion in text
Evolutionary basis of infant use	Group selection implied	Kin selection	Kin selection	Mutualism between paternal care and infant use	Kin selection	Reciprocal altruism between male and infant based on social strategies

account (Popp, 1978), infants younger then two months (who are in constant association with their mothers) are never observed being forcibly taken away from the mother's ventrum.

The greatest variation across the populations studied concerns (1) the relationship between the infant and the males involved in the interaction and (2) the contexts in which infants are used. In the majority of cases, the infants used are either cared for by the user (affiliated) or *presumed* to be related to him (Ransom and Ransom, 1971; Packer, 1980; Busse and Hamilton, 1981). Gilmore (1977) suggests that the infants being used are offspring of lower-ranking females, while Popp (1978) maintains that the infants are less likely to be related to the users than to the males being opposed.

With respect to the context of use, all studies agree that the proximity of another male acts to initiate an episode of infant use. In some studies (Popp, 1978; Packer, 1980), however, infants are regarded as a means to get resources, primarily estrous females or preferred foods, while in others (Ransom and Ransom, 1971; Gilmore, 1977; Busse and Hamilton, 1981) no necessary relationship is observed between using an infant and obtaining a resource.

For all populations, using infants is regarded as a successful strategy for males, particularly in agonistic situations. Males are less likely to threaten or attack a male with an infant than a male without one (Ransom and Ransom, 1971; Gilmore, 1977; Popp, 1978; Packer, 1980). Black infants seem more effective than brown ones, but detailed comparisons are not presented.

The cost of using an infant for the user is either not considered (Ransom and Ransom, 1971; Popp, 1978; Gilmore, 1977) or is linked to the cost of caring for the infant at other times (Packer, 1980) or protecting the infant during use (Busse and Hamilton, 1981). The cost to the infant, however, is assumed to be high in all studies, but the evidence for this is scanty (see details below). Only Busse and Hamilton (1981) suggest that using an infant reduces rather than increases its chances of injury.

The most agreed-upon benefit for the user is the thwarting of another male's aggression (Packer, 1980; Ransom and Ransom, 1971; Popp, 1978; Gilmore, 1977). Some studies also suggest that the use of an infant allows greater proximity to another male (Gilmore, 1977; Popp,, 1970; Packer, 1980), which in turn may permit appropriation of resources (Packer, 1980; Popp, 1978) and/or an increase in the dominance status of the user relative to the other male (Packer, 1980). The benefit to the infant is more problematic. Only Busse and Hamilton (1981) suggest that a male's carrying an infant is critical for its survival. Others propose a link between caretaking and using an infant (Ransom and Ransom, 1971; Packer, 1980) such that an infant benefits from the male's proximity and attention in either the immediate or some other context. Ransom and Ransom (1971) even suggest that the relationship between the male and the infant serves as a prototype for the infant's future supportive relationships.

In all these other baboon studies, then, where the relevant data are presented or discussed, the general pattern of who uses infants against whom is similar to that found in this study. Other features of infant use vary more across populations, but there is no strong pattern of findings that would contradict the data from the Pumphouse baboons: black and brown infants are used, affiliated infants are used, resources may or may not be involved, the success of the strategy varies, and observed costs to the infant are low (Table 7-7).

Comparative Advantages of the Present Interpretation

The present interpretation of male use of infants can be summarized in the following six points:

1. Males may derive contact comfort from the presence of infants, and the consequent change in these males' emotional state can potentially shift the asymmetry of their interaction with other males.
2. The behavior of the male who is being opposed will sometimes be modified in ways favorable to the user.
3. By importing an infant into a male-male interaction, the user provides an alternate point of focus which can act to diffuse tension and prevent an escalation of aggression.
4. The troop can be mobilized in defense of an infant in distress.
5. The success of infant use depends primarily on the cooperation of the infant with the user, and this cooperation is linked to a preexisting affiliative relationship between user and infant.
6. The actual cost of infant use for both the infant and the user is relatively low, while the immediate benefits for the user and the deferred benefits for the infant are high.

Studies from other baboon populations, where they present relevant and comparable data, pose no major contradictions to this interpretation. The reverse is not the case, however: the explanatory hypotheses on infant use that each of these other studies has offered cannot be applied broadly across populations. The hypothesis that infant coat color is an inhibitor of aggression falls in the face of evidence that brown infants and even females are used effectively as agonistic buffers. Hypotheses relating the inhibition of aggression to paternity, assessed either specifically or as a generalized probability, also do not apply across populations, especially where infants are frequently used against males to whom they could not possibly be related. The infanticide protection hypothesis may fit the Botswana data but seems not to apply to other populations where immigrant males, i.e., the very ones from whom infants are supposed to be protected, also carry and use infants. The fact that males sometimes carry infants who have already been weaned (and are thus past the age at which infanticide would yield any reproductive benefit for a male) and that immature males also carry infants (this study, Ransom and Ransom, 1971) poses additional problems for the infanticide protection hypothesis. If carrying an infant is meant to communicate the carrier's willingness to defend the infant, as Busse and Hamilton (1981) suggest, such behavior on the part of immature males seems suicidal since they could not possibly defend themselves or the infant against an adult male opponent.

Thus one advantage of the present interpretation over others is that it can be rendered compatible with more of the variations observed across populations. In addition, it does not require individuals to make complicated assessments of the likelihood of paternity; it explains why individuals other than black infants can be used successfully as agonistic buffers; and it is compatible with data suggesting that, under some circumstances, males may commit infanticide on just those infants whose supposed sign stimulus should inhibit such acts.

Finally, this interpretation helps to resolve certain anomalies in the data and hypotheses derived from other populations. Packer (1980), for example, was surprised to find that using infants correlated with a male's seniority in the troop; yet he also noticed, but did not explain, that an infant's cooperation was important in infant use. It is, of course, precisely because cooperation is important and dependent upon prior affiliation between users and infants that males of longer residency have many more opportunities (i.e., can establish affiliated relationships with) to use infants than do newer males.

One set of anomalies in particular has plagued every interpretation of infant use. These anomalies arise in connection with researchers' judgments about the costs of infant use to the infant and the possible biological relatedness between infants and their users. Three possible lines of interpretation have been suggested, but all three present serious puzzles: (1) Males use infants for whom they care and to whom they are related. The puzzle here is why should males risk injuring young relatives, and how are all those instances to be explained where males use infants to whom they could not be or are unlikely to be related? (2) Males use unrelated infants which they care for nonetheless. The puzzle here is why males should care for unrelated infants. (3) Males use infants whom they do not care for and to whom they are not related. This puzzle is why infants or their relatives should let the infants be exploited in this manner.

The present interpretation resolves the puzzles in each of these three scenarios in two simple steps: first, it denies the assumption that the risks to the infant are great on the grounds that there is very little empirical evidence of infant injury; and second, it dismisses the preoccupation with biological relatedness as irrelevant, or at least too narrow, by setting the behaviors of infant use in a much broader context of social strategies that do not depend on biological relatedness for their effectiveness.

As previously noted, despite references by almost all researchers (for a review, see Hrdy, 1976) to the high cost to the infant, direct evidence for such is exceedingly thin. In roughly 10,000 hours of observation in all the studies (see Table 7-7), only one injury to an infant was actually observed as a result of infant use (Packer, 1980). Five cases of injury were *presumed* to be the result of males using infants because the infants were later found to have canine bites (Packer, 1980; Gilmore, 1977). However, these injuries can just as easily be explained as the result of bites sustained during episodes of aggression between males not involving infant use or during punishment of infants undertaken by a single male.

Once the costs to the infant are recognized as minimal, it becomes fruitful to think of infant use as a subset of quite general and pervasive social strategies among baboons (Strum, ms. b, Fig. 7-7). A social strategy is defined here as an integrated set of tactics not directly dependent on aggressive behavior. These tactics involve the "manipulation" of others for an individual's own immediate or delayed benefit, where manipulation is taken to mean "the act of operating upon or managing persons or things with dexterity" *(Oxford English Dictionary).* For group-living primates, social strategies may play an important role in daily life by providing an individual with means to get what he wants without using aggression (social strategies of competition). In addition, when aggression is impending or unavoidable, these strategies may offer the individual methods of buffering the aggression and minimizing its costs (social strategies of defense).

From this perspective, male use of infants as buffers is just one of a number

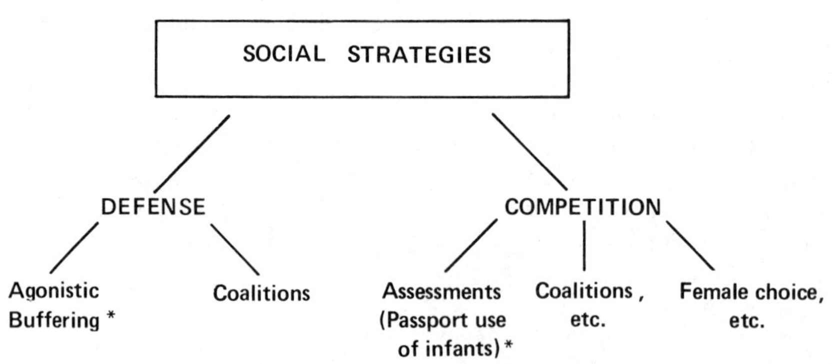

Fig. 7-7. The place of male use of infants in male social strategies. (*As defined in text in section, classification and terminology, p. 150-51.)

of social strategies of defense, which also includes the use of females as buffers (Strum, in pr. a) and may include the formation of coalitions with other males during aggressive episodes (Strum, in pr. a). Similarly, the use of infants as passports with which to approach other males is just one of a number of social strategies of competition (Strum, 1982). (Infants also use each other as passports to approach newcomer males, pers. obs.) These passport approaches are one way resident males assess newcomers and also familiar males at times when unsettling changes in troop membership make it necessary to acquire fresh information about male behavior. The process of approach and assessment may itself affect status relationships to a limited extent, but the primary benefit of assessment is to provide information useful for competitive interactions in other contexts and at other times (Strum, 1982).

These social strategies do not depend on biological relatedness for their effectiveness. Indeed, the whole question of relatedness becomes irrelevant in this framework. A male cares for infants (or females) because the bonds he thereby establishes can work to his benefit in defensive or competitive situations. "Care" represents an investment by the male, and the later cooperation of infants (or females) represents the return on the investment. The caring behavior of the male and the infant's cooperation during use constitute a "reciprocally altruistic" complex (Trivers, 1971).

There is, of course, the possibility of pure exploitation, where males use infants for whom they do not care and to whom they are not related. This possibility is the one that Popp sees as characteristically operative, but he himself calls this behavior "kidnapping" to indicate the opposition of others to the male's appropriation and use of the infant. The very terminology casts doubt on the effectiveness and pervasiveness of the strategy. Cases of pure exploitation from other populations seem rare, or rarely reported, and it is clear, at least from the Pumphouse data, that unaffiliated (i.e., uncared for) infants offer few if any benefits for males. It would be important to know whether or not Popp's (1978) "kidnappers" ever cared for their "victims." The data are lacking. Still, the possibility remains that males on occasion exploit, or at least try to exploit, infants.

The reverse of pure exploitation would be pure paternalism, which is Busse and Hamilton's (1981) interpretation of infant use. What looks like the use of an infant by a male for his own benefit is actually, according to these authors, protection of the infant against infanticide. That males are prepared to assume the high risks of protecting infants against infanticidal immigrants is explained in terms of the presumed close biological relationship between the infants and their protectors (i.e., kin selection). Several criticisms of this position have already been offered above, but it is worth noting additionally that the evidence for the infanticidal tendencies of immigrant males is scant.

It also seems relevant to ask why there should have been so relatively few cases of infants being carried for protection when the dangers of infanticide would have been fairly constant, and, additionally, why males should have picked up infants who were probably at some distance from immigrant males and then carried the infants to their potential killers. Although there may be cases of pure paternalism and kin selection among baboons, it seems difficult to reconcile these instances of male use of infants from this perspective.

CONCLUSIONS

What is the evolutionary basis of male use of infants? And what, if any, is the relationship between this set of behaviors and paternalistic behaviors?

Each of the interpretations considered here implies, and sometimes makes explicit, a mechanism by which infant use might have evolved (Table 7-7). The choice of a mechanism depends partly on how an author assesses the costs and benefits of the phenomenon to its various participants, partly on how he identifies these participants and their relationship to one another; and partly on how he understands the source of the effectiveness of infant use (e.g., sign stimulus, assessment of paternity, affiliation). In other words, the evolutionary explanation in each case is of a piece with the author's entire interpretative framework and cannot be evaluated independently of that framework. Thus Ransom and Ransom (1971) relying upon the concept of sign stimulus, imply that it is to the group's advantage not to have infants killed (group selection), while Gilmore (1977), and Popp (1978), relying on the ability of the animals to assess paternity, see kin selection at work when the use of an infant by one male inhibits the aggression of another. For Busse and Hamilton (1981), the evolutionary mechanism is also kin selection, which operates in this case through the willingness of resident males to defend presumably closely related infants against presumably dangerous immigrants. Packer (1980) sees the use of infants as simultaneously beneficial to the user and infant because the infant is thought to enjoy increased paternal care by being near the male. Further, because an infant may be at high risk of injury when aggressive males are nearby, the infant's safest option is presumed to be to remain near the male user or cling to his belly. Thus the evolutionary mechanism for Packer (1980) is best termed "mutualism" rather than reciprocal altruism, which involves helping someone else in return for past help and/or in expectation of future help. In light of the criticisms that have been presented here against all these interpretative frameworks, none of the above evolutionary explanations of infant use seems adequate.

This study has built its interpretative framework around the concept of social strategies that depend for their success on the skillful creation of social relationships. Male use of infants is one such strategy, and its success depends on the ability of males to win, or earn, the cooperation of infants by caring for them in a paternal manner. The evolutionary mechanism involved here is thus reciprocal altruism, implying minimal costs and *complementary* high benefits for both males and infants (Western and Strum, 1983.). This interpretation seems applicable to other baboon populations once the costs of infant use are appropriately assessed and the link between caring for and using an infant is clarified.

The social strategy framework forwarded here also helps to clarify the nature of paternalistic behaviors among nonhuman primates. Relying on the concept of kin selection, Hrdy (1976) has proposed a sharp dichotomy between caring for infants (paternalism) and exploiting them. She thinks that males care for related infants and exploit unrelated ones. For baboons and other promiscuously mating primates, this sharp dichotomy seems difficult to maintain, given the evidence of this and other studies that males do care for seemingly unrelated infants and may be "exploiting" infant relatives. For these species, most paternalistic behaviors are likely to occur as part of social strategies that link caring for infants and using them, regardless of biological relatedness. This does not mean, of course, that the possibility of real paternal care (Fig. 7-8) should be categorically ruled out. To be certain that paternalistic behavior is real paternal care, however, may require a situation where paternity can be ascertained and where an infant is cared for by a male but never used by him as part of a social strategy.

This distinction between real paternal care and paternalistic behaviors

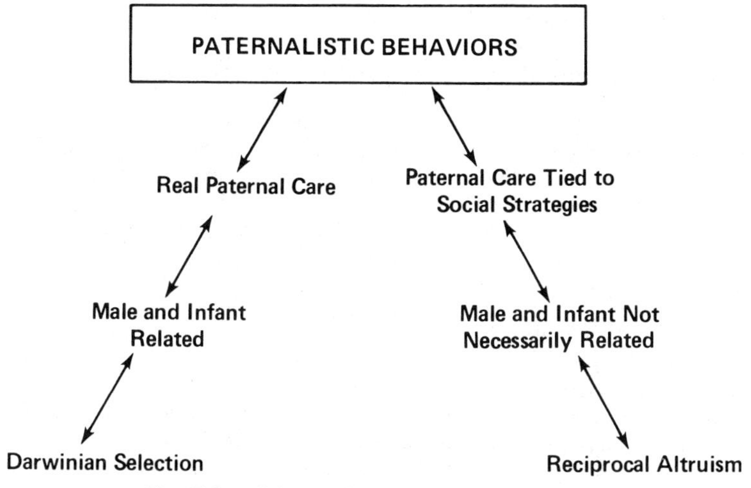

Fig. 7-8. Origins of paternalistic behavior.

that occur as part of social strategies is useful for understanding the evolution of social behavior. In particular, the distinction allows us to see that the close link between "care" and "use" with respect to infants also obtains with respect to other classes of individuals. As already noted, Pumphouse males, for example, care for and use females in much the same way that they care for and use infants. It makes little sense to call this "paternal care" because the female is an adult and obviously unrelated, but the relationship between the two sets of "care and use" behaviors is probably much closer in evolutionary terms than is the relationship between real paternal care and those paternalistic behaviors that are tied to social strategies.

ACKNOWLEDGMENTS

I am indebted to the University of Nairobi, the National Museums of Kenya, and the Institute of Primate Research for the sponsorship of my four field seasons in Kenya. I would also like to thank the government of Kenya for permission to conduct my research and Mr. and Mrs. Arthur Cole and Mr. Richard Dansie for their generosity and hospitality while I was at Gilgil. Drs. Dorothy Cheney, Robert Seyfarth, and David Western provided important comments on a related manuscript which are incorporated into this paper, and Dr. Charles Nathanson's editorial help was invaluable. Karen Savary is responsible for the beautiful illustrations and Barbara Boyer has my gratitude for her expert work at the typewriter.

REFERENCES

Alley, T. 1980. Infantile colouration as an elicitor of caretaking behaviour in Old World primates. *Primates* 2(3): 416-29.

Altmann, J. 1980. *Baboon Mothers and Infants.* Cambridge: Harvard University Press.

Blankenship, L., and Qvortrup, S. 1974. Resource management on a Kenya ranch. *Annals of the South African Wildlife Management Association* 4: 185-90.

Busse, C., and Hamilton, W. 1981. Infant carrying by male chacma baboons. *Science* 212: 1281-82.

Deag, J. 1980. Interactions between males and unweaned Barbary macaques: Testing the agonistic buffering hypothesis. *Behaviour* 56-81.

Deag, J., and Crook, J. H. 1971. Social behaviour and "agonistic buffering" in the wild Barbary macaque, *Macaca sylvana* L. *Folia Primatologica* 15: 183-280.

Gilmore, H. 1977. The evolution of agonistic buffering in baboons and macaques. Paper presented to the *American Association of Physical Anthropologists,* Seattle, Washington.

Harding, R. S. O. 1976. Ranging patterns of a troop of baboons *(Papio anubis)* in Kenya. *Folia Primatologica* 25: 143-85.

Harlow, H., and Harlow, M. 1965. The affectional systems. In A. Schrier; H. Harlow; and F. Stollnitz (eds.), *Behavior of Nonhuman Primates,* pp. 187-334. New York: Academic Press.

Hrdy, S. 1976. Care and exploitation of nonhuman primate infants by conspecifics other than the mother. In J. Rosenblatt; R. A. Hinde; C. Beer; and E. Shaw (eds.), *Advances in the Study of Behavior,* pp. 101-58. New York, Academic Press.

Itani, J. 1959. Paternal care in the wild Japanese monkey, *Macaca fuscata fuscata. Primates* 2: 61-93.

Manning, A. 1967. *An Introduction to Animal Behavior.* Menlo Park: Addison-Wesley.

Marler, P., and Hamilton, W. 1967. *Mechanisms of Animal Behavior.* New York: Wiley and Sons.

Packer, C. 1980. Male care and exploitation of infants in *Papio anubis. Animal Behavior* 28: 512-20.

Popp, J. 1978. Kidnapping among male *anubis* baboons in Masai Mara Reserve. Paper presented at *Werner-Gren Conference, Baboon Field Research: Myths and Models.* New York.

Ransom, T., and Ransom, B. 1971. Adult male-infant relations among baboons *(Papio anubis). Folia Primatologica* 16: 179-95.

Ransom, R., and Rowell, T. E. 1972. Early social development of feral baboons. In Frank Poirier (ed.), *Primate Socialization*, pp. 105-44. New York: Random House.

Strum, S. C. 1982. Agonistic dominance among male baboons: an alternative view. *International Journal of Primatology* 3:175-202.

———. In press a. Why males use females among olive baboons. *American Journal of Primatology.*

———. Manuscript a. Learning to exploit others: The ontogeny of infant use in immature male baboons.

———. Manuscript b. Social strategies and the evolutionary significance of social relationships.

Taub, D. 1978. Aspects of the biology of the wild Barbary macaque: biogeography, the mating system and male-infant associations. Ph.D. dissertation, University of California, Davis.

———. 1980. Testing the "agonistic buffering" hypothesis. *Behavioral Ecology and Sociobiology* 6: 187-97.

Trivers, R. 1971. The evolution of reciprocal altruism. *Quarterly Review of Biology* 46: 35-57.

Western, J., and Strum, S. 1983. Sex, kinship, and the evolution of social manipulation. *Ethology and Sociobiology* 4:19-28.

8

Triadic Interactions Among Male and Infant Chacma Baboons

Curt Busse

Division of Environmental Studies
University of California
Davis, California,
and
Yerkes Regional Primate Research Center
Emory University
Atlanta, Georgia

INTRODUCTION

A widely observed behavior among savanna baboons, in which one male holds or carries an infant in the presence of a second male, has attracted interest because of its possible relevance to theories on aggressive competition (Popp and DeVore, 1979), infant care (Blaffer-Hrdy, 1976), and infant killing (Blaffer-Hrdy, 1979) by male primates. Two consistent observations—that the male holding the infant usually has lower social rank than the second male and that these interactions rarely escalate to fights—have contributed to a widely accepted interpretation, referred to as the agonistic buffering hypothesis; this hypothesis asserts that by carrying an infant, a male reduces the probability of receiving aggression from a higher-ranking male (Ransom and Ransom, 1971; Gilmore, 1977; Popp, 1978; Packer, 1980; Stein, 1981). A recent, alternative interpretation, referred to as the infant protection hypothesis, suggests that carrying an infant functions instead to protect the infant from attack by a potentially infanticidal male (Busse and Hamilton, 1981). The present chapter examines these hypotheses in light of a nine-month study of male-infant interactions in two groups of chacma baboons *(Papio ursinus)* in the Moremi Wildlife Reserve, Botswana. This chapter amplifies the previous report (Busse and Hamilton, 1981) which set forth the infant protection hypothesis.

The hypotheses mentioned above overlap in several of their predictions. For example, the agonistic buffering hypothesis predicts the frequency at which males are threatened or attacked will be lower when they hold infants than when they do not. The infant protection hypothesis predicts that infants carried periodically by males will receive fewer wounds than will infants not carried by males (ultimately, survivorship will be higher for carried infants). Thus, both hypotheses suggest relatively low frequencies of aggression and wounding during triadic interactions, although for different reasons.

The predictions derived from these hypotheses differ, however, when concerned with the biological relatedness of the interacting participants to one another. The infant protection hypothesis suggests that the distribution of triadic interactions among males and infants will reflect paternity or other kinship relationships: infants will be carried by their possible fathers (or other relatives, in the presence of males that are not their fathers (nor other relatives). Predictions of the agonistic buffering hypothesis vary according to the proposed mechanism by which the presence of an infant inhibits aggression. One suggestion is that dominant males are inhibited from behaving aggressively because the infants carried in their presence could be their offspring and might be injured during a fight (Popp, 1978; note that this prediction that the recipient males are possible fathers of infants carried in their presence is just the opposite of that derived from the infant protection hypothesis). Another suggestion is that some characteristic of infants, such as their black coat coloration (see Ransom and Ransom, 1971; Alley, 1980), inhibits aggression in general. This possibility makes no inherent predictions about the relatedness of participants. Therefore, the following analysis of the distribution of interactions among males and infants and the possible paternity of infants is relevant primarily to the infant protection hypothesis and to one suggested mechanism for the agonistic buffering hypothesis.

METHODS

General

The study was conducted in the Moremi Wildlife Reserve (23°02′E, 19°31′S), 65km. NW of Maun, Botswana. This site is near the center of the Okavango Delta, between the Boro River and Chief's Island (Fig. 8-1). This area floods seasonally from May to September; floodplains reach an average depth of 1 m. during this time. The floodplains are interspersed with numerous small islands that are above the flood levels (Fig. 8-2). Islands typically consist of

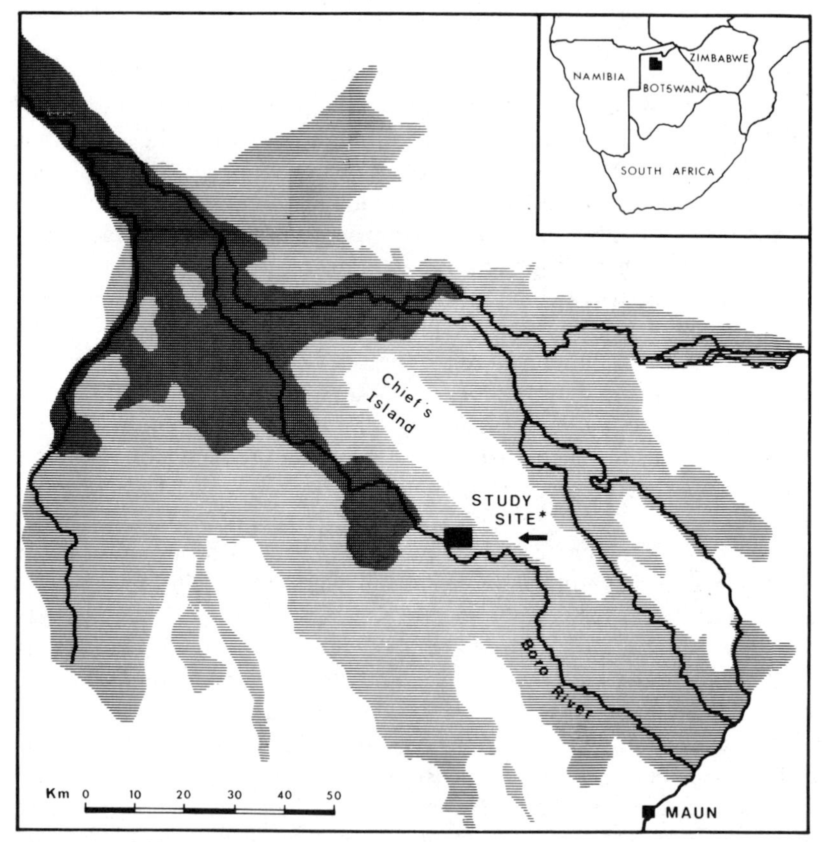

Fig. 8-1. Map of the Okavango Delta, Botswana, showing the location of the study site. Dark shading indicates perennial flooding; light shading indicates seasonal flooding (see Fig. 8-2 for a blowup of this area).

woodland with closed or broken canopies along the perimeter and grassland and open woodland in the interior (see Tinley, 1973, for detail).

Two chacma baboon *(Papio ursinus)* groups, group C and group W, were observed from August 1977 to June 1980. Each group averaged 70 members, including 6 to 10 adult males and 19 to 23 adult females plus their immature offspring. The present study of male-infant-male interactions spanned 270 days, from 13 September 1979 to 8 June 1980; demographic and behavioral records for each group had been maintained for over two years when the study began. During the study group C was censused on 138 days (51%) and group W was censused on 102 days (38%). During each census, births, deaths, disappearances, emigrations, immigrations, female reproductive

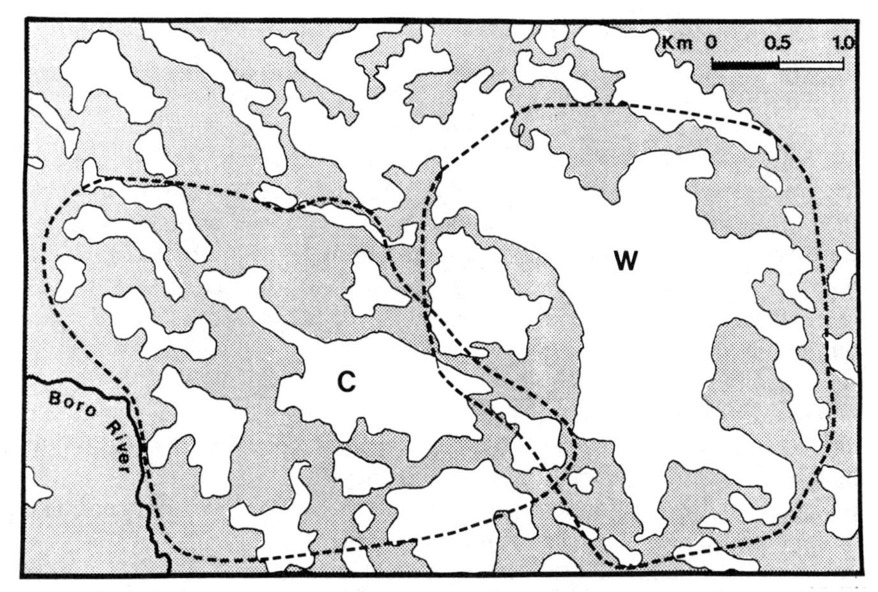

Fig. 8-2. Map of study site showing approximate home ranges of group C and group W. Shaded areas are seasonally flooded; open areas are permanently dry.

status, and consortships between males and females were recorded. Data collected from previous studies of these groups were also used in some of the demographic analyses. Dates of demographic events were estimated as the midpoint of the days immediately before and after the event had been detected.

Terminology

Male-infant-male interactions, in which one male holds or carries an infant in a ventral position in response to the behavior or proximity of a second male, are referred to for convenience as triadic interactions (after Taub, 1980). Other labels for this behavior pattern include agonistic buffering (Gilmore, 1977; after Deag and Crook, 1971), kidnapping (Popp, 1978), infant carrying (Busse and Hamilton, 1981), and countercarrying (Hamilton, chapter 12). The male holding or carrying the infant is referred to here as the carrier. Other labels include kidnapper (Popp, 1978) and carrying male (Busse and Hamilton, 1981). The male not holding or carrying the infant is referred to as the recipient, in that he is the focus of the carrier's behavior. Another label used elsewhere is the opponent (Popp, 1978; Busse and Hamilton, 1981).

Behavioral Observations

General. During the study interval, group C was observed for 335 hours and group W was observed for 240 hours. Subject animals were identified from distinctive physical characteristics and from eartags, which were applied to a majority of adults during the study. Observations were made on foot; all subjects allowed observers to approach within 1 to 3 m. of them. Subjects were usually observed from slightly greater distances (~5 m.) to minimize disturbance. When new males joined a group, they quickly became accustomed to humans, sometimes allowing observers to approach within 5 m. of them on the first day.

Male-Infant-Male Interactions. Because of their low frequency, triadic interactions were recorded ad libitum. Information scored for each interaction included (1) the identity of each participant; (2) how the interaction was initiated; (3) the duration that the carrier held the infant; (4) whether either male chased, displaced, or supplanted the other; (5) whether either male was consorting with a female; (6) whether the infant screamed or resisted being carried; and (7) whether the mother showed alarm or attempted to retrieve her infant during the interaction.

Dominance Measurements. Dominance relationships were determined from ad libitum observations of "who supplants whom" at food resources. From these dyadic interactions, dominance hierarchies were constructed separately for adult males and for adult females. Dominance relationships were known at the beginning of the study from 681 observations of supplanting that had been collected earlier. Supplanting records were maintained during the present study to monitor any changes in dominance relationships, especially when males migrated into the study groups.

Paternity Analysis

Assessment of the paternity of infants involved in triadic interactions is addressed in three ways. First, was the male in the group when the infant was conceived? This analysis permits exclusion of some males as possible fathers. Second, what was the male's social rank when the infant was conceived? High social rank may be one measure of a male's priority of access to ovulating females in some baboon populations (Hausfater, 1975; Popp, 1978; Packer, 1979 a,b; but see Smuts, 1982; Strum, 1982), including the population considered here (K. S. Smith, pers. comm.). Third, did the male copulate with the mother during the menstrual cycle in which the infant was conceived? For this last question, three categories of answer are considered: "yes" indicates that the male was observed copulating at least once with the mother; "possibly" indicates that the male was in the group but was not observed copulating with the mother; and "no" indicates that the male was not in the group.

RESULTS

Male Migration

Figure 8-3 shows adult and subadult male residency patterns in group C from January 1978 through June 1980. During the present study, male membership in group C changed 17 times as a result of nine immigrations, four emigrations, three disappearances, and one predation by lions. Four different males held

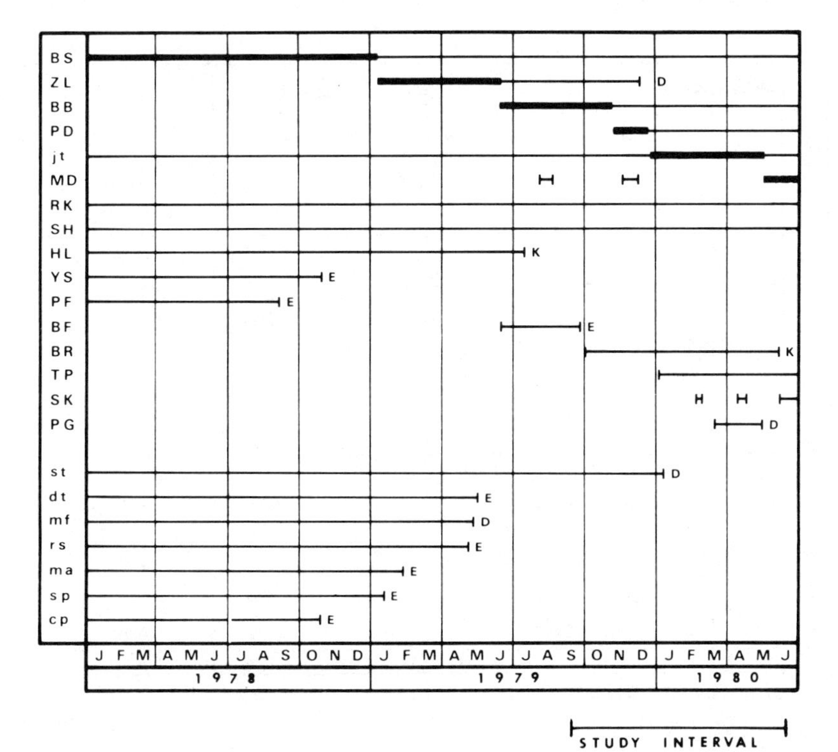

Fig. 8-3. Adult male membership in Group C from January 1978 through June 1980. Thin horizontal lines indicate presence in the group. Thick horizontal lines indicate that the male held alpha rank. Males that were subadults in January 1978 are shown in lower case letters. Disappearances are due to emigration (E), predation (K), or unknown causes (D).

alpha rank during the study. One of these was a resident male who could have been born in the group. The other three males gained alpha rank soon after immigrating.

Figure 8-4 shows adult and subadult male residency patterns in group W from October 1977 through June 1980. During the present study, male membership in group W changed six times as a result of one immigration, three emigrations, and two disappearances. One male, NR, held alpha rank throughout the study interval.

Considering both groups, 11 different males held alpha rank since studies began in 1977. Median tenure at alpha rank was 5 months, with a range in each group from 1.5 months to at least 12 months. In 8 of 10 rank changes, the new alpha male was a recent immigrant. In total, 8 of 14 immigrants held alpha rank. No males migrated between groups C and W since studies began.

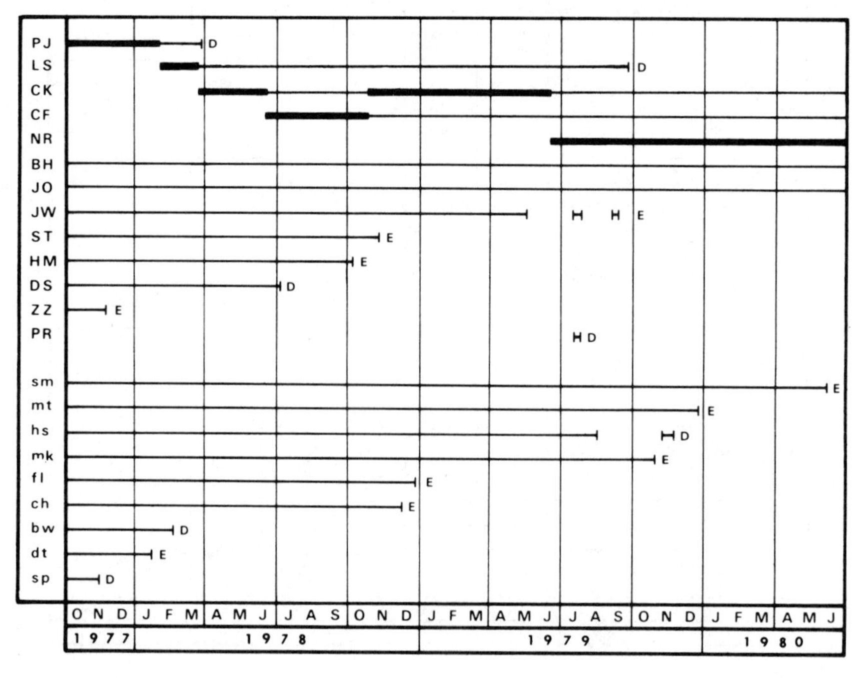

Fig. 8-4. Adult male membership in Group W from October 1977 through June 1980. Thin horizontal lines indicate presence in the group. Thick horizontal lines indicate that the male held alpha rank. Males that were subadults in October 1977 are shown in lower case letters. Disappearances are due to emigration (E) or unknown causes (D).

General Description of Triadic Interactions

Interactions involving two adult males and an infant were observed 51 times in group C and 61 times in group W. A typical interaction began as the recipient walked into the vicinity of the carrier and infant, who usually were frequent associates of each other (see Ransom and Ransom, 1971; Altmann, 1978, 1980; Packer, 1980; Stein, 1981; Smuts, 1982; Strum, chapter 7). The carrier then picked up the infant (in at least 12 episodes) or the infant ran to him (in at least 18 episodes). The carrier sometimes grunted repeatedly while the recipient was nearby. Infants often vocalized in unison with the carrier. Recipients usually were silent. After the recipient walked away, or the carrier took the infant away from the recipient, the infant broke contact with the carrier and the interaction was considered to be ended.

This generalized pattern had many variations, however (see Stein, 1981 and Smuts, 1982, for detailed descriptions of triadic interactions in *Papio cynocephalus* and *P. anubis,* respectively). In a recurring variation (13 episodes), infants incited the interaction by approaching and threatening the recipient while the carrier was nearby (see Packer, 1979a, description of "ambivalence"). The carrier then grabbed the infant, usually carrying it away from the recipient. One interaction occurred after a male chased a mother carrying an infant up a tree. The mother's screams attracted a second male, who ran up the tree and held the infant before the first male descended from the tree. Another interaction occurred after a male chased an infant through a large tree. The infant's screams attracted a second male, who ran up the tree, grabbed the infant, and chased the first male down the tree.

Three times resident males picked up and held infants while unknown males from adjacent groups were displaying through the group space and chasing group members. These interactions are not included in these analyses because the carriers never interacted with the outsiders. Nevertheless, the interval of carrying coincided with the outsiders' presence.

The median duration of infant contact with carriers during 112 interactions (both groups pooled) was 10 seconds, with a range from one to 360 seconds (mean = 36 seconds). Carriers chased or displaced recipients during four interactions (3.6%), and recipients chased or displaced carriers during 11 interactions (9.8%). Only one interaction escalated to a fight with canine fencing between the males. No males nor infants were wounded during any interaction.

Foods were involved twice (1.8%): a carrier supplanted a recipient once, and a recipient supplanted a carrier once. During nine interactions (8.0%), recipients were in consort with estrous females. By comparison, males consorted females during an average of 13% of the days males were observed during the study ($N = 14$ males censused during at least 50 days). Carriers never gained access to estrous females during or after interactions (see also Strum, chapter 7).

Infant-Carrier Combinations

Group C. Triadic interactions in group C involved 14 infants carried between 1 and 11 times each (mean = 3.6 interactions/infant, Table 8-1). Of the eight infants carried more than once, three had only one carrier, four had two carriers, and one had four carriers. The lowest degree of infant-carrier specificity observed in either group involved this last infant, CY. This infant, however, had been carried four times by male ZL before the male disappeared. Afterwards, three different males carried the infant one time each. Seven adult males in group C carried infants between 1 and 18 times each (mean = 7.3 interactions/male, Table 8-1). Three of these males, SH, JT, and RK, accounted for 82.4% of the carrying episodes. SH carried four infants a total of 18 times; one of these infants accounted for 9 of SH's interactions (50.0%) and another accounted for 7 of his interactions (38.9%). JT carried four infants a total of 13 times; one of these infants accounted for 8 of JT's interactions (61.5%). By contrast, male RK did not carry infants selectively: his 11 interactions were distributed among seven infants (mean = 1.6 interactions/infant), and the infant he carried most frequently accounted for only 27.3% of his carrying interactions.

For 20 to 21 infant-carrier combinations (95.2%) in group C, the carrier had been in the group when the infant was conceived (Table 8-1). Thus, all but one of the carriers were possible fathers of the infants they carried. The exception involved a relatively old, middle-ranking male, TP, who had immigrated 153 days before his infant-carrying interaction. During this interaction the mother closely followed TP and raised her tail (an alarm behavior exhibited by mothers to immigrant males; Busse, 1981). Long-term records of mating patterns were relatively incomplete for this group, and for only three combinations had the carrier been observed mating with the mother when the infant was conceived. Carriers usually had been high ranking at the time of conception, however; for 15 of the 21 combinations (71.4%) the carrier had ranked first, second or third in the male hierarchy.

Group W. Triadic interactions in group W involved 11 infants carried between 1 and 19 times each (mean = 5.5 interactions/infant, Table 8-2). Of the seven infants carried more than once, five had only one carrier and two had two carriers. The highest degree of infant-carrier specificity observed in either group involved infant FI, who was carried all 19 times by male CK.

Table 8-1. Distribution of Infant-Carrier Combinations and Evidence of Paternity in Group C.

Infant	Carrier	No. of Interactions	Carrier's Rank at Conception	Mated with Mother?
MX	SH	9	3 of 8	Possibly
	RK	2	4 of 8	Possibly
DN	JT	8	4 of 10	Possibly
	RK	1	3 of 10	Possibly
MY	SH	7	4 of 8	Possibly
CY	ZL	4	1 of 8	Yes
	BS	1	2 of 8	Possibly
	JT	1	3 of 8	Possibly
	TP	1	*	No
LJ	RK	3	4 of 10	Possibly
	BS	1	2 of 10	Possibly
NN	RK	2	2 of 10	Possibly
	SH	1	3 of 10	Possibly
BX	JT	2	1.5 of 8	Possibly
WE	JT	2	3 of 10	Possibly
TO	ZL	1	1 of 8	Yes
BU	BS	1	1 of 9	Yes
CB	RK	1	2 of 10	Possibly
SL	RK	1	2 of 9	Possibly
NS	RK	1	2 of 9	Possibly
OT	SH	1	4 of 9	Possibly
			% of Total	No. of Infants
Subtotals	SH	18	35.3	4
	JT	13	25.5	4
	RK	11	21.6	7
	ZL	5	9.8	2
	BS	3	5.9	3
	TP	1	2.0	1
Totals		51	100.1	21

*Not applicable; male was not a group member.

Four males in group W carried infants (Table 8-2). One of these males, CK, accounted for 77.0 percent of the carrying ($N = 47$). CK carried a total of seven infants, four of them seven or more times. The three other males also showed selectivity in which infants they carried, despite their low frequencies of participation in triadic interactions.

For all 13 infant-carrier combinations in group W, the carrier had been in the group when the infant was conceived (Table 8-2). For 12 of the 13 combinations (92.3%) the carrier had ranked first or second at the time of

Table 8-2. Distribution of Infant-Carrier Combinations and Evidence of Paternity in Group W.

Infant	Carrier	No. of Interactions	Carrier's Rank at Conception	Mated with Mother?
FI	CK	19	1 of 6	Yes
BL	CK	9	2 of 8	Possibly
SI	CK	7	1 of 9	Yes
	JO	2	8 of 9	Possibly
AY	CK	7	1 of 6	Yes
SL	CF	5	1 of 7	Possibly
LB	CK	3	1 of 9	Possibly
	CF	2	2 of 9	Possibly
BG	BH	3	2 of 10	Yes
GO	CK	1	1 of 8	Yes
FC	CK	1	1 of 8	Yes
LE	BH	1	2 of 10	Yes
BR	BH	1	2 of 10	Possibly
			% of Total	No. of Infants
Subtotals	CK	47	77.0	7
	CF	7	11.5	2
	BH	5	8.2	3
	JO	2	3.3	1
Totals		61	100.0	13

conception, and for seven combinations (53.8%) the carrier had been observed mating with the mother when the infant was conceived.

Infant-Recipient Combinations

Group C. Infant-recipient combinations ($N = 31$, Table 8-3) outnumbered infant-carrier combinations in group C. Of the eight infants carried more than once, six infants had three or more recipients (i.e., males in the presence of whom they were carried; see "Methods" section), one infant carried twice had two recipients, and one infant carried twice had one recipient. The highest frequency for any combination was seven interactions (infant DN carried in the presence of male BB).

Eight males were recipients of infant carrying in group C (Table 8-3). Two of these males, BB and MD, accounted for 58.9% of the interactions. BB was the recipient in 19 interactions involving 8 infants (mean = 2.4 interactions/infant). MD was the recipient in 11 interactions involving 8 infants (mean = 1.4 interactions/infant). For the other males as well, there was little specificity in terms of which infants were carried in their presence (Table 8-3).

Table 8-3. Distribution of Infant-Recipient Combinations and Evidence of Paternity in Group C.

Infant	Recipient	No. of Interactions	Recipient's Rank at Conception	Mated with Mother?
MX	BB	5	*	No
	SK	2	*	No
	BR	2	*	No
	MD	1	*	No
	TP	1	*	No
DN	BB	7	*	No
	MD	1	*	No
	PD	1	*	No
MY	PD	3	*	No
	BR	2	*	No
	BB	1	*	No
	MD	1	*	No
CY	MD	3	*	No
	BB	2	*	No
	SK	1	*	No
	TP	1	*	No
LJ	MD	2	*	No
	PD	1	*	No
	BR	1	*	No
NN	PD	1	*	No
	SK	1	*	No
	JT	1	>10 of 10	Possibly
BX	PG	2	*	No
WE	PD	1	*	No
	MD	1	*	No
TO	MD	1	*	No
BU	BB	1	*	No
CB	BB	1	*	No
SL	BB	1	*	No
NS	MD	1	*	No
OT	BB	1	*	No
			% of Total	No. of Infants
Subtotals	BB	19	37.3	8
	MD	11	21.6	8
	PD	7	13.7	5
	BR	5	9.8	3
	SK	4	7.8	3
	PG	2	3.9	1
	TP	2	3.9	2
	JT	1	2.0	1
Totals		51	100.0	31

*Not applicable; male was not a group member.

For 30 of 31 infant-recipient combinations (96.8%) in group C, the recipient had not been in the group when the infant was conceived (Table 8-3). Therefore, with the single possible exception of male JT (a low-ranking subadult at the time of the infant's conception who subsequently rose in rank; even in this instance he probably was not the infant's father), no recipient of infant-carrying could have sired any infants carried in their presence.

Group W. In this group, male NR was the recipient in 93.4% of the interactions ($N = 57$, Table 8-4). NR's interactions involved 10 of the 11 infants carried in the study. Of the seven infants carried more than once, three had a second recipient in addition to NR. The most frequently carried infant, FI, was carried all 19 times in NR's presence.

For 12 of 14 infant-recipient combinations (85.7%) in group W, the recipient had not been in the group when the infant was conceived (Table 8-4). In 10 of these combinations the recipient was NR; in two combinations it was male

Table 8-4. Distribution of Infant-Recipient Combinations and Evidence of Paternity in Group W.

Infant	Recipient	No. of Interactions	Recipient's Rank at Conception	Mated with Mother?
FI	NR	19	*	No
BL	NR	8	*	No
	SM	1	>8 of 8	Possibly
SI	NR	9	*	No
AY	NR	7	*	No
SL	NR	5	*	No
LB	NR	4	*	No
	BH	1	5 of 9	Possibly
BG	NR	2	*	No
	CK	1	*	No
GO	NR	1	*	No
FC	NR	1	*	No
LE	NR	1	*	No
BR	CK	1	*	No
			% of Total	No. of Infants
Subtotals	NR	57	93.4	10
	CK	2	3.3	2
	SM	1	1.6	1
	BH	1	1.6	1
Totals		61	99.9	14

*Not applicable: male was not a group member.

CK. Even though CK was the most frequent carrier in the group, he had immigrated after the conceptions of the two infants that were carried in his presence. Two recipients were resident males: one, SM, had been a low-ranking subadult and the other, BH, had been a middle-ranking adult when the infants carried in their presence had been conceived. As with group C, even in these instances in which the recipients were resident males, neither had a high probability of being the father of the carried infant.

Whether a male immigrated before or after the birth of an infant did not appear to influence his participation as a recipient in triadic interactions. For 55 interactions involving a total of 11 infant-recipient combinations in both groups, the recipient had immigrated during the six-month interval of pregnancy; whereas for 52 interactions involving 29 combinations, the recipient had immigrated while the infant was 0 to 12 months old. Binomial tests correcting for the longer duration of the post parturition interval reveal no significant difference from that expected by random when comparing the number of interactions ($z = 0.81, p = 0.21$, one tailed test) or the number of infant-recipient combinations ($z = -0.78, p = 0.22$, one-tailed test).

Carrier-Recipient Combinations

Group C. The six carriers and eight recipients in group C were paired in 22 combinations (Table 8-5). The most frequent carriers, SH, JT, and RK, interacted with 5, 4, and 7 recipients, respectively. The most frequent recipients, BB, MD, and PD, interacted with 5, 5, and 3 carriers respectively. Thus, specificity of interaction was low in both directions. Ten combinations of males interacted only once; the most frequent combination (JT versus BB) had eight interactions (15.7%).

Carriers ranked lower than recipients in 48 of the 51 interactions (94.1%) in group C (Table 8-5). For the two interactions involving JT and PG, no dominance relationships had been identified. In only a single interaction did the carrier rank higher than the recipient: JT carried an infant in the presence of BB after BB had dropped from alpha rank and JT had risen to alpha rank. (This interaction does not appear in Table 8-5 because the ranks shown are medians of several interactions.)

Carriers in group C usually not only were relatively lower in rank than recipients, but generally had low absolute rank: for 17 of the 22 combinations (77.3%), the carrier ranked in the lower half of the hierarchy. By contrast, recipients had high absolute rank: for 20 of 22 combinations (90.9%), the recipient ranked in the top half of the hierarchy. For all 51 interactions the mean difference in rank between carrier and recipient was 3.9 rank positions.

Table 8-5. Distribution of Carrier-Recipient Combinations and Social Ranks of Interactants in Group C.

No. of Interactions	Carrier	Median Rank of Carrier	Recipient	Median Rank of Recipient
7	SH	6 of 7	BB	1 of 7
4	SH	8 of 9	BR	6 of 9
3	SH	7 of 8	PD	2 of 8
2	SH	8 of 9	MD	1.5 of 9
2	SH	9 of 10	SK	4 of 10
8	JT	2 of 7	BB	1 of 7
2	JT	3 of 9	MD	1.5 of 9
2	JT	1.5 of 9	PG	1.5 of 9
1	JT	4 of 9	PD	1 of 9
3	RK	9 of 9	PD	1 of 9
2	RK	7 of 7	BB	1 of 7
2	RK	9 of 9	MD	1.5 of 9
1	RK	8 of 8	JT	1 of 8
1	RK	9 of 9	SK	4 of 9
1	RK	9 of 9	BR	7 of 9
1	RK	8 of 8	TP	3 of 8
4	ZL	7 of 9	MD	2 of 9
1	ZL	7 of 9	BB	4 of 9
1	BS	4 of 7	BB	1 of 7
1	BS	6 of 9	MD	2 of 9
1	BS	6 of 8	TP	3 of 8
1	TP	6 of 10	SK	4 of 10
Total: 51				

Group W. Carrier-recipient combinations in group W, by contrast to those in group C, were highly specific (Table 8-6). The combination of CK and NR accounted for 46 of the 61 interactions (75.4%). NR held alpha rank throughout the study, and CK held second rank until the last week of the study; and in all other interactions carriers ranked lower than recipients. The mean difference in rank between carriers and recipients was 1.6 rank positions.

Infant Age, Gender, and Mother's Rank

In both groups the infants carried most frequently were between two and seven months old. The monthly distributions of interactions as a function of infant age were unimodal, with peaks at 3 to 4 months in group C (median = 126 days, range = 39-603 days) and 5 to 6 months in group W (median = 168 days, range 53-642 days). Only 11 of the 112 interactions (9.8%) involved infants older than one year of age (which is the average age at which mothers' lactational amenorrhea ends, $\overline{X} \pm \text{SE} = 356 \pm 4$ days, $N = 21$ females).

Table 8-6. Distribution of Carrier-Recipient Combinations and Social Ranks of Interactants in Group W.

No. of Interactions	Carrier	Median Rank of Carrier	Recipient	Median Rank of Recipient
46	CK	2 of 6	NR	1 of 6
1	CK	3 of 6	SM	2 of 6
6	CF	6 of 6	NR	1 of 6
1	CF	6 of 6	BH	4 of 6
3	BH	4 of 6	NR	1 of 6
2	BH	4 of 6	CK	2 of 6
2	JO	5 of 6	NR	1 of 6
Total: 61				

In group C, only 2 of the 16 infants that were under two years of age at the end of the study (SA and WP) were not observed being carried (Table 8-7). For the mothers of both infants, lactational amenorrhea ended at the beginning of the study; presumably the infants were nearly weaned. The frequency at which the 16 infants were carried was inversely correlated with age (Pearson $r = -0.47$, $p < 0.05$, $df = 15$, one-tailed test).

Table 8-7. Characteristics of Infants Carried During Triadic Interactions in Group C[1].

Infant I.D.	No. of Interactions	Gender	Age Range (Days)	Mean age When Carried	I.D. of Mother	Rank of Mother
MX	11	F	80–350	205	MN	22 of 22
DN	9	M	95–365	142	DS	15 of 22
MY	7	F	0–213	156	MS	19 of 22
CY	7	M	26–296	149	CH	10 of 22
LJ	4	M	33–303	204	LL	3 of 22
NN	3	F	255–525	422	NT	9 of 22
BX	2	M	0–91	39	BN	17 of 22
WE	2	F	26–296	107	WD	12 of 22
TO	1	M	1–271	78	TB	4 of 22
BU	1	M	122–392	126	BT	1 of 22
CB	1	M	297–567	329	CL	14 of 23[2]
SL	1	F	332–602	345	SS	16 of 22
NS	1	M	353–623	603	NL	8 of 22
OT	1	M	381–651	398	OO	5 of 22
SA	0	F	313–583	*	SD	6 of 22
WP	0	M	434–704	*	WM	2 of 22

*Not applicable: infant not carried during any interactions.
[1]Included are all individuals that were less than two years old at the end of the study.
[2]Mother disappeared at the beginning of the study.

Table 8-8. Characteristics of Infants Carried During Triadic Interactions in Group W[1].

Infant I.D.	No. of Interactions	Gender	Age Range (Days)	Mean Age When Carried	I. D. of Mother	Rank of Mother
FI	19	F	51-321	145	FR	5 of 19
BL	9	F	161-431	261	BA	7 of 19
SI	9	M	41-311	230	CD	4 of 19
AY	7	M	86-356	178	AN	14 of 19
SL	5	M	150-420	158	SH	2 of 19
LB	5	M	20-290	144	LR	9 of 19
BG	3	M	383-653	490	BB	6 of 19
GO	1	F	335-605	356	GR	12 of 19
FC	1	F	313-583	382	FN	11 of 19
LE	1	M	356-626	579	LL	18 of 19
BR	1	F	397-667	642	BN	13 of 19
PN	0	M	0-64	*	PG	15 of 19
SZ	0	M	0-110	*	SF	17 of 19
EN	0	M	0-164	*	EM	1 of 19
AM	0	F	61-331	*	AC	3 of 19
WK	0	F	185-455	*	WW	8 of 19
TF	0	M	317-587	*	BE	16 of 19
LN	0	F	394-664	*	MD	10 of 19

*Not applicable: infant not carried during any interactions.
[1]Included are all individuals that were less than two years old at the end of the study.

In group W, 7 of 18 infants younger than two years old (38.9%) were not observed being carried (Table 8-8). For the mothers of two of these infants (TF and LN), lactational amenorrhea ended at the beginning of the study. A third infant, WK, had been carried before the study by male LS, who had mated with the mother when the infant was conceived, but who disappeared 11 days after the study began. After LS disappeared the infant received no further male care. A fourth infant, AM, had a male associate, BH, but she was not carried by this male in any triadic interactions reported here.

The other three infants that were not carried were born during the study: two, PN and EN, were possible offspring of the alpha male, NR. The third infant, SZ, disappeared when it was 110 days old. Partly because these young infants were not carried, the frequency that infants were carried in group W was not significantly correlated with age (Pearson $r = -0.16$, n.s., $df = 17$, one-tailed test).

Infant gender was unrelated to frequency of participation in triadic interactions. In group C, 6 female infants were carried a total of 24 times, and 10 male infants were carried 27 times. In group W, 8 female infants were carried 31 times, and 10 male infants were carried 30 times. After controlling for infant age, the distributions of interactions were unrelated to infant gender in group C (point biserial $r = -0.26$, n.s., $df = 15$, two-tailed test) and in group W (point biserial $r = -0.17$, n.s. $df = 17$, two-tailed test).

Infants carried most frequently in group C had low-ranking mothers, and the distribution of interactions was correlated significantly with mother's rank (Pearson $r = -0.86$, $p < 0.01$, $df = 15$, two-tailed test). By contrast, infants carried most frequently in group W had high-ranking mothers, but the distribution of carrying in this group was not significantly correlated with mother's rank (Pearson $r = 0.39$, n.s., $df = 17$, two-tailed test).

Infant Carrying by Juvenile and Subadult Males

In 31 additional triadic interactions in both groups, the carrier was a juvenile or subadult male between four and seven years old. In 23 of these interactions (74.2%), the recipient was an adult male that had immigrated after the carried infant was conceived, in 3 interactions (9.7%), the recipient was an unidentified male from an adjacent group during an intergroup encounter; and in 5 interactions (16.1%), the recipient was an older subadult male that had been present but had held low social rank when the infant was conceived. Thus, recipients probably were never the fathers of infants carried by immature males, as has been shown to be the case for infants carried by adult males.

All immature males that carried infants had been group members when the infants they carried were conceived, but were probably not the fathers because of their low social ranks. Also, some were not sexually mature. Unfortunately, the groups had not been studied long enough to determine whether immature males could have been siblings or other relatives of the infants they carried.

Because interactions involving immature males were not scored every time they were observed, their frequency can not be compared with that for adult males. Durations of interactions and ages of infants carried, however, differed significantly between these two age categories of carriers. The median durations of interactions involving immature males was 5 seconds; for those involving adult males it was 10 seconds (median test, $\chi^2 = 15.9, p < 0.01$). The median age of infants carried by immature males was 317 days versus 168 days by adult males (median test, $\chi^2 = 15.9, p < 0.01$). Juvenile and subadult males also differed from adult males by frequently carrying infants outside the context of triadic interactions and by occasionally adopting infants whose mothers died (Hamilton, et. al., 1982).

DISCUSSION

Does carrying an infant function to protect the carrier, the infant, or both? Although patterns of male-infant-male interactions are similar across savanna baboon populations, this question remains controversial (see also Strum, chapter 7). Despite the many similarities, however, triadic interactions do not necessarily function identically in each population that has been studied; hence, the interpretations offered here are meant to apply only to the two groups studied at Moremi. In addressing the alternative hypotheses, the discussion first considers the descriptive evidence; then considers the distribution of triadic interactions among group members and the possible relatedness of participants; and finally considers possible proximate and ultimate mechanisms by which carrying an infant could inhibit aggression.

The two study groups provide a marked contrast for the analysis of triadic interactions because of the opposite patterns of male composition and stability of dominance ranks during the study. Group W was more stable: it consisted of the same six adult males during almost the entire study (95%). The alpha male, NR, a recent immigrant, held his rank for the entire interval. The only dominance reversal, in which SM took second rank from CK, occurred at the end of the study. By contrast, group C was less stable: six males immigrated and four males held alpha rank at different times during the study. Interestingly, no males migrated between these two groups during the study nor for at least two years before the study; thus, there were no dual male memberships in the two study groups.

One commonality of triadic interactions in both groups was the absence of distress between carriers and infants. Carriers did not handle infants roughly, nor did infants scream or otherwise resist being carried. Only once did a mother attempt to retrieve her infant before a triadic interaction ended; another time a mother raised her tail (a sign of alarm; Busse, 1981) during the only interaction (of 112) in which an immigrant male carried an infant. This descriptive evidence provides little support for the facet of the agonistic buffering hypothesis, which holds that males are exploiting infants they carry.

After the present study was completed, Hamilton (pers. comm.) observed several interactions in which young infants screamed and resisted being carried by males that handled them roughly. Observations of distress involving young infants (also see Ransom and Ransom, 1971), who may have had little prior experience being carried by males, suggest that distress may be associated with the ontogeny of carrying involving some infants (Hamilton, pers. comm.; see Strum, chapter 7). Far more often, however, infants and mothers exhibit no distress toward carriers (also see Packer, 1980; Stein, 1981; Smuts, 1982).

By contrast, infants and mothers persistently exhibit alarm toward males that are frequent recipients of triadic interactions. Mothers carrying infants raise their tails and sometimes avoid and scream when in close proximity to immigrant males (Busse, 1981), who are the most frequent recipients of triadic interactions. Infants exhibit alarm toward these males by approaching them within a few meters, threatening, and sometimes screaming at them (Packer, 1979a; Busse, 1981). Some of the triadic interactions reported here were initiated when infants threatened the recipients in this manner.

Analyses of which males carry which infants support a primary prediction of the infant protection hypothesis that males may be fathers or other relatives of infants they carry. Almost always carriers had been members of the group, usually they had held high social rank, and, where data are available, they had mated with the mother during the menstrual cycle in which the infants they carried were conceived.

The specificity in the distribution of infant-carrier combinations agrees with the pattern predicted where confidence of paternity is relatively high. Infants that were carried frequently usually had an exclusive male carrier, or a primary carrier and a secondary carrier (terminology follows Taub, 1980). The one exception to this pattern involved an infant carried by three different males after its primary carrier (and probable father) had disappeared. One prediction, untestable from available data, is that exclusive or primary carriers mated with the mother near the optimal time for conception, and that secondary carriers, if any, may have mated with the mother near but not at the optimal time.

Whereas all but one infant was carried by one or two males, some males carried several different infants. This observation also is consistent with a paternity interpretation: if alpha males sire a majority of infants, then they may have a cohort of infant offspring they could carry.

The agonistic buffering hypothesis predicts that males who carry infants most frequently will be those that receive the most aggression. One might therefore expect an inverse correlation between male social rank and frequency of carrying infants (see Taub, 1980), although this is not necessarily so (see Bernstein and Gordon, 1974). Two of the four most frequent carriers, SH and RK, fit this prediction: they were the lowest-ranking adult males in group C. RK's pattern of carrying also fits a second prediction of the agonistic buffering theory that males will carry infants at random rather than selectively (see Taub, 1980): during a total of 11 interactions, RK carried seven infants. Four of the infants were relatively old, however, and had been conceived when RK had held high rank. Thus, the pattern of carrying by this male is also partly consistent with the infant protection hypothesis. Nevertheless, RK's pattern of carrying suggests that some triadic interactions in this population could function to protect the carrier. The other two most frequent carriers, CK and JT, did not fit the agonistic buffering pattern: they ranked high and they carried infants selectively. Male SH also carried infants selectively, even though he held low rank.

By what proximate mechanisms might males interact differentially with infants according to the probability of paternity? One possibility identified in studies of mice is mating experience with the mother at the time the young are conceived (Labov, 1980; vom Saal and Howard, 1982). Baboon reproductive physiology could allow a similar mechanism to operate. In particular, the perineal skin of adult females swells during each menstrual cycle, reaching maximum turgescence during the periovulatory phase (Hendrickx, 1967). Furthermore, this skin does not swell during pregnancy; instead, the paracallosal skin turns bright red (S. A. Altmann, 1970). Thus, external characteristics appear to signal ovulation and subsequent pregnancy; mating experience during the most recent interval of perineal swelling before birth could act as one mechanism upon which subsequent behaviors are based (Hamilton, chapter 12).

Although this idea remains speculative, it is supported by the observation that carriers almost always had immigrated before conception but not during intervals of pregnancy nor lactation of the infants they carried. Recipients, by contrast, had immigrated during pregnancy or lactation, but rarely before conception of infants carried in their presence. Thus, a male's presence in the group and his interactions with a female near the time of conception may influence his subsequent interactions with her infant.

A limitation to the analysis of carrier-infant relationships is that paternity has not been measured directly in any study including the present one; instead, paternity has been estimated or inferred from incomplete records of copulatory patterns and by reference to studies of caged baboons which suggest that ovulation usually occurs 2 to 3 days before the onset of rapid deturgescence of the female's perineal skin (Gillman and Gilbert, 1946; Hendrickx and Kraemer, 1969; MacLennan and Wynn, 1971; but see Wildt, et al., 1977). Needed are studies of triadic interactions in captive baboon groups, in which paternity can be precisely measured and related to infant carrying patterns.

Although an imprecise understanding of paternity is a limitation in the analysis of infant-carrier relationships, it is not so for infant-recipient relationships. Nonpaternity of most recipients is certain: recipient males had not been group members when the infants carried in their presence had been conceived. Two of three recipients that had been group members when the carried infants were conceived were unlikely to be fathers because they had been subadults with low social ranks. The results of the distribution of infant-recipient episodes lend support to the infant protection hypothesis and run counter to Popp's (1978) proposed mechanism by which agonistic buffering could develop in the population.

Could recipient males be related to infants in other ways, perhaps through male relatives that had immigrated to the group earlier and sired infants? As yet there is little information to support or reject this possibility. Even if supported, however, one would need to explain why, if this mechanism works in the case of distant relatives, are infants not also carried in the presence of possible fathers?

Specificity of recipient-infant combinations reflected the number of high-ranking immigrant males in each group. Several such males resided in group C during the study, and the overall specificity of infant-recipient combinations was low. Only one such male resided in group W, and the specificity of infant-recipient combinations was high, almost always involving this male. By contrast, high-ranking resident males rarely were recipients; the agonistic buffering hypothesis predicts that they should be, however. In group C, long-term resident male JT was observed as a recipient only once during his five-month tenure at alpha rank. In group W, the second-ranking male, CK, was a recipient only twice, each time involving a relatively old infant conceived before he joined the group.

The infant protection hypothesis suggests a specific benefit to high-ranking immigrant males that kill infants: a higher probability of subsequently impregnating the mother, since females conceive significantly sooner when an infant offspring dies than when it survives to weaning (see Busse and Hamilton, 1981, for results from the Moremi groups). Top-ranking males may benefit most because they have priority of access to ovulating females, at least in the population under study (K.S. Smith, pers. com.) Also, since the average tenure of males at alpha rank in the study groups was relatively short, the value to a top-ranking immigrant male of a female conceiving again as soon as possible would be greater than if average alpha tenure were longer.

The hypothesis also suggests that when males join a new group they could benefit from the death of infants born for at least six months after their arrival, since they cannot be fathers of infants born during this interval. In agreement with this prediction, observations of triadic interactions show that some males were recipients long after they joined a group. The most frequent recipients, NR and BB, had joined the study groups three months before the study began. Throughout the study, these males continued to be recipients of triadic interactions involving infants conceived before their arrivals.

The conceptual model for the infant-killing hypothesis is based on studies of langur monkeys *(Presbytis entellus)* in India (Blaffer-Hrdy, 1974, 1977, 1979; Chapman and Hausfater, 1979). Evidence of infant-killing by male baboons in the study groups and in other populations (reviewed by Busse, 1982; Collins, et al., in press) does not as yet confirm an adaptive basis, if any, to infant killing, but it does show that infants receive a relatively high frequency of wounds, including fatal ones. The accumulating evidence of male baboons biting infants also undermines a premise of the agonistic buffering hypothesis that there is a general inhibition against directing aggression toward infants.

What would happen if there were no potentially infanticidal males in the group; for example, when a male holds alpha rank long enough to be a possible father of all the infants. The infant protection hypothesis predicts that rates of triadic interactions will drop for two reasons. First, the simple proximity of an alpha male might deter lower-ranking males in the group from attacking infants, if they were inclined to do so. Second, lower-ranking males probably would not have priority of mating access to females whose infants die.

In groups C and W, male tenure at alpha rank was too short to test this prediction. In a study of three chacma baboon groups in Namibia, however, no males immigrated to any group for the first eight months (Hamilton and Tilson, 1982) and no triadic interactions were observed during this period (Buskirk, et al., 1974). Then, after two males joined one of the groups, the resident males in this group began to carry infants frequently in the presence of these newcomers (W. J. Hamilton III, pers. comm.). This observation from another chacma baboon population agrees with the above prediction for the Moremi population, although conceivably some of the males that resided in the group when the study began could have immigrated recently themselves. Also, in a study of yellow baboons *(Papio cynocephalus)* at Amboseli, Kenya, Stein and Stacey (1981) observed no triadic interactions in a group with only one adult male; this result is consistent with both hypotheses, however.

In the present study, as in others (Ransom and Ransom, 1971; Popp, 1978; Packer, 1980; Stein, 1981), males that carry infants usually are lower ranking than the recipients. This finding is one basis for the agonistic buffering hypothesis. Yet, it is also consistent with the hypothesis presented here. Males that carry infants rank high at the time the infants were conceived, but have dropped significantly in rank at the time of the interaction, at least six months later (Busse and Hamilton, 1981). The recipients usually are high-ranking immigrants, who would have a relatively high probability of impregnating any female whose infant dies (see Strum, chapter 7 for a different residency-rank pattern for baboons).

The age distribution of carried infants is also consistent with both hypotheses. Ninety percent of the interactions involved unweaned infants, infants who are at risk of attack by recipients, according to the theory on infanticide. The youngest infants are at greatest risk, theoretically, because their mothers would conceive a year sooner if they die than if they survive. Thus, one might expect the youngest infants to be carried most frequently.

Although a majority of infants carried were less than six months old, they were rarely carried before they were two months old, perhaps because they were almost continuously in contact with their mothers during this time (see Altmann, 1980). The analyses of infant gender and mother's rank reveal no consistent patterns, although neither hypothesis explicitly predicts any.

Patterns of carrying by juvenile and subadult males were similar to those of adult males in that recipients could not have been fathers of the carried infants. Like adult males, immature males may be protecting relatives (but not offspring) from potential harm. At Gombe, Tanzania, Packer (1980) observed four juvenile and subadult male olive baboons *(Papio anubis)* carrying infants during interactions with subadult and immigrant males; all four males carried their siblings more frequently than they carried other infants. The present study was too short to determine possible relatedness of immature males to the infants they carried, although the infant protection hypothesis predicts that they are related.

Why do high-ranking males rarely attack or threaten lower-ranking males carrying infants? The infant protection hypothesis suggests that the costs to a recipient of attacking an infant are greater when an infant is being carried by a male than when it is not. Carriers may be willing to protect infants, despite the higher rank of recipients, because of their possible relatedness to the infants they carry. By this reasoning, the payoff to the male carrying the infant, survival of a possible offspring, exceeds the payoff to the recipient, a higher probability of impregnating the mother if the infant dies (Busse and Hamilton, 1981). Interactions therefore rarely escalate.

Until recently, no ultimate explanation for the agonistic buffering hypothesis had been advanced. The idea that neonatal characteristics, such as black coat coloration, might inhibit attack in general is a proximate explanation that requires further explanation for how such characteristics originally developed to function this way (see Strum, chapter 7). Popp's (1978) hypothesis, that recipients might be inhibited from attack because the infants could be their offspring, is a plausible and testable ultimate explanation for agonistic buffering. This hypothesis, however, can be rejected for baboon groups studied at Moremi and elsewhere (Packer, 1980; Stein, 1981; Smuts, 1982).

The most recent and most plausible suggestion of a mechanism for agonistic buffering is that other group members, including adult females, might mob the recipient if a triadic interaction escalates and the infant exhibits distress (Stein, 1981; Smuts, 1982; Strum, chapter 7). Being mobbed could be costly to a recipient because he might be injured and because the infant's mother, and other females as well, might subsequently be reluctant to mate with the male (Stein, 1981; and especially Smuts, 1982). Mobbing was observed only once during the present study (during an intergroup encounter several resident males chased and screamed at an intruder male as he hit a subadult male holding an infant), but it need not be frequent to effectively inhibit aggression by a recipient male (Stein, 1981). The possibility that the recipient might be mobbed, regardless of whether the carrier is acting on behalf of the infant or himself, has contributed to the interpretation for one baboon *(Papio anubis)* population that some triadic interactions might function to protect the infant, whereas other interactions might function to protect the carrier (Smuts, 1982).

The infant protection hypothesis, in summary, is a robust one. It makes specific predictions, upheld for the groups examined here, that infants and carriers are related to each other, but that infants and recipients are not. Also, it identifies a specific reason dominant, recipient males might behave aggressively in the first place. Finally, it integrates a diverse set of other observations: alarm by mothers toward immigrant males; alarm by infants toward immigrant males; associations between mothers, infants, and possible fathers; and evidence of infrequent killing of infants by immigrant males.

Patterns of intergroup transfer, rank change, and infant carrying among male baboons at Moremi suggest a generalized (though by no means invariant) male life history pattern in this population. As in other populations (e.g., Packer, 1979a) males reside in their group of birth until they reach adult size and their canine teeth approach prime condition. Transfer out of the natal group may be a relatively long and selective process, in which a male samples various groups and eventually resides in a group in which he can outcompete most or all of the other males. During the post transfer interval, coinciding with top or high rank, he sires a majority of his offspring, is a frequent recipient of triadic interactions, and may even kill infants if opportunities arise. After he has resided in a new group for at least six months, infants born in the group could be his own offspring. Eventually he loses high rank to even more recent immigrants and he starts carrying infants during triadic interactions with these males. Thus, the period in which a male is a frequent recipient of triadic interactions is followed by an interval of paternalistic care in which he is a frequent carrier of infants. The subsequent behavior of males as they age and their offspring become weaned, remains an unknown aspect of the life history pattern.

ACKNOWLEDGMENTS

I am indebted to Mr. K. T. Ngwamotsoko, Director of the Department of Wildlife, National Parks, and Tourism, and to the Office of the President, Botswana, for permission to conduct studies in the Moremi Wildlife Reserve. Much of the credit for this work belongs to William J. Hamilton III, who provided ideas and guidance at all stages of the project. Kenneth S. Smith, my collegue in the field, was instrumental in establishing and maintaining the field research program. I am especially indebted to the book's editor, David M. Taub, for suggesting new approaches to analyzing and interpreting the results. Sarah Blaffer-Hrdy, Anne B. Clark, Russell O. Davis, Barbara E. Kus, Steve Neudecker, Peter J. Richerson, Peter S. Rodman, Matthew P. Rowe, Euclid O. Smith, and Ronald L. Tilson provided valuable comments on drafts of the manuscript. Carolyn Krause and Louise Wright provided clerical assistance. This research was funded by NIH grant 5 RO1 RR01078 (W. J. Hamilton III, P. I.).

REFERENCES

Alley, T. R. 1980. Infantile coloration as an elicitor of caretaking behavior in Old World primates. *Primates* 21: 416-29.

Altmann, J. 1978. Infant independence in yellow baboons. In G. M. Burghardt, and M. Bekoff (eds.), *The Development of Behavior: Comparative and Evolutionary Aspects,* pp. 253-77. New York: Garland Press.

———. 1980. *Baboon Mothers and Infants.* Cambridge: Harvard University Press.

Altmann, S. A. 1970. The pregnancy sign in savannah baboons. *Lab. Anim. Dig.* 6: 7-10.

Bernstein, I. S., and Gordon, T. P. 1974. The function of aggression in primate societies. *Am. Sci.* 62: 304-11.

Blaffer-Hrdy, S. 1974. Male-male competition and infanticide among the langurs *(Presbytis entellus)* of Abu, Rajasthan, *Folia Primatol.* 22: 19-58.

———. 1976. The care and exploitation of non-human primate infants by conspecifics other than the mother. *Adv. Stud. Behav.* 6: 101-58.

———. 1977. *The Langurs of Abu: Female and Male Strategies of Reproduction.* Cambridge: Harvard University Press.

———. 1979. Infanticide among animals: A review, classification and examination of the reproductive strategies of females. *Ethol. Sociobiol.* 1: 13-40.

Buskirk, W. H.; Buskirk, R.E.; and Hamilton, W. J., III. 1974. Troop-mobilizing behavior of adult male chacma baboons. *Folia Primatol.* 22: 9-18.

Busse, C. 1981. Infanticide and parental care by male chacma baboons, *Papio ursinus.* Ph. D. dissertation, University of California, Davis.

———. 1982. Migrant males, infant killing, and the social organization of baboons. Paper presented at the *Wenner-Gren Conference on Infanticide in Animals and Man,* Ithaca, New York.

Busse, C., and Hamilton, W. J., III. 1981. Infant carrying by male chacma baboons. *Science* 212: 1282-83.

Chapman, M., and Hausfater, G. 1979. The reproductive consequences of infanticide in langurs: A mathematical model. *Behav. Ecol. Sociobiol.* 5: 227-40.

Collins, D. A.; Busse, C. D.; and Goodall, J. In press. Infanticide in two populations of savanna baboons. In G. Hausfater, and S. B. Hrdy (eds.), *Infanticide: Comparative and Evolutionary Perspectives,* New York: Aldine.

Deag, J. M., and Crook, J. H. 1971. Social behaviour and "agonistic buffering" in the wild barbary macaque, *Macaca sylvana* L. *Folia Primatol.* 15: 183-200.

Gillman, J., and Gilbert C. 1946. The reproductive cycle of the chacma baboon *(Papio ursinus)* with special reference to the problems of menstrual irregularities as assessed by the behaviour of the sexual skin. *S. Afr. J. Med. Sci.* 11: 1-54.

Gilmore, H. B. 1977. The evolution of agonistic buffering in baboons and macaques. Paper presented at the *46th Annual Meeting of the American Association of Physical Anthropologists,* Seattle, Washington.

Hamilton, W. J., III, and Tilson, R. L. 1982. Solitary male chacma baboons in a desert canyon. *Am. J. Primatol,* 2 (2): 149-58.

Hamilton, W. J., III; Busse, C.; and Smith, K. S. 1982. Adoption of infant orphan chacma baboons. *Anim. Behav.* 30: 29-34.

Hausfater, G. 1975. Dominance and reproduction in baboons *(Papio cynocephalus). Contrib. Primatol.,* vol. 7. Basel: Karger.

Hendrickx, A.G. 1967. The menstrual cycle of the baboon as determined by the vaginal smear, vaginal biopsy, and perineal swelling. In H. Vagtbord (ed.), *The Baboon in Medical Research,* vol. 2, Austin: University of Texas Press. pp. 437-60.

Hendrickx, A. G., and Kraemer, D. C. 1969. Observations on the menstrual cycle, optimal mating time and pre-implantation embryos of the baboon, *Papio anubis* and *Papio cynocephalus. J. Reprod. Fert., Suppl.* 6: 119-28.

Labov, J. 1980. Factors influencing infanticidal behavior in wild male house mice *(Mus musculus). Behav. Ecol. Sociobiol.,* 6: 297-303.

MacLennan, A. H., and Wynn, R. M. 1971. Menstrual cycle of the baboon. I. Clinical features, vaginal cytology, and endometrial histology. *Obstet. Gynecol.* 38: 350-58.

Packer, C. 1979a. Inter-troop transfer and inbreeding avoidance in *Papio anubis. Anim. Behav.* 27: 1-36.

———. 1979b. Male dominance and reproductive activity in *Papio anubis. Anim. Behav.* 27: 37-45.

———. 1980. Male care and exploitation of infants in *Papio anubis. Anim. Behav.,* 28: 512-20.

Popp, J. L. 1978. Male baboons and evolutionary principles. Ph. D. dissertation, Harvard University.

Popp, J. L., and DeVore, I. 1979. Aggressive competition and social dominance theory: Synopsis. In D. A. Hamburg, and E. R. McCown (eds.), *The Great Apes,* pp. 317-40. Menlo Park: Benjamin/Cummings.

Ransom, T. W., and Ransom, B. S. 1971. Adult male-infant relations among baboons *(Papio anubis). Folia Primatol.* 16: 179-95.

Smuts, B. B. 1982. Special relationships between adult male and female olive baboons *(Papio anubis).* Ph. D. dissertation, Stanford University.

Stein, D. M. 1981. The nature and function of social interactions between infant and adult male yellow baboons *(Papio cynocephalus).* Ph. D. dissertation, University of Chicago.

Stein, D. M., and Stacey, P. B. 1981. A comparison of infant-adult male relations in a one-male group with those in a multi-male group for yellow baboons *(Papio cynocephalus). Folia Primatol.* 36: 264-76.

Strum, S. C. 1982. Agonistic dominance in male baboons: An alternative view. *Int. J. Primatol.* 3(2): 175-202.

Taub, D. M., 1980. Testing the "agonistic buffering" hypothesis. I. The dynamics of participation in the triadic interaction. *Behav. Ecol. Sociobiol.* 6: 187-97.

Tinkley, K. L. 1973. *An Ecological Reconnaissance of the Moremi Wildlife Reserve, Botswana.* Gaborones, Botswana: Okavango Wildlife Society.

vom Saal, F. S., and Howard, L. S. 1982. The regulation of infanticide and parental behavior: Implications for reproductive success in male mice. *Science* 215: 1270-72.

Wildt, D. E.; Doyle, L. L.; Stone, S. C.; and Harrison, R. M. 1977. Correlation of perineal swelling with serum ovarian hormone levels, vaginal cytology and ovarian follicular development during the baboon reproductive cycle. *Primates* 18: 261-70.

9

Ontogeny of Infant-Adult Male Relationships during the First Year of Life for Yellow Baboons *(Papio cynocephalus)*

David M. Stein
Department of Psychiatry
John A. Burns School of Medicine
University of Hawaii
Honolulu, Hawaii

INTRODUCTION

During the past decade there has been a surge of interest in interactions between primate infants and adult males. Inquiries into this area have included attempts to understand the relationship of infant-male interactions to phylogeny (Mitchell and Brandt, 1972), to mating system (Redican and Taub, 1981; Bales, 1980), to demographic factors (Bogess, 1979; Stein and Stacey, 1981), to biosocial factors (Hrdy, 1976), and to socioecological and geographic influences (Stein, 1981). Unfortunately, most of these efforts have been severely hampered by a lack of systematically collected data and have had to base their conclusions on scattered and often anecdotal observations.

This chapter is an attempt to help remedy this shortcoming for interactions between yellow baboon infants and adult males. I do not endeavor to test any hypotheses or falsify any predictions, but simply provide a quantitative description of infant-adult male relations during the first year of life in such a manner that it can be used to test a wide variety of hypotheses concerning all of the above topics. The emphasis in this presentation is on changes in the interactions across time and on the relative role of the adult male, the infant, and the infant's mother in these interactions. A cursory look is also taken at the correspondence between certain key sociobiological variables and the closeness of the relationship between an infant and an adult male.

METHODS

A thorough exposition of the rationale, study site, subjects, sampling methods, behavioral categories, and data analysis can be found in Stein (1981).

Study Site and Study Group

The field observations for this research were conducted in Amboseli National Park and adjacent lands in Kenya, East Africa. This semi arid savannah habitat is described in detail in Struhsaker (1967) and Western (1973). Systematic samples were taken of Alto's Group, a group of approximately 45 baboons studied extensively since 1971 (e.g., J. Altmann, 1980; Hausfater, 1975; Post, 1978; Walters, 1980).

The present Alto's Group was formed in 1972 by the fusion of two smaller groups (J. Altmann et al., 1977). Since then, two fairly distinct and stable subgroups have been maintained, with female membership in the subgroups corresponding largely to membership in the original groups (see J. Altmann, 1980; Stein, 1984). Most mating, grooming, and other affiliative interactions took place within subgroups rather than between them. Four of the 12 reproductively active adult males in Alto's Group during my study were first observed as young juveniles in the group and were almost certainly born there. All other adult males immigrated into the group as adults. During the present study, Alto's Group ranged in size from a high of 53 to a low of 39 members.

Focal Samples

Focal sampling (J. Altmann, 1974) of infants in their first year of life was carried out from 1 January 1978 to 31 December 1978. Samples were 20 minutes long, were scheduled to begin once each hour on the hour between 0800 and 1755, and were generally conducted five days per week. Beginning with the day an infant was born, subsequent time was divided into four-week blocks, henceforth referred to as "months." Infants were sampled throughout their first eight months and again in their thirteenth month. Younger infants were sampled more than once per day if the number of other animals being sampled permitted it. The actual amount of time infants were in sight during focal samples at each age is indicated in Table 9-1.

Table 9-1. Number of Subjects and Amount of In-Sight Time in Focal Samples at Each Infant Age.

Infant Age	Number of Different Infants	Minutes of Sample Time	Number of Adult Males	Dyad-Minutes of Sample Time	Number of Infant-Male Dyads*	Average Number of Adult Males Per Infant
Month 1	5	5281.82	14	47901.96	49	9.8
Month 2	4	3677.66	12	33585.91	38	9.5
Month 3	6	3741.82	13	35773.93	59	9.8
Month 4	7	3838.95	13	36542.27	70	10.0
Month 5	7	3622.09	13	34022.30	67	9.6
Month 6	6	2838.96	12	27482.27	60	10.0
Month 7	8	3036.01	13	28513.41	82	10.3
Month 8	4	1368.47	12	12573.59	41	10.3
Month 13	6	1839.67	14	15159.96	52	8.7
Total	12	29245.45	16	271555.60	136	11.3

*Because of male migration, not all males were present for each infant at each age, and therefore the number of dyads is often less than the product of the number of infants and the number of adult males.

Specified Behaviors

During focal samples on infants, all events or transitions within the following categories were recorded, together with the time, the specific behavior, and the identities of all relevant participants:

Five-Foot Transitions. All movements that brought the center of mass of the focal infant and that of an adult male within 5 feet of each other were considered approaches. All movements that took the center of mass of the focal infant and that of an adult male beyond 5 feet from one another were termed departures. These transitions could be effected by the infant, the adult male, the infant's mother, or two of these at once. The state of an infant and adult male being within 5 feet of one another is referred to as proximity.

Contact Transitions. All movements that brought any part of the focal infant's body into sustained contact (more than .02 minute) with any part of an adult male's body were considered making contact. Momentary touching, hitting, and brushing were not included. All movements that terminated sustained contact between the infant and an adult male were termed breaking contact. These transitions could be effected by the infant, the adult male, the infant's mother, another animal, or two of these at once.

Connection Transitions. A connection was defined as a subcategory of contact in which either (1) the infant or adult male held onto the other, or (2) the adult male completely supported the weight of the infant. Thus connection subsumed such behaviors as embracing, riding dorsally, clinging ventrally, climbing on, holding against substrate, or grasping a body part. All connections were scored as being initiated by either the infant or the adult male, depending on who first performed one of the following "initiating acts" (regardless of who was actually holding whom): (1) the onset of a connection, as defined above; (2) an incipient form of a connection, as defined above; or (3) palming of the infant by the adult male, i.e., the male contacted the infant on its dorsum with the palm of his hand and guided it toward his ventrum.

Grooming Transitions. When the focal infant or an adult male began systematically parting the fur of the other with visual attention focused on the act, this was considered the onset of grooming. Only the infant or adult male were considered as actors, with the one who was parting the fur being designated as both grooming and stopping grooming, regardless of who initiated or terminated the interaction.

Harassments. Any behavior toward the focal infant by a baboon other than the infant's mother was considered harassing the infant if, within −1 minute, either (1) the infant responded by screaming, squealing, distress-geckering, moaning, raising its tail, cowering, grimacing, running away, or hiding; or (2) the infant's mother (when she was in contact with it) responded by screaming, squealing, cackling, raising her tail, cowering, grimacing, running away with the infant, restraining the infant, shielding the infant, threatening, or attacking. The relevant actors were considered to be adult males and other animals besides the mother. An adult male was considered to have defended the focal infant if he punished (i.e., threatened or attacked) an animal who had just harassed the infant.

Social Contexts. The predominant activity of the group was considered to be feeding when at least 70% of the animals in the group were feeding or foraging, that is, their visual attention was directed toward procuring specific food items or water. The predominant activity of the group was considered to be not feeding when 30% or fewer of the animals in the group were feeding or foraging. At other times, if more than 30% but fewer than 70% of the animals were feeding or foraging, the predominant activity of the group was considered to be mixed. These three categories were not considered to be scored reliable by me until 17 February 1978.

If an adult male was known to be involved in an agonistic interaction with another adult male at the time he interacted with the focal infant, it was so indicated for all of his interactions with the infant. Since I was focusing on the infants in these samples and not on the adult males, this procedure undoubtedly underestimated the rate of infant-adult male interactions during inter-male agonistic encounters.

Food-sharing. Any time the focal infant fed within 5 feet of an adult male on food on which that male expended energy to process, the event was recorded along with the identity of the adult male. These events were only scored after 14 June 1978. Since I could not reliably estimate the amount of nourishment an infant was receiving, and only wanted to assess the prevalence and distribution of the phenomenon, the event was scored only once per feeding site, regardless of the duration or the magnitude of the food sharing. If, however, the infant went beyond 5 feet from the adult male and then returned to the site and fed again, it was scored as a new event.

Scan Samples

Census. Scan sampling (J. Altmann, 1974) of all individuals in the group was performed each morning as soon as possible after their descent from the sleeping trees. Each adult male and infant was scored as present or absent. For mature females, I recorded the size of the sexual skin swelling and the day of detumescence, as in Hausfater (1975).

Sleeping-grove Subgroups. On many mornings when Alto's Group slept in noncontiguous clusters of trees (almost always two), I recorded the identities of the individuals in each of the separate clusters. Only if the baboons were in a grove where visibility was good, if all the baboons were in the trees when I arrived, and if I thought that it would not interfere with my scheduled samples did I attempt to record such data. The percentage of samples in which a given infant and adult male were in the same cluster was the quantitative measure that I used for assessing the strength of subgroup affiliation of that dyad.

Ad Libitum Samples

Paternity. In order to ascertain probable paternity of infants, I collected data concerning all observed consortships and mountings on an ad libitum basis (J. Altmann, 1974). Using the data on consortships and mountings together with the census records permitted me to dichotomize all infant-adult male dyads along each of the following dimensions (1) whether the male was present in the group when the infant was born; (2) whether the male was present in the group when the infant was conceived; (3) whether the male was seen copulating with the mother during the days of most probable conception (the four days prior to the day of detumescence of the female's sexual skin swelling; See Gillman and Gilbert, 1946; Hendrickx and Kraemer, 1971); and (4) whether the male monopolized the consortship record of the mother during the days of most probable conception.

Dominance. In order to determine the relative dominance ranks among mothers and among adult males, all observed supplantations and signs of submission between mothers and between males were recorded. Individual A was considered dominant to B for any days in which it supplanted B or received signs of submission from B without any reversals. For days on which reversals occurred in the dominance relation between two animals, or on which they were involved in a circular dominance relationship, they were considered to have the same rank. For days on which no dominance interactions were observed, the relative ranks were inferred from the nearest days with observed interactions before and after the day in question.

Data Analysis and Presentation

Interactions. During each month, interactions with all adult males from all samples were pooled for each infant. For each month, then, a simple average was taken of the results for each infant, regardless of the number of adults present at each infant age. Since there was little variability from month to month in the average number of adult males available to infants, this procedure had little impact on my results. In comparisons with other studies, however, the number of adult males ought to be taken into consideration, and this number is given in each figure for each month. In Figures 9-1, 9-3 through 9-9, 9-11, 9-12, and 9-16, data are summed over all males for each infant and the number of infants observed and the average number of adult males available to interact with each infant are indicated at the top of the figure for each infant age.

Biosocial Variables. At each infant age, all infant-adult male dyads were classified into one of two categories. If the adult male individually accounted for more than 20% of the infant's total time in proximity to adult males that month, the dyad was said to have a "preferred" relationship. Otherwise, the dyad was designated as having a "non preferred" relationship. At each month for each sociobiological variable and for each category, a simple mean was taken of the value of that variable over all the dyads in that category.

RESULTS

Static Spatial Relationships

A simple overview of infant-adult male relationships during the first year of life can be obtained from Figure 9-1, where I have presented the percentage of time spent by infants in each of three spatial states—proximity, contact, and connection—with adult males, as a function of infant age. During their first week of life, infants spent, on the average, about one-third of their time between the hours of 0800 and 1800 within 5 feet of an adult male. This level was maintained throughout the first seven weeks and then dropped sharply,

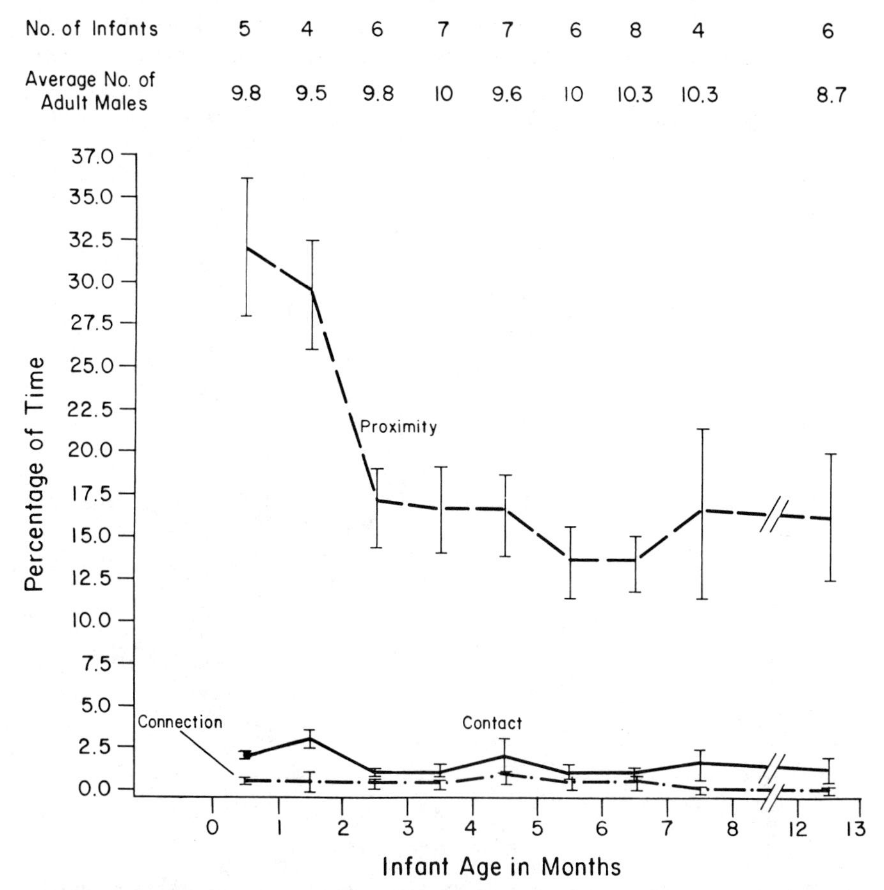

Fig. 9-1. Percentage of time spent by infants in each of three spatial states with adult males, as a function of infant age (simple mean over all infants and one standard error).

so that from the third month through the end of the first year, infants spent about one-sixth of their day time in proximity to adult males.

The amount of time infants spent in contact with adult males rose steadily from 1% in their first week of life to 3% in their eighth week. Again there was a marked decline at the end of the second month, and the level remained at about 1% from months three through thirteen, except for a brief doubling in the fifth month. Infants spent roughly .5% of their time connected with adult males during their first half-year of life, with a peak of almost 1% in the fifth month; but they spent less than .2% of their time connected with adult males during the second half of the year.

Figure 9-2 illustrates the changes in the bout lengths of the above three spatial states as a function of infant age. Bout lengths were estimated for each infant by dividing the total amount of time spent in a given state by the total number of times that state was initiated during focal samples on that

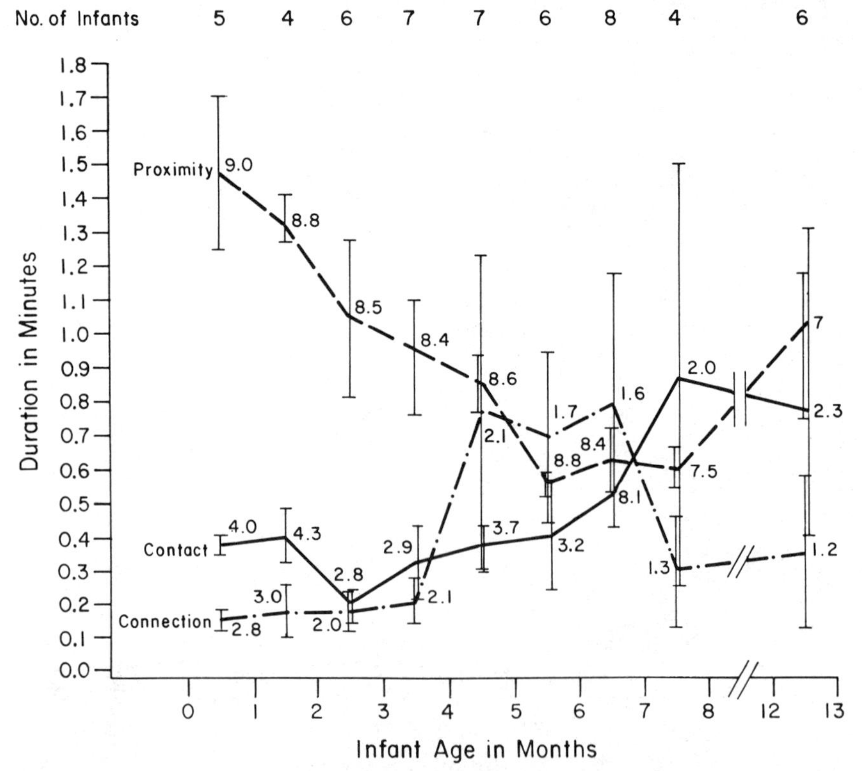

Fig. 9-2. Mean bout length of three spatial states, as a function of infant age (simple mean over all infants and one standard error). The estimate for each infant was calculated by pooling all its interactions with all adult males. The number of infants observed at each age is indicated at the top of the figure. The average number of adult males engaging in each type of bout with each infant is indicated next to the plotted points.

infant. Proximity bouts were longest during an infant's first month of life, and diminished fairly steadily throughout the first six months. At the end of the first year they were considerably longer once again. Contact bouts were fairly long for the first eight weeks, but then shortened substantially at the end of the second month. Thereafter, the length of contact bouts tended to increase with infant age, at least through the eighth month of life. Connection bouts were quite short when the infants were young, and tended to lengthen very gradually over the course of the first year. During months 5, 6, and 7, however, the average duration of connection bouts was profoundly longer, related, no doubt, to the increased incidence of agonistic buffering during these months (see below).

Dynamic Spatial Relationships

The frequency of connections (clinging, embracing, etc.) between infants and adult males, as a function of infant age, is portrayed in Figure 9-3. During the first seven weeks, adult males connected with each infant between 9 and 17 times per 600 minutes, on the average, but this rate decreased markedly at

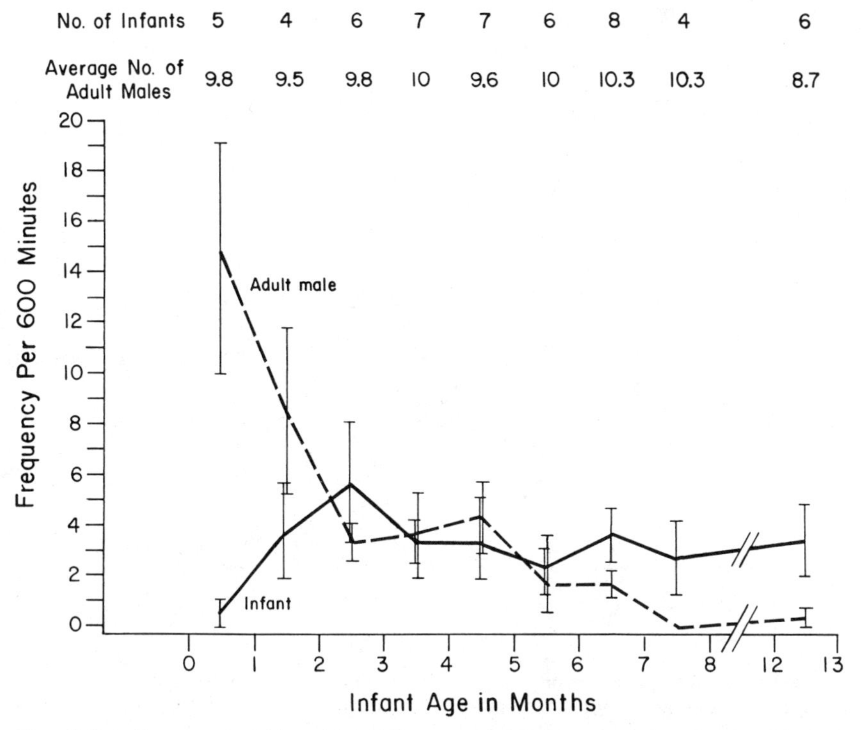

Fig. 9-3. Frequency with which infants and adult males connected with one another, as a function of infant age (simple mean over all infants and one standard error).

the end of the second month. From then on the rate of male-initiated connections declined gradually, with a secondary peak in the fifth month.

Infants were not seen connecting with adult males until their third week of life, but then the incidence of this behavior rose steadily until the third month. Through the remainder of their first year, the rate of infant-initiated connections fluctuated between two and five times per 600 minutes. During the first seven weeks, adult males connected with infants much more often than vice versa, but after about the fifth month the majority of connections were effected by the infants.

Figure 9-4 reveals the frequency with which infants and adult males contacted each other, as a function of infant age. During the first six weeks, adult males initiated contacts between 10 and 18 times per 600 minutes, on the average, but at the end of the second month this rate fell dramatically to about three times per 600 minutes. The frequency of adult male contacting

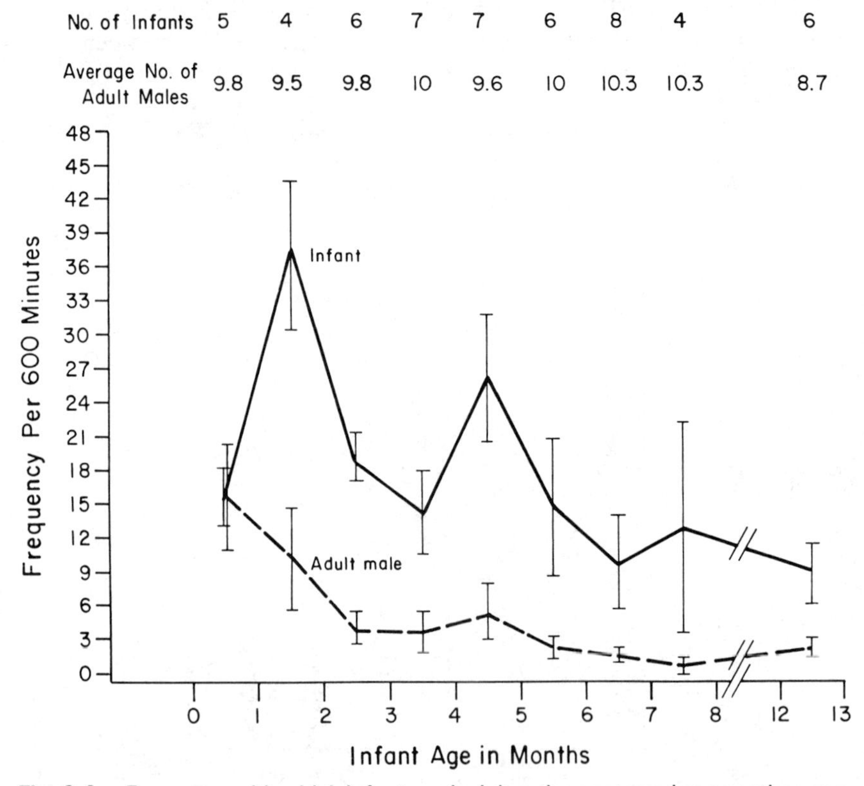

Fig. 9-4. Frequency with which infants and adult males contacted one another, as a function of infant age (simple mean over all infants and one standard error). This was summed over all males for each infant.

remained at about this level for the remainder of the first year, with a slight secondary peak in the fifth month.

The frequency of infant-initiated contacts climbed sharply from weeks 1 through 6, as infants became ever more mobile, and then dropped off abruptly at the end of the second month. Except for a substantial secondary peak in the fifth month, the frequency of contact effected by the infant declined gradually from month 3 through 13. From the fourth week on, infants consistently made 2 to 12 times as many contacts with adult males as the adult males did with the infants.

The frequency with which infants and adult males contacted each other while the adult male was known to be involved in an agonistic interaction with another adult male is plotted in Figure 9-5. This behavior is a good index of the prevalence of agonistic buffering, any context-dependent interaction between infants and adult males that decreases the expected level of aggression against the adult male in that context (see Stein, 1984). Since the actual frequency of inter-male aggression, and hence of agonistic buffering, is largely a function of varying social and demographic conditions in the group, comparisons between different ages in Figure 9-5 are not liable to be very meaningful. This graph can be used, however, for assessing the relative role of the infant and adult male in these interactions at each age, and for seeing the relationship between agonistic buffering and other aspects of infant-adult male relationships.

In these samples, adult males contacted infants during male-male fights most often in months 1 and 2 and again in months 5 and 6. The frequency with which infants contacted an adult male during his fights with another male increased fairly steadily to a maximum in the sixth month and then declined again steadily thereafter. During the first two months, adult males played a much more active role than the infant in establishing these contacts, but after the third month the infant was the primary instigator of these contacts.

Comparing Figures 9-4 and 9-5, it is clear that in the first two months only about 1 to 3% of the contacts by infants occurred during the males' agonistic interactions, but during month 6 about 14% did. In contrast, 10 to 30% of the contacts by adult males in the first two months occurred while the male was engaged in an agonistic interaction, and for infants in their sixth month roughly 50% did.

Figure 9-6 displays the frequency with which adult males approached infants and departed from them, as a function of infant age. During their first week of life, infants were approached by adult males an average of 98 times per 600 minutes. The rate remained high throughout the first six weeks and then fell precipitously at the end of the second month. The frequency of adult male approaches continued to decline, for the most part, until by the

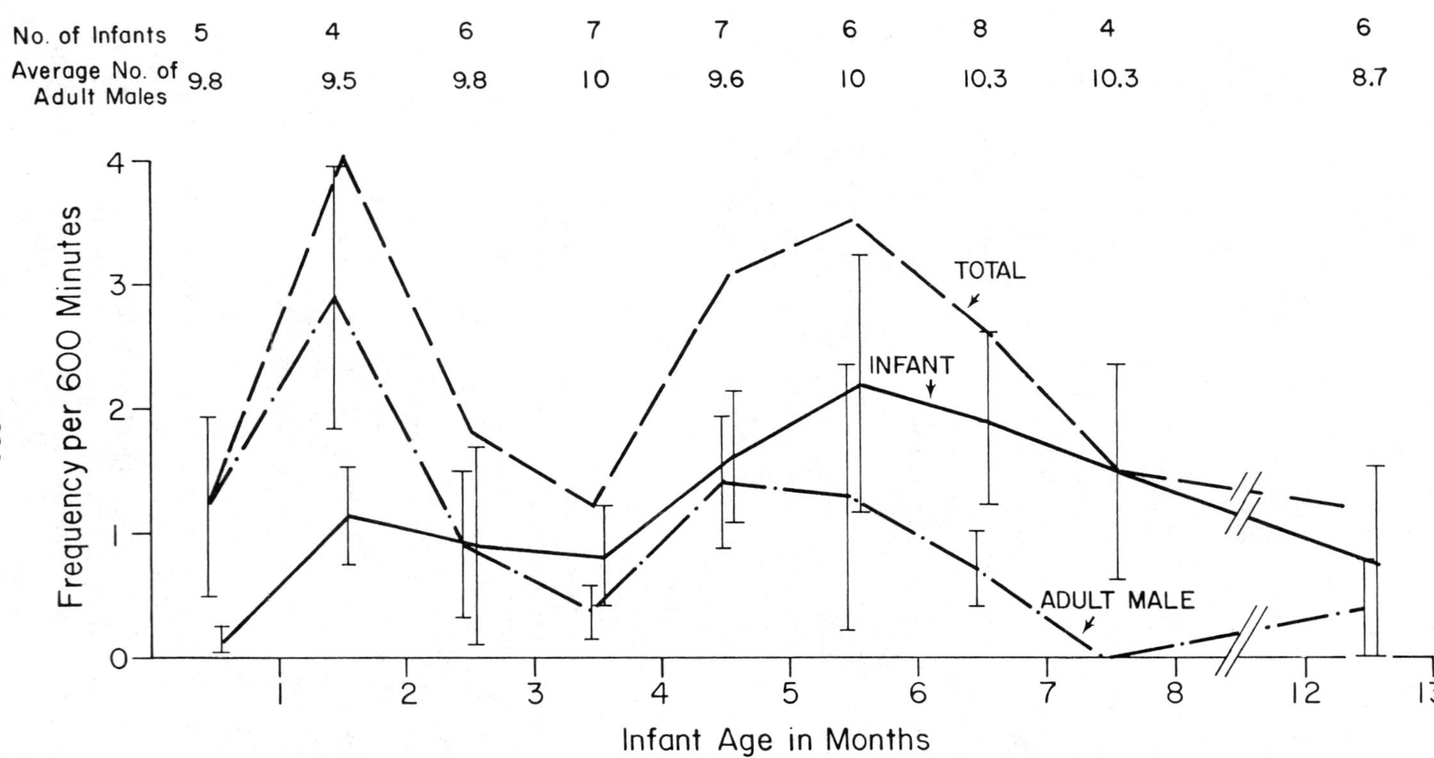

Fig. 9-5. Frequency with which infants and adult males contacted one another while male was engaged in agonistic interaction with another male, as a function of infant age (simple mean over all infants and one standard error). This was summed over all males for each infant.

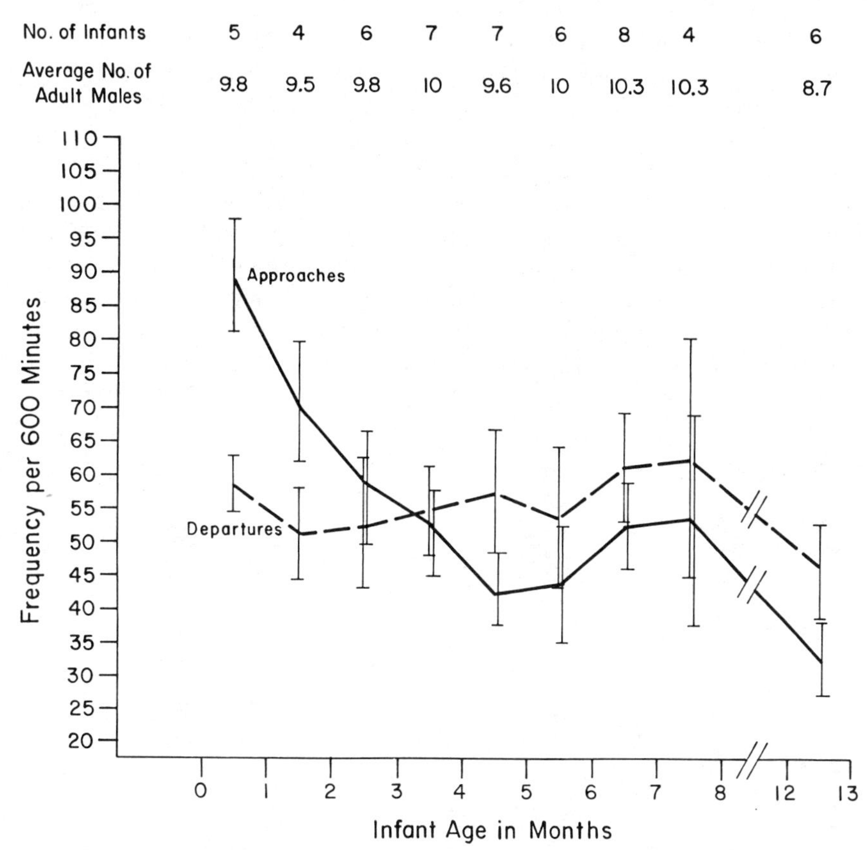

Fig. 9-6. Frequency with which adult males approached and departed from infants, as a function of infant age (simple mean over all infants and one standard error).

end of the first year adult males moved to within 5 feet of each infant an average of 32 times per 600 minutes.

The frequency of adult male departures from infants, in contrast, was fairly constant throughout the first year, ranging between 45 and 60 times per 600 minutes. Thus during an infant's first three months, adult males tended to approach it more often than they left it, whereas from month 4 on, adult males departed from an infant more frequently than they approached it.

Mothers carried their infants to within 5 feet of adult males 40 to 50 times per 600 minutes during the first eight weeks after parturition (Fig. 9-7). Thereafter the rate dropped steadily until, by the end of the first year, there were virtually no maternally effected approaches. At nearly every infant age, mothers carried their infants away from adult males more frequently than they carried them toward the males. This difference was more pronounced the younger the infant was, and during the first few weeks maternal departures were nearly twice as common as maternal approaches.

Infants were not seen moving to within 5 feet of adult males on their own until their second week of life (Fig. 9-8), but then the frequency of this behavior increased rapidly until infants were approaching adult males an average of 95 times per 600 minutes during their fifth month. Thereafter the frequency of infant approaches remained fairly constant through the rest of the first year. Changes in the frequency of infant departures from adult males followed essentially the same time course. Unlike their mothers, however, infants over three months of age exhibited a consistent tendency to approach adult males more frequently than they departed from the males.

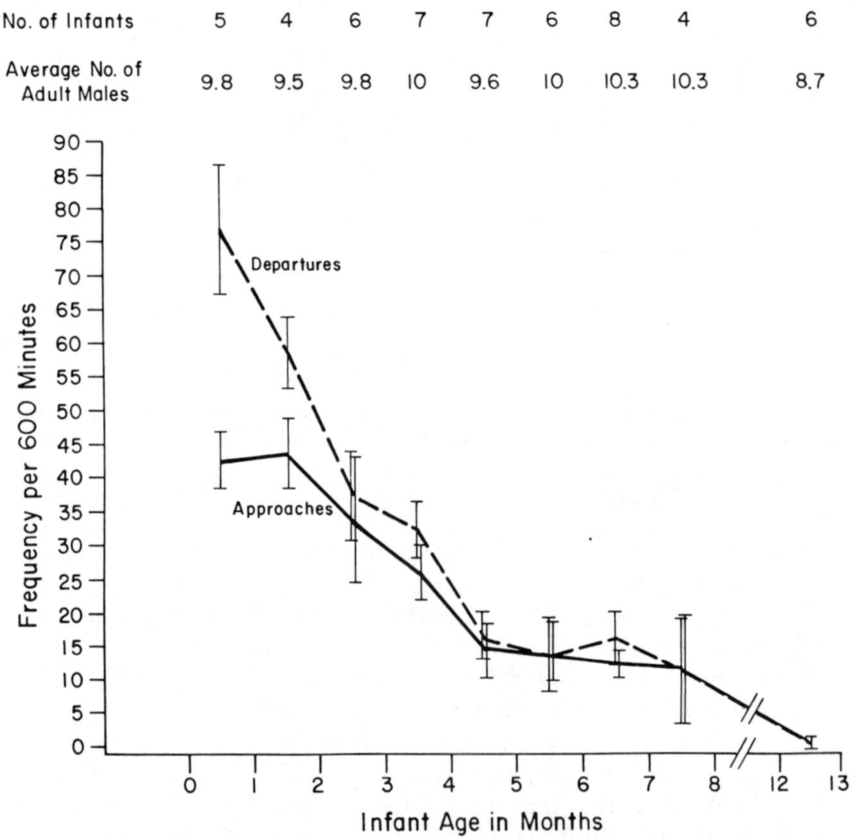

Fig. 9-7. Frequency with which mothers carried their infants to within 5 feet of, or to beyond 5 feet from, adult males, as a function of infant age (simple mean over all infants and one standard error).

No. of Infants	5	4	6	7	7	6	8	4	6
Average No. of Adult Males	9.8	9.5	9.8	10	9.6	10	10.3	10.3	8.7

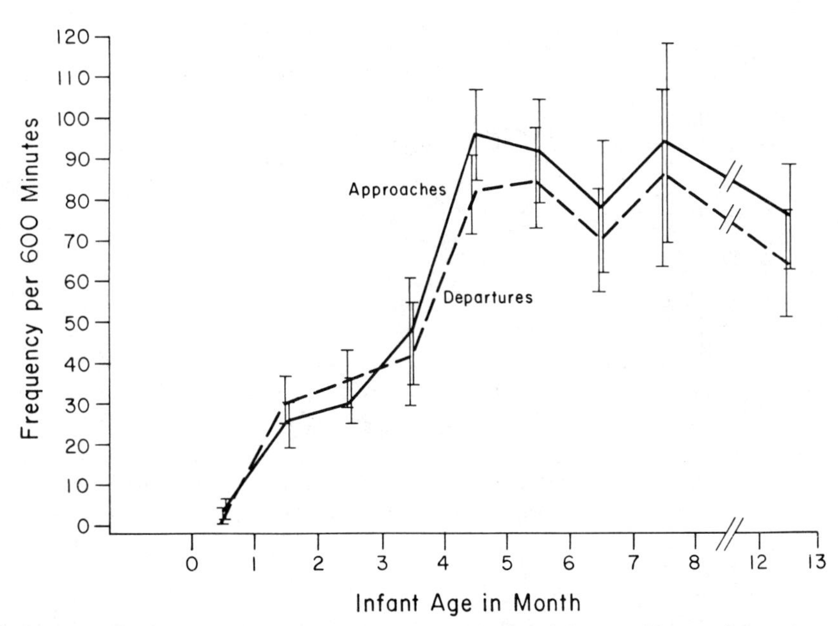

Fig. 9-8. Frequency with which infants approached and departed from adult males, as a function of infant age (simple mean over all infants and one standard error).

Relative Roles

A partial integration of Figures 9-6 through 9-8 is achieved in Figure 9-9. The first week after an infant was born, adult males were responsible for two-thirds of infant-male approaches, the mother was responsible for one-third, and the infant effected none. One year later the infant was responsible for two-thirds of all approaches, the adult males were responsible for one-third, and the mother effected virtually none. The percentage of infant-male approaches effected by the mother diminished fairly steadily throughout the first year, while the percentage made by adult males dropped steeply for the first five months and then underwent little change. The percentage of infant-male approaches made by the infant rose sharply for the first five months, until the infant was effecting more approaches than its mother and the adult males combined, and then the percentage increased gradually for the rest of the year.

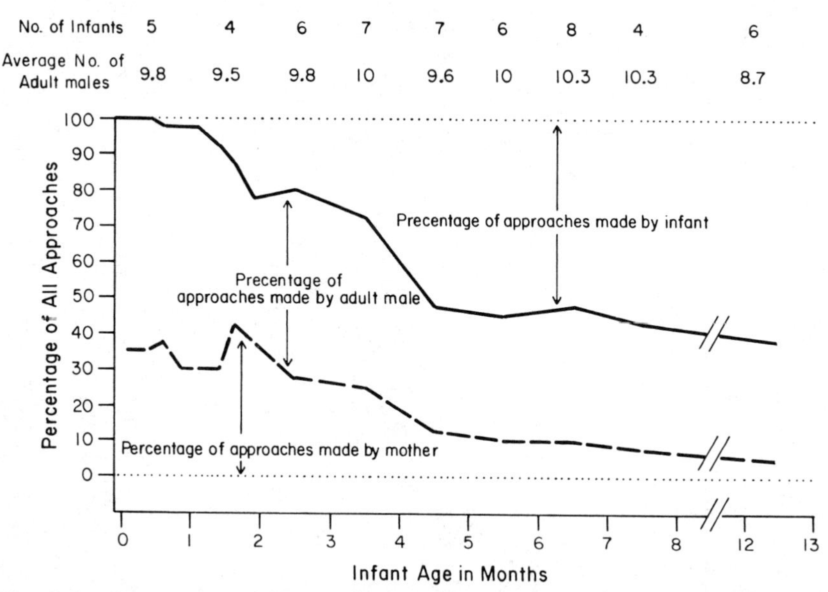

Fig. 9-9. Percentage of infant-male approaches effected by the infant, its mother, and adult males, as a function of infant age (simple mean over all infants). The estimate for each infant was calculated by pooling all its interactions with all adult males.

In order to summarize the information on approaches and departures as simply as possible, I constructed a measure called "the relative role of adult males in maintaining proximity to the infant." It is defined as:

$$\frac{n\,(\text{approach by male}) + n\,(\text{departure by infant}) + n\,(\text{departure by mother})}{n\,(\text{approach}) + n\,(\text{departure})}$$

where $n(x)$ indicates the frequency of x. This relative role measure is an estimate of the probability that a 5-foot transition is either an approach by an adult male or a departure by the infant or its mother.

The relative role of adult males in maintaining proximity to infants dropped fairly steadily from about 60% at birth to about 45% at one year of age. Comparing the relative role of adult males when the group was feeding and when the group was not feeding (Fig. 9-10) revealed that after the third month, infants and their mothers played a relatively greater role in maintaining proximity to adult males when the predominant group activity was feeding. This difference was most pronounced in months 5 through 8, the ages at which J. Altmann (1980) predicted baboon infants would have to begin

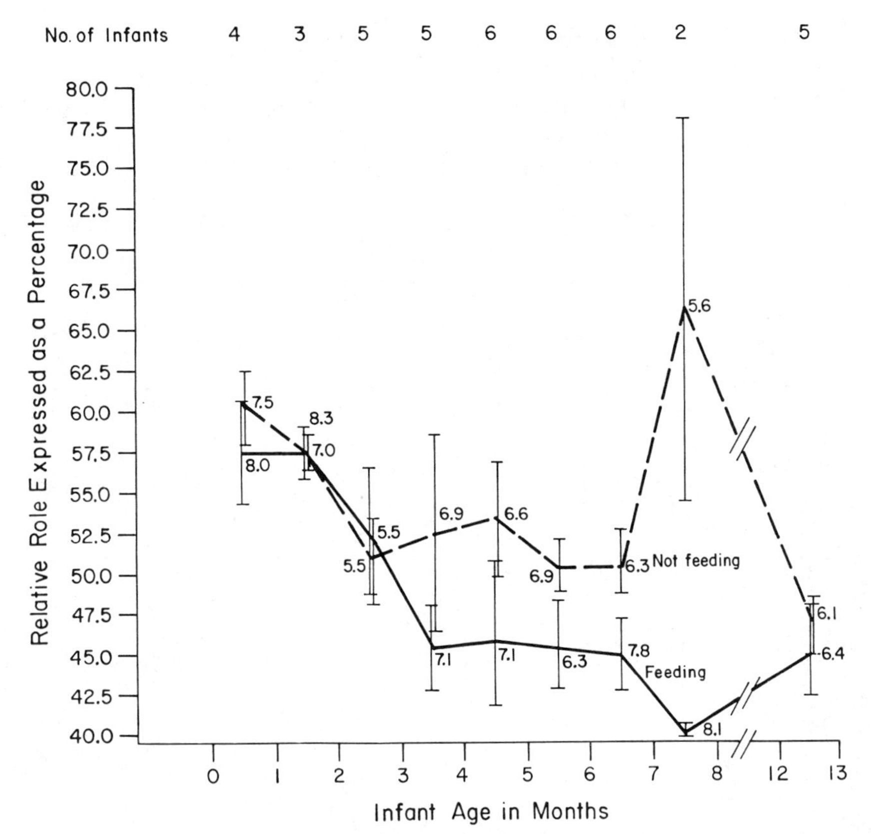

Fig. 9-10. Effect of group activity—feeding versus not feeding—on relative role of adult males in maintaining proximity to infants, as a function of infant age (simple mean over all infants and one standard error). This was pooled over all males for each infant. The number of infants observed at each age, after I began scoring group activity reliably, is indicated at the top of the figure for each age. The average number of adult males engaging in 5-foot transitions with each infant during each group activity is indicated next to the plotted points.

providing a sizable share of their own nutrition lest their mothers incur a nutritional deficit.

Potential Benefits

The greater relative role of infants and their mothers in maintaining proximity to adult males when the group is feeding suggests that the presence of the male enables them to obtain more or higher-quality food. Additionally, beginning in their second month of life, infants fed periodically within 5 feet of an adult male on food that he expended energy processing (Fig. 9-11). In all instances observed during my samples, the food items were grass corms, the primary food of Amboseli baboons during the dry season (Altmann and Altmann, 1970). Infants do not have the strength to pull these corms out of

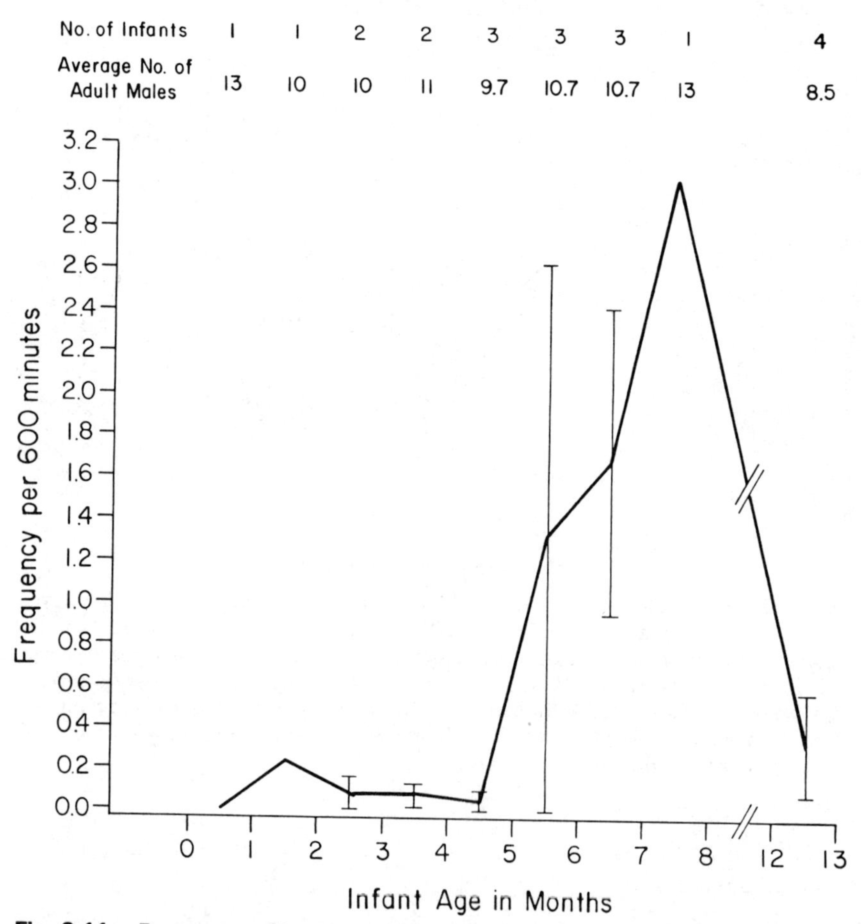

Fig. 9-11. Frequency with which infants fed within 5 feet of adult males on food processed by that male, as a function of infant age (simple mean over all infants and one standard error). This was summed over all males for each infant. The number of infants observed, after I began recording this behavior, and the average number of adult males available to each infant are indicated at the top of the figure for each age.

the dry soil and relied heavily on the scraps left by other baboons. This overt food sharing by adult males with infants was most pronounced in months 6 through 8, again, the critical period in terms of the mother's nutritional balance (J. Altmann, 1980).

As can be seen in Figure 9-12, adult males actively punished other group members for harassing infants, while the infants were young. Both the absolute frequency of this punishing and the percentage of harassments within 5 feet of an adult male that were punished by that male decreased steadily as the infants matured. This overt defense by adult males did not, however, have any obvious impact on the frequency with which others harassed the infant (Fig. 9-13). On the average, infants were harassed just as often in proximity to an adult male as when no adult males were within 5 feet.

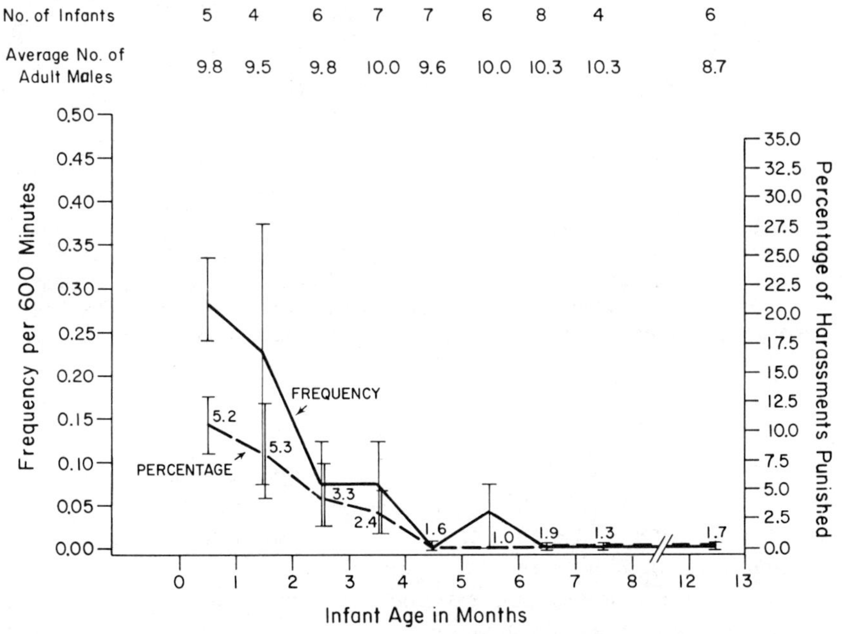

Fig. 9-12. Frequency with which adult males defended infants, and percentage of harassments by others within 5 feet of adult males that were punished by the male, as a function of infant age (simple mean over all infants and one standard error). The estimates for each infant were calculated by pooling all its interactions with all adult males. The number of infants observed and the average number of adult males available to each infant are indicated at the top of the figure for each age. The average number of adult males within 5 feet of whom each infant was harassed by others is indicated next to each plotted percentage.

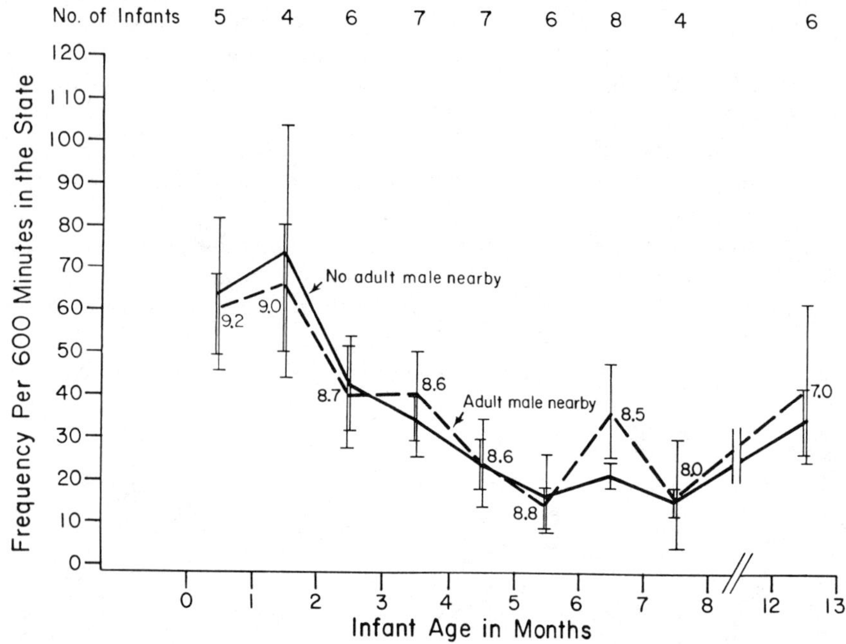

Fig. 9-13. Effect of proximity to adult males on frequency with which others harassed infants, as a function of infant age (simple mean over all infants and one standard error). The estimate for each infant was calculated by pooling over all adult males. The number of infants observed at each age is indicated at the top of the figure. The average number of adult males seen in proximity to each infant is indicated next to the appropriate plotted points.

Since the adult males themselves harassed infants much more frequently when the infants were nearby (Fig. 9-14), the net result is that infants received harassments at a much higher rate when an adult male was within 5 feet than when no adult male was (Fig. 9-15).

Infants and adult males spent extremely little time grooming one another (Fig. 9-16). Adult males groomed infants only during the first six weeks and again in months 5 and 6. Infants were first seen grooming adult males during their third month and tended to spend more time doing so as they got older.

Biosocial Variables

As one would suspect from the large standard errors of the mean for most behaviors, there was considerable variance among the different infants in my

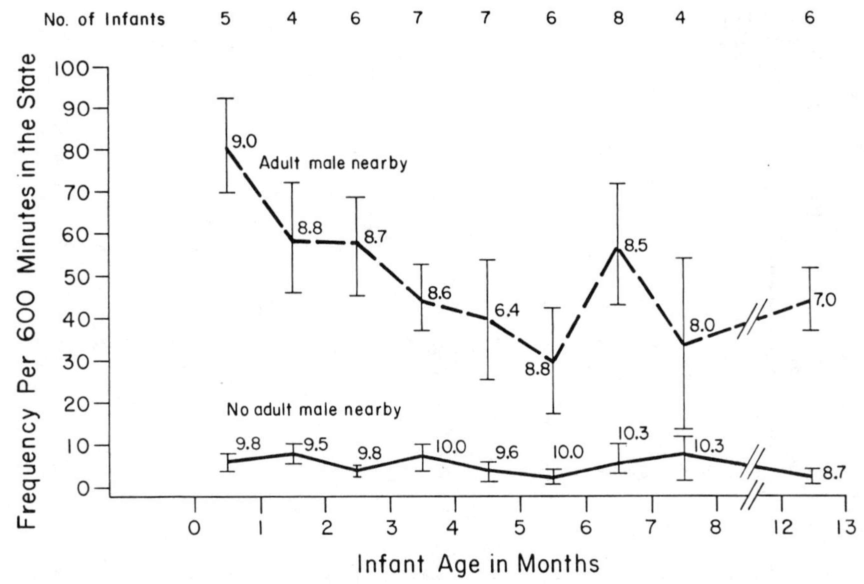

Fig. 9-14. Effect of proximity to adult males on frequency with which adult males harassed infants, as a function of infant age. (simple mean over all infants and one standard error). The estimate for each infant was calculated by pooling over all adult males. The number of infants observed at each age is indicated at the top of the figure. The average number of adult males participating in each spatial state for each infant is indicated next to plotted points.

study. Not discernible from these graphs is the even larger variance among the adult males. Most adult males made little or no contribution to the total score received by an infant, while one or two adult males accounted for nearly all of it. In order to help understand some of the sociobiological determinants of this variance, I distinguished the few infant-adult male dyads having these "preferred" relationships from the majority having "nonpreferred" relationships.

Figure 9-17 displays the average dominance rank of the adult males in preferred dyads and in nonpreferred dyads, at each infant age. Since dominance ranks are not measured on an interval scale, the actual values plotted may have no biological meaning. They do indicate, however, that through the entire first year of life, adult males in preferred dyads tended to be higher ranking than the adult males in non preferred dyads. They also show that it was not the very highest-ranking males that were preferred, but rather males in the range of third to fourth in rank, on the average.

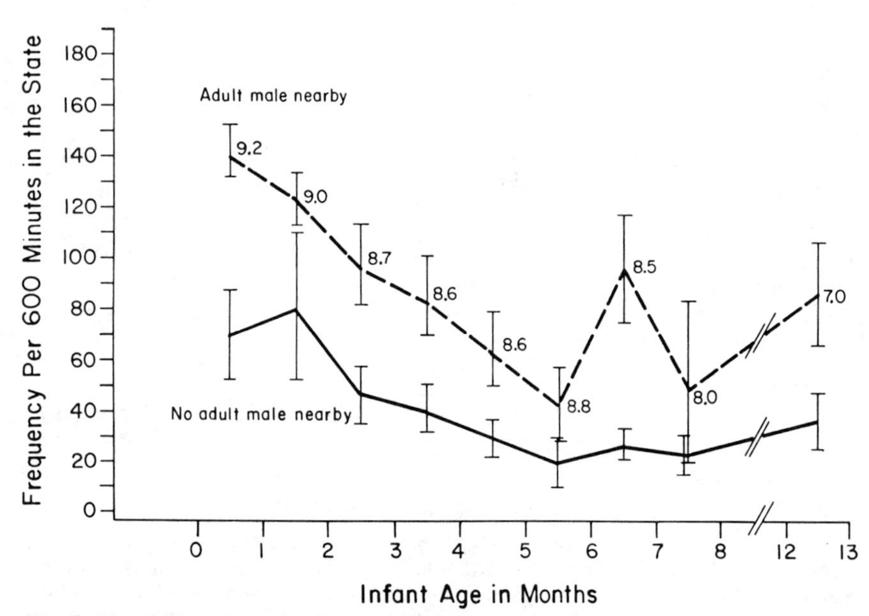

Fig. 9-15. Effect of proximity to adult males on total frequency with which infants were harassed, as a function of infant age (simple mean over all infants and one standard error). The estimate for each infant was calculated by pooling over all adult males. The number of infants observed at each age is indicated at the top of the figure. The average number of adult males seen in proximity to each infant is indicated next to the appropriate plotted points.

Next I examine four indicators of the likelihood of paternity. For each indicator, dyads were given a score of "1" if the criterion of that indicator was satisfied, and a score of "0" otherwise. Thus the average score over all dyads in a given class is an estimate of the probability that the dyads in that class meet the criterion of the indicator.

1. The probability that the adult male was in Alto's Group when the infant was born is presented in Figure 9-18 for preferred and for nonpreferred dyads. As is apparent, the adult male in preferred dyads was nearly always present in the group at the infant's birth, although this was not the case for nonpreferred dyads. At all ages during the first year, moreover, the probability that the male was present at the infant's birth was higher in preferred dyads than in other dyads.

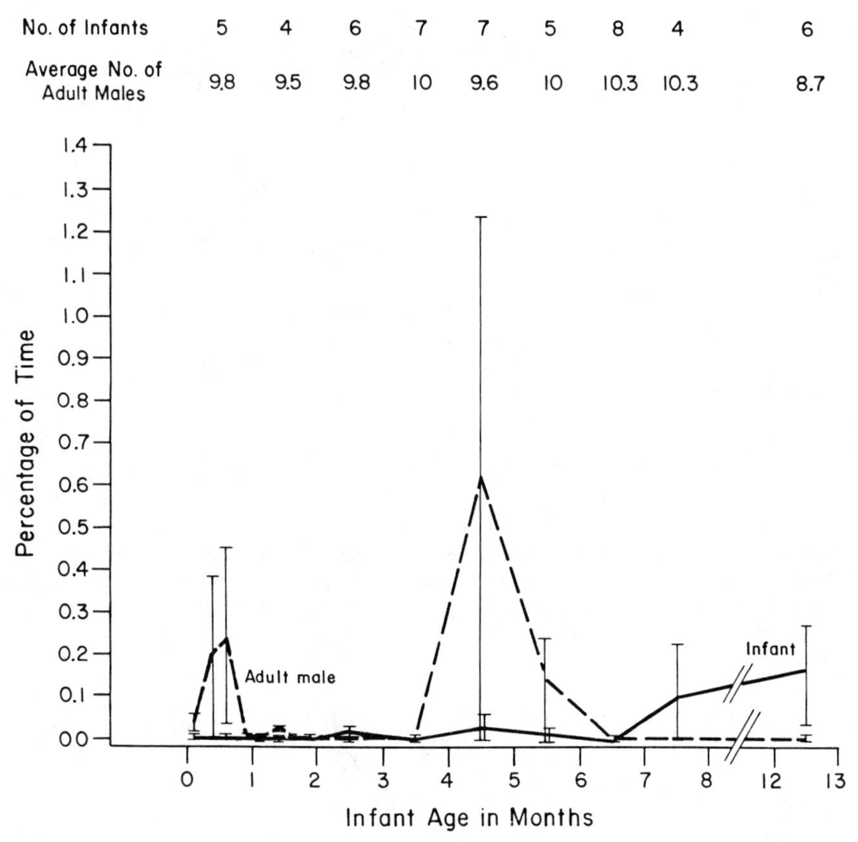

Fig. 9-16. Percentage of time infants and adult males spent grooming one another, as a function of infant age (simple mean over all infants and one standard error).

2. In Figure 9-19, I have recorded the probability that the adult male was in Alto's Group at the time the infant was conceived, for preferred and non-preferred dyads. Males absent from the group at the time of conception are highly unlikely to have sired the infant. Again, the adult male in preferred dyads was almost always present at the infant's conception, and at every age the probability that he was present at conception was higher than that for adult males in nonpreferred dyads.

3. The probability that the adult male copulated with the infant's mother at the time of conception, and thus is a potential father of the infant, is depicted in Figure 9-20 for the two types of dyad. Throughout the first year,

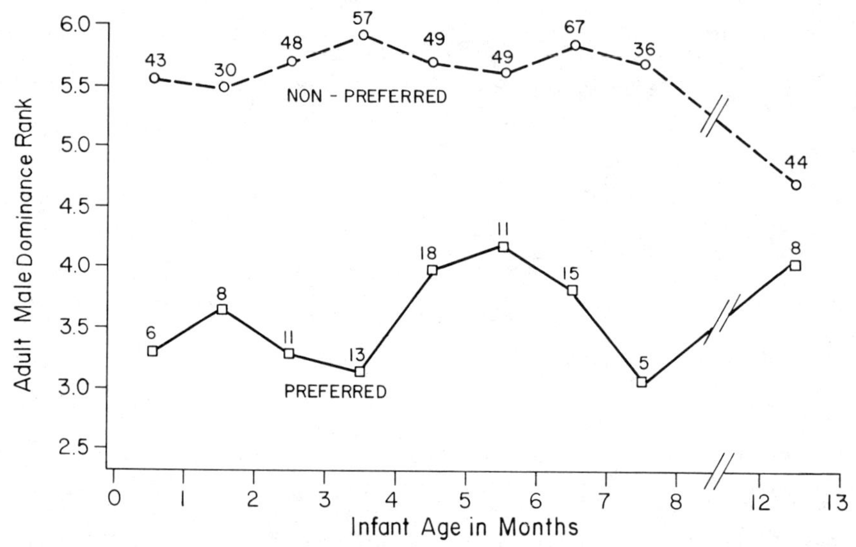

Fig. 9-17. Relative dominance rank of adult males in preferred and nonpreferred dyads, as a function of infant age (simple mean over all dyads in each class). The highest rank was assigned a value of "1". The number of dyads in each class at each age is indicated next to plotted points.

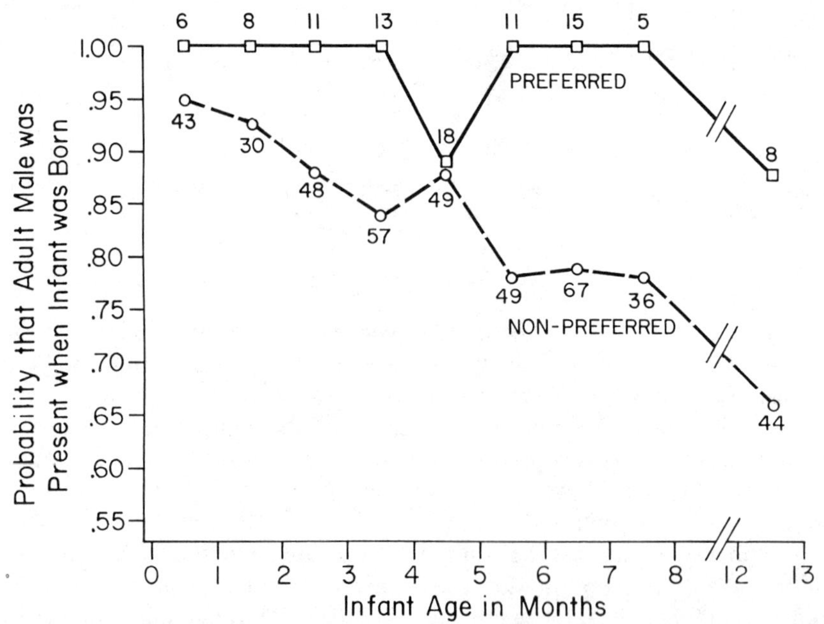

Fig. 9-18. Probability that adult male was present in group at infant's birth, for preferred and nonpreferred dyads, as a function of infants age (simple mean over all dyads in each class). The number of dyads in each class at each age is indicated next to plotted points.

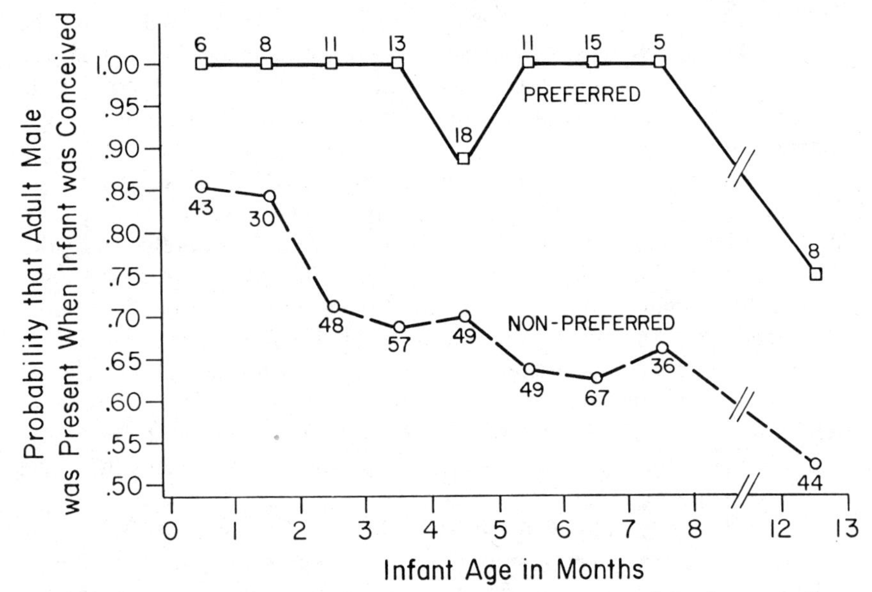

Fig. 9-19. Probability that adult male was present in group at infant's conception, for preferred and nonpreferred dyads, as a function of infant age (simple mean over all dyads in each class). The number of dyads in each class at each age is indicated next to plotted points.

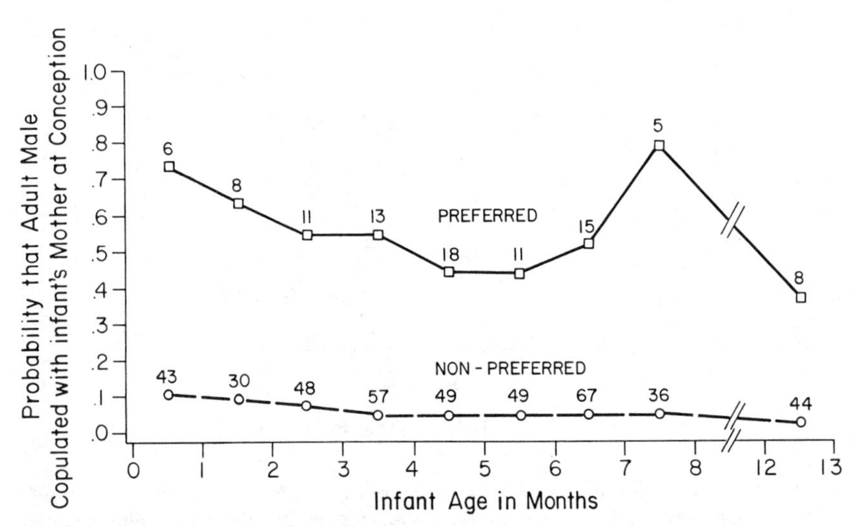

Fig. 9-20. Probability that adult male copulated with infant's mother at conception, for preferred and nonpreferred dyads, as a function of infant age (simple mean over all dyads in each class). The number of dyads in each class at each age is indicated next to plotted points.

adult males in preferred dyads were more likely to have copulated with the mother when the infant was conceived than were adult males in nonpreferred dyads. Indeed it was quite rare for any nonpreferred dyads to contain an adult male that had copulated with the mother at the time of the infant's conception.

4. The probability that the adult male monopolized the consortship record of the infant's mother at the time of conception and thus is almost certainly the father is indicated in Figure 9-21 for preferred and nonpreferred dyads. For none of the infants that I observed during their first two months was there an adult male who had monopolized the mother at conception, but at all other ages adult males in preferred dyads had a greater likelihood of having monopolized the infant's mother than did males in nonpreferred dyads. Although the preferred dyads often contained males who had not monopolized the mother's consortship record during the estrus of conception, all males who had done so had a preferred relationship with the offspring, with one exception in month 13.

During my study there were only six adult males in Alto's Group whose relative ages were known for sure (G. Hausfater, pers. comm.). Considering only dyads containing these six males, I give in Figure 9-22 the average

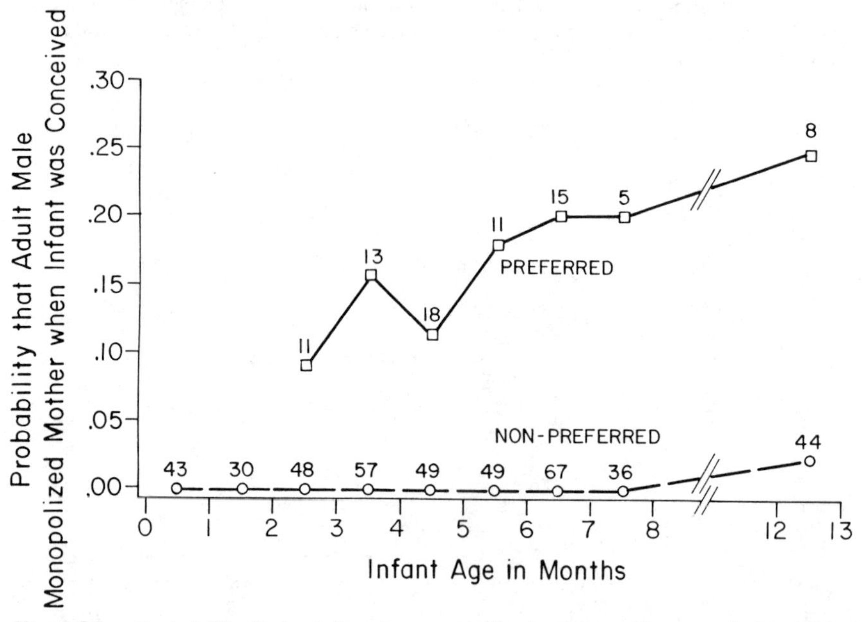

Fig. 9-21. Probability that adult male monopolized consortship record of infant's mother at conception, for preferred and nonpreferred dyads, as a function of infant age (simple mean over all dyads in each class). The number of dyads in each class at each age is indicated next to plotted points.

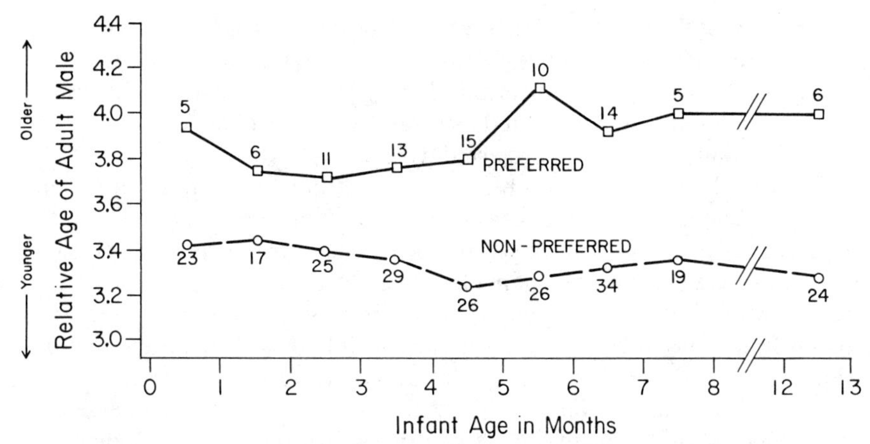

Fig. 9-22. Relative age of adult males in preferred and nonpreferred dyads, as a function of infant age (simple mean over all dyads in each class). Only the six adult males whose relative ages were known for certain are included. The youngest male was assigned a value of "1." The number of dyads in each class at each age is indicated next to plotted points.

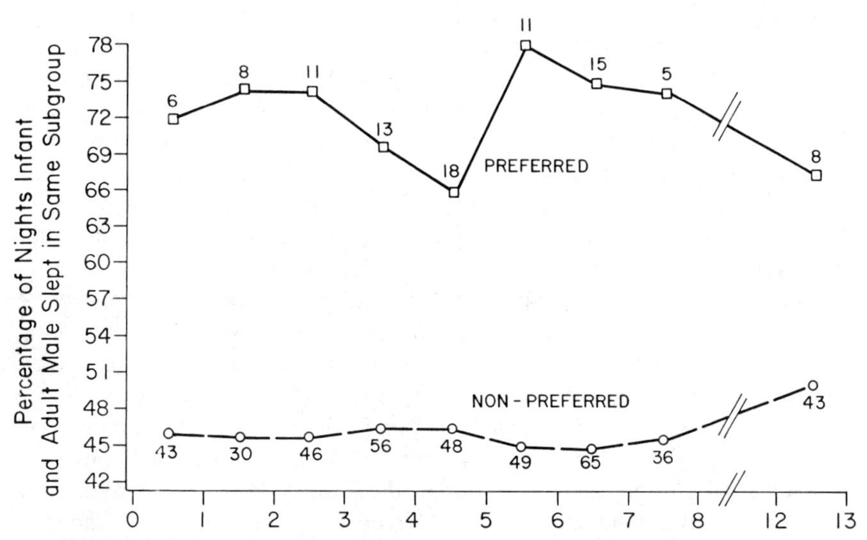

Fig. 9-23. Strength of subgroup affiliation in preferred and nonpreferred dyads, as a function of infant age (simple mean over all dyads in each class). The number of dyads in each class at each age is indicated next to plotted points. One adult male was not present during any samples of sleeping-grove subgroups.

relative age of the adult males in preferred dyads and in nonpreferred dyads. Since the actual age differences between the males are not known, the exact meaning of this graph is not certain. It suggests, nevertheless, a consistent trend for adult males in preferred dyads to be slightly older, on the average, than the adult males in nonpreferred dyads.

Figure 9-23 displays the average strength of subgroup affiliation for preferred and nonpreferred dyads, as a function of infant age. Throughout the first year of life, dyads having a preferred relationship slept in the same cluster of trees considerably more often than did dyads having a nonpreferred relationship. In fact, the strength of subgroup affiliation accounted for a much larger proportion of the variance in infant-adult male interactions than any of the other biosocial parameters that I examined.

Infant gender did not appear to have an appreciable effect on infant-adult male relationships, except for the fact that all grooming of adult males by infants during focal samples was done by female infants. Likewise, the relative dominance of the infant's mother was not a useful predictor of any aspect of infant-adult male social interactions.

DISCUSSION

All of the above observations can best be understood, perhaps, with the following view of infant-adult male relationships (elaborated upon considerably in Stein, 1984), although there may be other schemata with which the data are also compatible:

It seems that adult males in a multi-male yellow baboon group seek to be near and to handle newborn infants mainly for two reasons. First, newborn infants are very effective agonistic buffers. Second, infants develop a selective attachment to the adult males to whom they are most exposed as neonates, and by increasing his contact with an infant during its first few weeks of life, an adult male increases that infant's cooperation with him in future agonistic buffering episodes.

Due to the ability of dominant males to supplant subordinate males, the males that succeed in spending much time near the infants and in establishing special relationships with them tend to be higher ranking than those who do not. Because mothers are more tolerant of males with whom they are more familiar, these males tend to have been present in the group for a longer time (both when the infant was born and when it was conceived), tend to be older, and tend to be in the same subgroup as the mother. For the above reasons, the same males are the ones most likely to consort with the mother, and thus males having preferred relationships with an infant also have a higher probability of being the infant's father than other males.

After about the first six weeks of life, adult males interact with infants much less frequently and for shorter periods of time, possibly for three different reasons. First, the period of special sensitivity of infants to exposure to adult males has presumably passed. Second, older infants may be less effective as agonistic buffers. Third, the infants take an increasingly active role in the relationship, and after about six weeks, when needed in an agonistic encounter they can go to the adult male.

This trend continues for most of the first year, with the adult males playing an ever-smaller role in the relationship and the infants taking more and more of the initiative. When agonistic tensions rise, however, adult males do not hesitate to renew their attentions to infants, as was seen in my study for infants in their fifth and sixth months. Indeed, except during the early period of intense exposure to males, this was the only time when adult males groomed infants.

Infants probably cooperate with specific adult males and serve as agonistic buffers for them "in exchange" for unrelated benefits that they receive from their relationship with these males. Some adult males actively defend certain infants from the harassment of other group members, and some adult males share food with certain infants. Infants and their mothers play a more active relative role in maintaining proximity to adult males when the group is predominantly feeding than when it is predominantly not feeding, suggesting that proximity to the adult male enhances their acquisition of nutrition. More specifically, infants feed near males on food processed by those males, primarily at the ages when additional nutrition is expected to be most critical to their survival.

By the end of an infant's first year of life, its mother is probably cycling again (J. Altmann, et al., 1978). The mother is apt to be fairly inattentive to her offspring at this time, and the infant may more actively seek out adult males. When the infant is with its mother, an adult male is more likely to be nearby than had previously been the case. Because of the relatively high turnover of adult male dominance ranks and the relatively high migration rates of adult males, this will often be an adult male with whom the infant had had no prior special relationship. Thus infants may be spending more time with adult males at this age and having longer proximity bouts, even though they are harassed more frequently by the adult males.

This scenario is only one possible interpretation, albeit the one that seems most plausible to me, of the patterns observed during my limited study. These ideas and other ones suggested by my data must be tested in many other groups, habitats, and species. More importantly, these data should be used for testing other hypotheses that arose independently of my results, and for comparison with infant-adult male interactions at other study sites having different demographic compositions, different ecological environments, different social structures, and even different primate species.

ACKNOWLEDGMENTS

I would like to thank the people of the United States for their support of this research through Medical Scientist Training Grant PHS 5 T32 GM07281 to the author and National Institutes of Mental Health grant MH19617 to Stuart Altmann. The cooperation of the Office of the President, the Ministry of Tourism and Wildlife, and the Institute of African Primatology of Kenya is also greatly appreciated. I especially wish to express my gratitude to Stuart Altmann, whose advice and assistance has enhanced all aspects of this study from its inception through this final manuscript.

REFERENCES

Altmann, J. 1974. Observational study of behavior: Sampling methods. *Behaviour* 49: 227-67.

———. 1980. *Baboon Mothers and Infants.* Cambridge: Harvard University Press.

Altmann, J.; Altmann, S. A.; Hausfater, G.; and McCuskey, S. A. 1977. Life history of yellow baboons: Physical development, reproductive parameters, and infant mortality. *Primates* 18: 315-30.

Altmann, J.; Altmann, S. A.; and Hausfater, G. 1978. Primate infant's effects on mother's future reproduction. *Science* 201: 1028-30.

Altmann, S. A., and Altmann, J. 1970. *Baboon Ecology.* Chicago: University of Chicago Press.

Bales, K. B. 1980. Cumulative scaling of paternalistic behavior in primates. *American Naturalist* 116: 454-61.

Bogess, J. 1979. Troop male membership changes and infant killing in langurs *(Presbytis entellus). Folia Primatologica* 32: 65-107.

Gillman, J., and Gilbert, C. 1946. The reproductive cycles of the chacma baboon with special reference to the problems of menstrual irregularities as assessed by the behaviour of the sex skin. *South African Journal of Medical Science* 11: 1-54.

Hausfater, G. 1975. Dominance and reproduction in baboons. *(Papio cynocephalus). Contributions to Primatology,* vol. 7. Basel: S. Karger.

Hendrickx, A. G., and Kraemer, D. G. 1971. Reproduction. In A. G. Hendrickx (ed.), *Embryology of the Baboon,* pp. 1-30. Chicago: University of Chicago Press.

Hrdy, S. B. 1976. Care and exploitation of nonhuman primate infants by conspecifics other than the mother. *Advances in the Study of Behavior* 6: 101-58.

Mitchell, G. D., and Brandt, E. M. 1972. Paternal behavior in primates. In F. E. Poirier (ed.), *Primate Socialization,* pp. 173-205. New York: Random House.

Post, D. 1978. Feeding and ranging behavior of the yellow baboon *(Papio cynocephalus).* Doctoral dissertation, Yale University.

Redican, W. K., and Taub, D. M. 1981. Adult male-infant interactions in nonhuman primates. In M. E. Lamb (ed.), *The Role of the Father in Child Development,* 2nd ed. New York: John Wiley and Sons.

Stein, D. M. 1981. The nature and function of social interactions between infant and adult male yellow baboons *(Papio cynocephalus).* Doctoral dissertation, University of Chicago.

——— 1984. The sociobiology of infant and adult male baboons. *Monographs on Infancy,* vol. 5. Norwood: Ablex.

Stein, D. M., and Stacey, P. B. 1981. A comparison of infant-adult male relations in a one-male group with those in a multi-male group for yellow baboons *(Papio cynocephalus). Folia Primatologica* 36: 264-76.

Struhsaker, T. 1967. Behavior of vervet monkeys and other cercopithecines. *Science* 156: 1197-1203.

Walters, J. 1980. Interventions and the development of dominance relationships in female baboons. *Folia Primatologica* 34: 61-89.

Western, D. 1973. The structure, dynamics and changes of the Amboseli ecosytem. Doctoral dissertation, University of Nairobi, Kenya, East Africa.

10

Adult Male-Infant Interactions in the Chimpanzee *(Pan troglodytes)*

Diana Davis
Institute for Primate Studies
Norman, Oklahoma

INTRODUCTION

Paternal behavior as an area of study has, until recently, received minimal attention. This is true for both human and nonhuman primates. The purpose of the present study is to examine adult male-infant interactions in the chimpanzee *(Pan troglodytes)*. As in the human literature, a great deal of attention has been given to the mother-infant bond in the chimpanzee and very little to the role of the adult male chimpanzee and its significance to the development of the infant.

A study of this nature addresses several issues that are important to the field of primatology. Since chimpanzees are considered one of our closest living relatives, such a study addresses the issue of the evolution of human paternal behavior. In addition to evolutionary implications, the generalizability of data across primate species must be considered; that is, the feasibility and validity of utilizing these animals as a model for human paternal behavior should be addressed. This type of research also has implications for the species preservation of the chimpanzee. With the dramatic reduction in numbers of chimpanzees, proper rearing of these animals in captivity is crucial for successful breeding to occur. Thus, an important question to ask is what extent the presence or absence of an adult male will have on the normal social development of the immature chimpanzee in captivity. Should contact be minimal, maximal, or nonexistent? These questions and issues are far from being answered and resolved. Nevertheless, studies that focus upon adult male-infant interactions in nonhuman primates provide some degree of clarity and resolution and should generate subsequent research.

A growing interest has developed concerning paternal behavior in nonhuman primates, as evidenced in general reviews by Raphael (1969), Mitchell (1969, 1977), Mitchell and Brandt (1972), Spencer-Booth (1970), Redican and Taub (1981), and Blaffer-Hrdy (1976). Yet, very little information has been documented on male-infant interactions in the chimpanzee (see chapter 15). The available information typically is found in general studies of social behavior in feral chimpanzees (e.g. van Lawick-Goodall, 1968) and in captive chimpanzees (King et al., 1980).

Even though male-infant interactions were not the focal point of these studies, the information they provide is valuable, and it is desirable to summarize the patterns of male-infant interactions as a framework for interpreting the present study. Under feral conditions, the opportunity for adult male-infant interactions is far less than for mother-infant interactions. This is due to the solitary nature of the female with dependent offspring (Goodall, 1977) and the patrolling behavior of adult males, who form a kinship group and move about in the group's home range (Pusey, 1979; Nishida, 1979). Nevertheless, these interactions do occur and change in quantity and quality as the infant matures.

As in the case of all the pongids, females appear to be most restrictive with their infants during the first several months of the infants' lives; the female chimpanzee inhibits almost all of its contact with other group members. However, an adult male may achieve contact by reaching out to the infant and touching it on its head or face (van Lawick-Goodall, 1968).

During the second half of the first year, when continuous mother-infant contact has broken, the males have greater opportunity to interact with infants. The adult male may greet the infant by reaching out to touch or pat the infant and offer comfort and protection by picking up and embracing infants who are temporarily separated from their mothers. Also, they have been observed to assist infants in their uncoordinated endeavors and to exhibit a degree of tolerance when infants interfere with their activities (van Lawick-Goodall, 1968).

It is during the latter part of the second year that adult males appear to become less tolerant of the infants' activities. They push the infants away more roughly when they interfere with feeding or copulation. It is during this time that an infant appears to learn when to approach or avoid an adult male and to recognize signs of potential excitement displayed by a male. It is noteworthy that the infant's aggressive displays increase during the second year of life. Another type of behavior that adult males sometimes direct toward the infant of this age is mounting and thrusting movements made against the infant's back in the context of play or sometimes during greeting. Likewise, analogous infantile sexual patterns are observed during this period (van Lawick-Goodall, 1968).

While play interactions between the males and infants increase during the second year, they begin to wane during the third year. During the earlier part of the third year, infants are still approached and embraced by mature males and are allowed to climb into their laps. However, by the end of this year, the interactions consist mainly of grooming, greeting, aggressive-submissive encounters, and sometimes play. Although males still tolerate and protect infants during the third year, they direct threatening gestures toward them more frequently, and the infants begin to direct a submissive behavior, bobbing, at the adult male (van Lawick-Goodall, 1968).

It appears that the adult male chimpanzee is capable of positive interactions with younger chimpanzees. Yet a socially excited adult male poses some degree of danger to infants, who may be used as display objects. In addition, adult males as well as adult females have been known to engage in infanticide and cannibalism (Suzuki, 1971; Bygott, 1972; Goodall, 1977). However, Goodall (1977) suggested that inter-group infanticide, associated with males, can be explained in terms of inter-group aggression directed at the infant's mother, rather than at the infant itself. She also noted that adult males are known to defend new mothers from intragroup attacks that are associated with adult females.

In summary, most adult male-infant interactions in feral conditions are positive in nature and typically occur in the presence of the infant's mother. In the rare cases involving infanticide on the part of the male, the infant and its mother are from another group.

Positive male-infant interactions have also been observed in a study of the social behavior of a group of captive chimpanzees, comprised of one adult male and three mother-infant pairs (King et al., 1980). It is important to note that unlike most captive conditions, an adult male was housed with the mother-infant pairs, and the group had lived together for many years. In this study, the adult male preferred to interact with not only the male infant, but also with the male infant's mother. Likewise, the male infant interacted as frequently with the adult male as he did with the female infants. The male-infant interactions of these captive chimpanzees were similar to those reported by van Lawick-Goodall (1968) for feral animals.

The present study examines adult male-infant interactions in captive chimpanzees, where the mother, infant, and male were periodically grouped in the same cage. It was therefore possible to determine the degree to which female restrictiveness affected the quantity of male-infant interactions and the extent to which this restrictiveness changed over time. Lastly, this study is designed to identify the degree to which the male was interested in the infant and vice versa.

METHOD

Subjects

Initially, the study was planned to involve three subjects: the biological mother (Washoe), the biological father (Ally), and the infant at the Institute for Primate Studies in Norman, Oklahoma. However, the infant died two months after birth, and the original study was modified to include an adopted infant.

Fifteen days after the infant's death, a 10½-month-old male infant (Loulis), who was received on loan from Yerkes Regional Primate Center, was introduced to Washoe for the purpose of adoption. While this infant had been with its biological mother for its first 10½ months of life, it had neither visual nor physical access to an adult male chimpanzee. Thus, his first exposure occurred upon his arrival in Norman. The adult male, from the onset of Loulis's arrival, was housed in the adjacent cage and was allowed access to Washoe and Loulis's cage only during the data collection process. This housing arrangement did not preclude male-infant interactions through the cage.

The cages were centered in a large room and each cage was 8½′ × 7½′ × 7½′ and was divided by a sliding door that was secured and locked, except during the actual study. The cage material consisted of expanded metal; the subjects therefore had visual access to each other, but when the sliding door was closed, actual physical contact was limited to touching or poking with the fingers.

At the beginning of the study, the adult male was 10 years old, and the adult female, approximately 14. Both subjects were home-reared, the male for 4 years and the female for 5½ years. Since their arrival in Norman, Oklahoma, they have had extensive social interactions with other chimpanzees and with each other. The female at the time of the study was still clearly dominant to Ally, although during the course of the study, the dominance relationship was undergoing transition.

Both adults had previous exposure to infants; they had been housed adjacent to a mother-infant group. Washoe, also had had direct access to infants; her first infant died several hours after birth, and her second infant died two months after birth. While the second infant was alive, Ally had direct access to the mother-infant pair on several occasions but this access was limited and carefully monitored by experimenters.

Data Collection

Data collection for the triadic interactions began approximately six months after Loulis's arrival in Norman and continued for the following four months.

Loulis was 15 months old at the beginning of the study. In total, there were nine episodes in which the male was allowed to enter Washoe and Loulis's cage. Each of these episodes was filmed and subsequently analyzed for the following variables: (1) maintenance of proximity between adults; (2) adult male-female interactions and (3) adult male-infant interactions. An interaction was demarcated as an initiation and any termination or attempt to terminate.

Analysis

The data were recorded as frequencies for each episode, reported as percentages (Tables 10-1 through 10-6) and plotted over time. The following variables were examined: (1) Maintenance of adult male-female proximity; (2) adult withdrawals preceded by male-infant interactions; (3) male-infant contact following adult interactions; (4) adult male versus infant initiations; and (5) terminator of male-infant interactions—male initiated versus infant initiated.

Inter-observer reliability measures for the occurrence of behaviors were obtained by randomly selecting 10-second intervals from each video tape and utilizing Pearson's r correlation coefficient ($r = 0.94$).

RESULTS

Maintenance of Adult Male-Female Proximity

Table 10-1 indicates that overall, the male was principally responsible for the maintenance of proximity. In all but three of the episodes, the male both approached and withdrew from the female over 69% of the time. In contrast, the female rarely both approached and withdrew from the male, accounting for zero percent of the interactions except in episode 7 (16.7% of total). Because the infant was usually near the female until the later episodes, the male had to approach the female in order to gain access to the infant. The female's failure to approach the male could be interpreted as maternally restrictive behavior. Consequently, since she did not typically make the infant accessible to him, it was the male who actively attempted to overcome the restrictiveness by approaching her.

Adult Withdrawals Preceded by Male-Infant Interactions

This variable was examined to determine whether the female used withdrawal as a strategy for making the infant inaccessible to the male, and

Table 10-1. Maintenance of Adult Male-Female Proximity.

	Approach/Withdraw			
Episodes	Male-Male	Male-Female	Female-Female	Female-Male
1	77.8	22.2	0	0
2	80	20	0	0
3	100	0	0	0
4	25	75	0	0
5	69.2	15.4	0	15.4
6	37.5	62.5	0	0
7	33.3	33.3	16.7	16.7
8	87	4.3	0	8.7
9	87.5	0	0	12.5

Table 10-2. Adult Withdrawals Preceded by Male-Infant Interactions.

	Withdrawals			
Episodes	Male preceded by male-infant interactions	Male not preceded by male-infant interactions	Female preceded by male-infant interactions	Female not preceded by male-infant interactions
1	77.8	0	11.1	11.1
2	50	30	10	10
3	75	25	0	0
4	25	0	50	25
5	53.8	30.8	15.4	0
6	12.5	25	25	37.5
7	33	16.7	33	16.7
8	87	8.7	4.3	0
9	87.5	12.5	0	0

whether the male's withdrawals were related to prior terminations of interactions with the infant. Of the female's withdrawals (Table 10-2), there does not appear to be much difference between those that involved a prior adult male-infant interaction and those that did not. Thus, the female did not typically utilize withdrawal as a strategy for inhibiting or terminating interactions between the male and the infant. In contrast, a high incidence of the male's withdrawals were associated with prior male-infant interactions, particularly in the earlier (e.g., Episodes 1, 2, and 3) and later (e.g., Episodes 8 and 9) Episodes of the study. It therefore appears that upon termination of a male-infant interaction, the male typically increased the distance not only between himself and the infant, but also between himself and the female.

Male-Infant Contact Following Adult Interactions

This variable was examined to determine whether the male, in an attempt to reduce female restrictiveness, used the strategy of first interacting with the female before attempting to interact with the infant, and if the attempt resulted in actual contact with the infant. In the earlier Episodes (Table 10-3), there is little difference between those that involved prior interactions with the female and those that did not. For the last four episodes, however, prior interactions markedly decreased, with a corresponding increase in situations involving no prior interactions with the female. For example, there is a dramatic increase in the frequency with which the male achieves contact with the infant without first interacting with the female (28% in Episode 1 versus 70% in Episode 9) and a dramatic decrease with prior interactions (36% in Episode 1 versus 5.5% in Episode 9). The lowest percentages across episodes typically occur for situations where the male first interacts with the female but does not achieve contact with the infant. These data suggest that the strategy of first interacting with the female to gain access to the infant was not a salient feature of the triadic interactions. While the male may have initially utilized this strategy to some degree, he did so with less frequency as the study progressed, and by the end of the study his success in achieving contact with the infant was not contingent upon a prior interaction with the female.

Adult Male Versus Infant Initiations

This variable was examined to determine whether the male or the infant was more frequently the initiator of their interactions, and from Table 10-4 it can be seen that the adult male consistently initiated a great majority (70+%) of the interactions, until the last two episodes. Thus by Episode 8, the infant initiated 44.4% of the interactions, and in Episode 9 the infant initiated a majority of the interactions (66.9%). This reversal in initiations is significant since it indicates that, over time, the infant became more assertive while in the presence of the male; that is, the infant switched from being a passive recipient of the male's attention to being an active solicitor of the male's attention. The male's initiations, in contrast, decreased drastically, possibly indicating a waning of interest on his part.

Terminator of Male-Initiated Male-Infant Interactions

This variable was examined to determine which subject most frequently terminated the adult male-initiated infant contacts. In the first Episode

Table 10-3. Male-initiated Infant Contact Following Adult Interactions.

	No Adult Interaction		Adult Interaction	
Episodes	% Contact	% No Contact	%Contact	% No Contact
1	36	16	28	20
2	37.5	12.5	33.3	16.1
3	32.7	23.6	16.4	27.3
4	5.3	9.2	47.4	38.1
5	32.5	13.5	21.6	32.4
6	17.2	8.6	55.2	19
7	9.3	7.4	66.7	16.6
8	11.1	.7	56.3	31.9
9	5.5	.9	70	23.6

Table 10-4. Adult Male Versus Infant Initiations.

	Initiator	
Episodes	% Male	% Infant
1	69.6	30.4
2	92	8
3	74	26
4	96.2	3.8
5	75.7	24.3
6	71.2	28.8
7	86	14
8	55.6	44.4
9	33.1	66.9

(Table 10-5), the female terminated 80% of the male-infant interactions, whereas the infant alone terminated only 5%. By the last Episode, however, the infant terminated 63% of the interactions, and the female accounted for 14.8%. Thus, over time, the female became less restrictive, putting the infant in a position of having greater control in the management of its interactions with the male. In addition, Table 10-5 indicates that the male rarely terminated interactions that he had initiated with the infant.

Terminator of Infant-Initiated Male-Infant Interactions

This variable was examined to determine which subject most frequently terminated the infant-initiated male-infant interactions. Table 10-6 reveals a pattern of interaction terminations similar to that seen in Table 10-5. In the first Episode, the infant and female were the sole terminators of the male-infant interactions (42.9% each), and in the second Episode the female terminated all the interactions. For the last three Episodes, this pattern was reversed, with the infant being responsible for most or all the terminations of male-infant interactions initiated by the infant. This further supports the notion that, over time, the female was less restrictive on the infant's behavior. It should also be pointed out that until the last two Episodes (8.1% and 30.6%, respectively), the male never terminated interactions initiated by the infant. This sudden increase in male terminations may indicate again a waning interest in the infant on the part of the male.

Table 10-5. Terminator of Male-Infant Interactions (Male Initiated).

	Interaction Terminated by			
Episodes	% Female only	% Infant only	%Both female and infant	% Male only
1	80	5	5	10
2	44	24	25	12
3	58.9	12.5	14.3	14.3
4	63.2	9.2	21	6.5
5	41.1	25	26.8	7.1
6	25	39.6	27.1	6.3
7	2.1	83.3	12.5	2.1
8	8.2	69.1	19.1	3.6
9	14.8	63	14.8	7.4

Table 10-6. Terminator of Male-Infant Interactions (Infant Initiated).

	Interaction Terminated by			
Episodes	% Female only	% Infant only	%Both female and infant	% Male only
1	42.9	42.9	14.2	0
2	100	0	0	0
3	25	75	0	0
4	33.3	33.3	33.3	0
5	16.7	77.7	5.6	0
6	50	38.9	11.1	0
7	0	100	0	0
8	1.1	89.7	1.1	8.1
9	.9	63.1	5.4	30.6

DISCUSSION

The above results provide data on adult male-infant interactions, with emphasis placed upon change over time. The infant initially was hesitant, but with time began to seek social contact with the male. The female initially was restrictive of both the infant and the male, but with time she began to be much more permissive. In addition, the male, being principal monitor of spatial relationships between himself and the female, did not always use the strategy of first interacting with the female as a mechanism for gaining access to the infant. The female did not utilize the strategy of increasing the spatial distance as a means for decreasing the infant's accessibility to the male. Thus, early adult negotiations regarding the infant necessarily involved close proximity and contact.

When the male would approach, the female would typically orient away from him and toward the infant and would frequently pull the infant to her. The infant would typically move toward the female upon the males approach, assuming a ventral-ventral position. If the male first interacted with the female before attempting to interact with the infant, he would typically groom her or initiate play. Playing and grooming, initiated by both the male and female, also occurred if the male first attempted to interact with the infant and was inhibited by the female. Thus, it appears that adult male-female initiations were used to distract the male from the infant, to reduce female restrictiveness, and to mollify any ensuing aggression or excitement. As the study progressed, the infant attempted to participate in the adult play sessions. Thus, the interadult interactions also served to provide the infant with a situation where proximity or contact with the male was less aversive. In fact, it was during a vigorous adult play session that the infant initiated its first contact with the male by cautiously placing its wrist in the male's mouth. It should also be pointed out that the infant was often ignored during vigorous play, which resulted in the infant's pouting, whining, and once having a tantrum.

A decrease in female restrictiveness became apparent when she began to withdraw from the infant and male. In one episode, she entered the adjacent cage and closed the sliding door, but continued to pay attention to the activities of the male and infant. When she did interfere with the interactions, she would typically block the male's reach with her arm or hand, gently push his hand away, or hold his hand if he was too persistent. Often she would use her body as a means of interference, by orienting away from the male (perhaps to present for grooming), leaning over or toward the infant, and pulling the infant closer to her. In the last two Episodes, her interference was minimal and her interest in their activity was rare. A large portion of these last episodes primarily involved interactions between the male and the infant. It is suggested that her waning interest was a function of her confidence in the social bond that had formed between the male and infant.They had obviously become play partners and her protectiveness was minimally needed.

The male-infant interactions at the beginning of the study can be characterized as one of approach-avoidance. The infant exhibited anxiety and fear at the male's solicitations and generally utilized the female as a site of refuge. In the beginning, the male's contact with the infant would simply involve a quick touch, since neither the infant nor the female permitted easy access. With time, the male began to pat the infant or attempt to groom it while it was in contact with or near the female.

As the study progressed, contact lasted for extended periods of time although it initially took place alongside the female. Eventually, these dyadic interactions took place some distance from the female, since the infant had begun to withdraw from the female to solicit the male's attention. The infant's initial initiations usually served no other purpose than to make the infant accessible to the male. With time, however, the infant's initiations involved actual contact such as play slapping or play kicking or ever assuming a ventral-ventral embrace with the male.

Male-infant interactions typically involved play, and, with time, the play became more vigorous. If the infant became frightened, the male, offering comfort, would stop playing and embrace and pat the infant. It should be pointed out that the male never exhibited any signs of aggression toward the infant. Also, whenever the male attempted to groom the infant, the infant would withdraw. The infant was never observed attempting to groom the male.

The last two Episodes are of particular interest since it was at these times that dramatic changes in the direction of the interactions began to appear. As noted earlier, most of these interactions were dyadic, involving only the male and infant; the female showed little interest in their activities. During these Episodes, it appeared that the approach-avoidance condition noted earlier was reversed, with the male now avoiding the solicitations of the infant. Not only did the male fail to respond in some situations, but he also began to withdraw from the infant, who would follow and hold onto the male. When the male sat down, the infant attempted to sit in his lap. If the male discouraged this position, the infant would sit beside or behind the male with his arms around him and, very frequently, with his head resting on the male's body. The male was even observed to gently push the infant away when the latter attempted to gain contact. It should be pointed out that the infant never gave up in his attempts to prolong these interactions. For example, in the last Episode, the male entered the adjacent cage and closed the door, blocking the infant's entrance. The infant attempted to pull the door open and, upon failure, stomped his feet on the door and toward the male.

It can be concluded that a strong social bond can form between an adult male and infant chimpanzee in the presence of an initially restrictive mother. While many of the activities observed in this study have also been observed in feral conditions, the male nevertheless does not act as principal caregiver. What must be addressed then is the role that the adult male plays in the enhancement of the infant's survival and its development. Certainly, the infant who can utilize both adult males and adult females will be at a greater advantage than the infant who utilizes only the mother.

Clearly, all group members, including infants, benefit by the male's patrolling of the group's territory, which serves to locate strangers, predators, and rich food sources. It is suggested that another role of the male in development concerns the ontogeny of social skills. Without this input, it is doubtful that normal integration into the group would occur; an individual as it matures must learn through contact and observation the importance and dominance of the adult male, the appropriate social signals for submission, and when the exhibition of these signals is appropriate. This may be even more important for the infant male chimpanzee, since at some point in the future, he will begin to integrate into the adult male group. The mature male chimpanzee, therefore, as a salient and focal member of the group, provides a social stimulus to infants, who, as a result, learn their own status as well as the status of other group members relative to the adult male. Particularly for the infant males, he is someone to emulate. It is also suggested that in this learning process, quality of interaction or observation may be more important than quantity.

If indeed strong social bonds can and do form between mature males and infant chimpanzees, then there must be a selective advantage. What advantage, then, does the male receive in expending energy that enhances an infant's survival, particulary where a genealogical relationship, or paternity confidence (Alexander, 1974), is unknown? According to parental investment theory (Trivers, 1972), when paternity confidence is minimal, as in the chimpanzee, paternal investment is predicted to be minimal. This prediction appears to be met, in that the male chimpanzee does not typically remain with one particular female and aid her in the rearing of a particular infant. Yet, caregiving behaviors do occur, even though not in the context of apparent paternity. What evolutionary advantage does the male receive by exhibiting these behaviors? One explanation may involve the concept of shared genes (i.e., inclusive fitness). Since adult males form a kinship group, they share common genes. From a sociobiological perspective (see chapter 11), any behavior on the part of a male that enhances an infant's survival will simultaneously enhance the probability that those shared genes will be passed on in future generations. Thus, a male need not sire an infant to gain genetically as long as there is a genealogical relationship with the actual father of the infant. This explanation can also be applied to the male's interactions with any offspring that his mother may produce.

Another explanation, offered by McGrew (pers. comm., 1979), involves the phenomenon of consortship, in which the male chimpanzee singles out a female in estrus and remains with her, isolated from the group, until her tumescence subsides (McGinnis, 1979; Tutin, 1974). In these cases, paternity confidence is believed to be high. However, to persuade the female to accompany him, the male needs the necessary social skills to interact not only with the female, but also with her still dependent offspring. Should he intimidate either the female or her offspring, their resistance might attract all nearby males. Such a failure to form a consortship results in a dramatic reduction in paternity confidence. Thus, it is to the male's advantage to have in his repertoire of social skills those behaviors that will be attractive and least intimidating to the infant as well as to its mother. The males who are most successful in their interactions with immature members of their group will gain genetically. In turn, these unaversive behaviors will be selected for in subsequent generations.

Perhaps it is through this mechanism that paternal behavior evolved in the human species. Tutin (1974) found that the adult male, who is most likely to form a successful consortship, is especially generous with a specific female during meat-sharing activities. This suggests the incipience of strong pair bonding between the adult male and adult female. If food sharing is an important variable to consider in the phylogeny of familial ties (Lancaster, in pr.), then it is reasonable to consider the chimpanzee as an appropriate model for the evolution of human paternal behavior. Furthermore, it is suggested that the chimpanzee can be utilized as an ontogenetic model to provide a clearer understanding of the role of the human male as a father or caregiver. For example, the social development of infant chimpanzees exposed only to adult females (or to adult males) can be compared to that of infants reared in the presence of both sexes. Critical periods and sex differences during development can be examined. These studies may indeed have important implications for the development of fatherless and motherless infants in the human species.

Since the chimpanzee is now an endangered species, it is crucial that these animals receive the most humane treatment in captive environments. It is suggested that an infant chimpanzee reared without the benefit of contact with adult males may be deficient in some social skills. Such social aberrations enhance the problems associated with captivity. As in the King et al. (1980) study, it is suggested that the social environment imposed by captivity should be carefully considered. To summarize, there is currently enough data on captive chimpanzees that indicate the adult male chimpanzee plays an important role in the maintenance of normal group behavior and normal social development of immature group members.

ACKNOWLEDGMENTS

I would like to thank Jane Lancaster, Roger Fouts, George Kimball, and Bill McGrew for their consultation in this project. I would also like to express gratitude to Yerkes Regional Primate Center for the loan of the infant used in this study and to Bill Lemmon for contributing the adult male. Finally, I would like to acknowledge a dedicated graduate student and friend, Jill Camp. Her outstanding support, tenacity, promptness, and remarkable willingness to learn were invaluable to this project as well as to those of us who had the good fortune to work with her.

REFERENCES

Alexander, R. D. 1974. The evolution of social behavior. *Ann. Rev. Ecol. Sys.* 5: 325-83.

Blaffer-Hrdy, S. 1976. Care and exploitation of nonhuman primate infants by conspecifics other than the mother. In J. S. Rosenblatt; R. A. Hinde; E. Shaw; and C. Beer (eds.), *Advances in the Study of Behavior,* vol. 2. pp. 101-58. New York: Academic Press.

Bygott, J. D. 1972. Cannibalism among wild chimpanzees. *Nature* 238: 410-11.

Goodall, J. 1977. Infant killing and cannibalism in free-living chimpanzees. *Folia Primatologica* 28: 259-82.

King, N. E.; Stevens, V. J.; and Mellen, J. D. 1980. Social behavior in a captive chimpanzee *(Pan troglodytes)* group. *Primates* 21: 198-210.

Lancaster, J. B. In press. Primate sex role and the evolution of the division of labor in humans. In R. Hall, and G. Bauer (eds.), *Sexual Dimorphism in Homo sapiens.* New York: Praeger.

McGinnis, P. 1979. Sexual behavior in free-living chimpanzees: Consort relationships. In D. Hamburg, and E. McCown (eds.), *Perspectives on Human Evolution: The Great Apes,* vol. 5. Menlo Park, California: Benjamin/Cummings.

Mitchell, G. D. 1969. Paternalistic behavior in primates. *Psychological Bulletin* 71: 399-417.

———. 1977. Parental behavior in nonhuman primates. In J. Money, and H. Musaph (eds.), *Handbook of Sexology.* Elsevier, North Holland: Biomedical Press.

Mitchell, G., and Brandt, E. M. 1972. Paternal behavior in primates. In F. Poirier (ed.), *Primate Socialization.* New York: Random House.

Nishida, T. 1979. The Social structure of chimpanzees of the Mahale Mountains. In D. Hamburg and E. McCown (eds.), *Perspectives on Human Evolution: The Great Apes,* vol. 5. Menlo Park, Cal.: Benjamin/Cummings.

Pusey, A. 1979. Intercommunity transfer of chimpanzees in Gombe National Park. In D. A. Hamburg, and E. McCown (eds.), *Perspectives on Human Evolution: The Great Apes,* vol. 5. Menlo Park, Ca.: Benjamin/Cummings.

Raphael, D. 1969. Uncle rhesus, auntie pachyderm, and mom: All sorts and kinds of mothering. *Perspectives in Biology and Medicine* 12: 290-97.

Redican, W. K., and Taub, D. M. 1981. Adult male-infant interactions in nonhuman primates. In M. E. Lamb (ed.), *The Role of the Father in Child Development,* 2nd ed. New York: John Wiley & Sons.

Spencer-Booth, Y. 1970. The relationship between mammalian young and conspecifics other than mothers and peers: A review. In D. S. Lehrman; R. A. Hinde; and E. Shaw (eds.), *Advances in the Study of Behavior,* vol. 3. New York: Academic Press.

Suzuki, A. 1971. Carnivority and cannibalism observed among forest-living chimpanzees. *Journal of the Anthropological Society of Nippon* 79: 30-48.

Trivers, R. L. 1972. Parental investment and sexual selection. In B. Campbell (ed.), *Sexual Selection and the Descent of Man.* Chicago: Aldine.

Tutin, C. E. G. 1974. Exceptions to promiscuity in a feral chimpanzee community. *Proceedings of the 5th Congress of the International Primatological Society, Nagoya, Japan.* Basal: Karger.

van Lawick-Goodall, J. 1968. The behaviour of free-living chimpanzees in the Gombe Stream Reserve. *Animal Behaviour Monographs* 1: 161-311.

11

The Evolution of Male Parental Investment: Effects of Genetic Relatedness and Feeding Ecology on the Allocation of Reproductive Effort

Jeffrey A. Kurland
Department of Anthropology
The Pennsylvania State University
University Park, Pennsylvania

Steven J. C. Gaulin
Department of Anthropology
The University of Pittsburgh
Pittsburgh, Pennsylvania

One can, in effect, treat the sexes as if they were different species, the opposite sex being a resource relevant to producing maximum surviving offspring.

R. L. Trivers, 1972

To breed or not to breed; the answer is a parent.

Anonymous

INTRODUCTION: SELECTION AND THE COST OF REPRODUCTION

Reproduction can be usefully partitioned into sexual and parental processes. Sexual processes include any and all attempts to increase the probability of syngamy, the successful fusion of gametes. Parental processes include any and all attempts to increase the reproductive and survival prospects of the resultant zygotes. Here we focus on the evolution and maintenance of parental processes and especially male parental traits although, as we will show below, there are critical interrelationships between sexual and parental processes. Even a cursory survey of the animal kingdom would demonstrate that male parental behavior is considerably less conspicuous than female parental behavior. There are, however, certain notable exceptions, including teleost fish with external fertilization (Breder and Rosen, 1966; Blumer, 1979), the so-called polyandrous birds (Jenni, 1974; Graul et al., 1977) and humans (Campbell, 1966). Neither a review of the rather extensive literature on parental behavior nor definitive tests of the proposed hypotheses will be presented in this paper. Instead, we provide a general logical framework within which to analyze the evolution of parental traits.

Typically discussions of parental adaptations have been confined to an examination of *parental care.* This, however, is misleading in that it unnecessarily restricts the range of parental traits, yielding an incomplete analysis of parental adaptations, in particular, and reproductive systems, in general. A more useful analytic concept is Trivers's (1972, 1974) notion of *parental investment,* "any investment by the parent in an individual offspring that increases the offspring's chances of surviving (and hence reproductive success) at the cost of the parent's ability to invest in other offspring" (Trivers, 1972, p. 139).[1] Thus parental investment includes not only such obvious characters as feeding or carrying; but also defense of a territory in which the young may harvest resources relatively unmolested (e.g. red, but not sage, grouse; Wiley, 1974); guarding against the infanticidal tactics of reproductive competitors (e.g., several mammals, Hrdy, 1979); or even "prezygotic" investments such as courtship feeding (e.g., insects, Thornhill 1976a; and birds, Nisbet, 1973), nest building (e.g., many birds, Welty, 1975),

[1]Low (1978) distinguishes between "parental effort" and "parental investment," where the former term refers to the sum total of parental investment made in one's offspring. Low is primarily concerned with analyzing how parental effort is best allocated among various potential recipients. We, however, are more concerned with the fact that parental effort is just one component of reproductive effort, and that selection may sometimes favor the shifting of reproductive effort from one component to another. Thus we emphasize alternative allocations of reproductive effort rather than alternative allocations within a given category, and therefore we use the terms "parental effort," "parental investment," and "parental expenditure" interchangeably.

and the production of nutrient-rich spermatophores (e.g., in insects, Thornhill, 1976b; Mullins and Keil, 1980; Boggs and Gilbert, 1979).

Parental investment is a cost function that is only adaptively expended when there is a sufficiently large compensating benefit (Williams, 1966a, 1966b; Trivers, 1972, 1974). That is, the present increment in offspring survival or reproduction must be greater than the decrement in production of future offspring resulting from increased parental mortality or decreased parental fecundity. This can be simply represented as:

$$r_o W_1 > r_o W_2 \quad (1)$$

where W_1 = the fitness gained from investment (benefit), W_2 = the fitness sacrificed by not investing in future offspring (cost), and r_o = the genetic relatedness between parent and offspring, here assumed to be constant.

Students of life-history strategy have defined the concept of *reproductive effort* as the cost of attempting to reproduce (e.g., Fisher, 1958; Williams 1966a, 1966b; Gadgil and Bossert, 1970; Goodman, 1974; Stearns, 1976, 1977; Horn, 1978). As defined by Trivers, parental investment is a subset of reproductive effort, the other principal subset being *mating effort* (Trivers, 1972; Arnold, 1976; Daly, 1978; Kurland and Gaulin, 1978; Low, 1978; Wade, 1979). Thus, one can begin to describe the reproductive strategy of an individual organism in terms of its allocation of reproductive effort between these two major components. Because at any point in time reproductive effort is a finite commodity, reproductive effort expended in parenting cannot be applied to mating (Bateman, 1948; Williams, 1966a; Trivers, 1972; Kurland and Gaulin, 1978; Wade, 1979). This is why, as stated above, an evolutionary analysis of parental processes requires consideration of sexual processes. As in the case of parental investment, a particular allocation of reproductive effort is only adaptive to the extent that it satisfies equation (1), where the right-hand side refers to the reproductive effort expended, and the left-hand side represents the fitness thereby gained (cf. residual reproductive value, Williams, 1966b). Because mating and parental effort are usually mutually exclusive, Trivers's (1972) statement that the relative parental investment of the sexes governs the course of sexual selection is well taken. However, as Bateman (1948) emphasizes, the higher variance in reproductive success within one sex is the "sign"of sexual selection, and the stronger correlation between number of mates and reproductive success in that sex is the actual "cause" of such sexual selection. Therefore, relative parental investment will drive sexual selection only to the extent that if some members of one sex can increase their reproductive success by increasing the availability of mates, then those same individuals will not be able also to make available additional amounts of time, energy, or risk for investment in the resulting zygotes. Bateman's two principles are really the more general logical foundation for the theory of sexual selection (cf. Wade, 1979, and see below "A Brief Survey of Parental Adaptation and Relatedness").

Although Trivers's (1972) paper is central to the *evolutionary* analysis of reproductive systems, others have provided *ecological* or *phylogenetic* analyses. In general, these three approaches remain separate, with each focusing on different problems: evolutionary analysis deals with the emergence of distinct male and female characters as a function of the different ways in which the sexes maximize reproductive success (e.g., Bateman, 1948; Trivers, 1972; Alexander et al., 1979; Wade, 1979); ecological approaches consider the association between environmental variables and "mating systems" (e.g., Orians, 1969; Eisenberg et al., 1972; Altmann et al., 1977; Emlen and Oring, 1977); and phylogenetic approaches attempt to reconstruct the evolutionary history of male and female reproductive traits within particular lineages (e.g. Darwin, 1871; Lack, 1968; Crook, 1972; Wells, 1977).

In examining male parental investment, we will emphasize both an evolutionary determinant, genetic relatedness, and an ecological factor, the metabolic consequences of body size.[2] We hope to show that interspecific variation in these two factors can account for much of the interspecific variability in parental investment tactics. In order to facilitate such cross-taxa comparisons, and in order to avoid some of the terminological confusion associated with the traditional classification of "mating systems," our analysis will focus exclusively on how the evolutionary and ecological factors mentioned above shape the reproductive effort allocations of the sexes.

THE COMPONENTS OF REPRODUCTIVE EFFORT

As already described, reproductive effort can be usefully partitioned into two major components: namely, mating and parental effort. Mating effort is defined as all expenditures or investments of resources and risk on mating attempts, that is, the cost of attempting to increase the syngamic success of gametes. In contrast, parental effort includes all expenditures or investments of resources and risk in the bearing or rearing of offspring. These two major components of reproductive effort can be further divided in a manner that can facilitate both evolutionary and ecological analyses of reproductive strategies (Table 11-1).

---

[2]Since we do not covet either the Hugo or Nebula awards, we refrain from attempting to reconstruct the phylogeny of parental behavior—nevertheless, the relevant review panels may find sufficient material here to warrant some consideration.

Table 11-1. The Components of Reproductive Effort.

I. Mating Effort: cost of attempting to increase success of gametes
   A. Exploitative Competition:
      1. Sexual: increasing the probability of syngamy
   B. Interference Competition:
      1. Mating Offense: decreasing competitors' sexual success
      2. Mating Defense: preventing others from lowering one's sexual success
II. Parental Effort: cost of attempting to bear/rear offspring*
   A. Prezygotic Effort:
      1. Gametic: sex cell production
      2. Nongametic: increasing unformed offspring's fitness
   B. Postzygotic Effort:
      1. Preparturitional: increasing offspring's fitness prior to separation from the parental body
      2. Postparturitional: increasing offspring's fitness after separation from the parental body

*Extragametic parental effort includes nongametic and postzygotic expenditures.

Mating effort can itself take two forms: *exploitative* and *interference* competition. Exploitative mating effort refers to the decrease in a competitor's reproductive success due to the prior utilization of reproductively limiting resources (I A 1 in Table 11-1). It is therefore a form of mate competition without direct conflict. In most species, this simply entails the successful exhaustion of female sexual and parental functions by means of successful insemination. On the other hand, interference competition occurs whenever an individual increases its relative fitness by effectively lowering the reproductive success of conspecific competitors (Arnold, 1976; Kurland and Gaulin, 1978). Interference mate competition has two aspects: *offensive* and *defensive* mating effort. Offensive interference competition (I B 1 in Table 11-1) represents a direct form of mating effort in which the individual increases its potential reproductive success by lowering a competitor's sexual success. Copulatory harassment in baboons (e.g., DeVore, 1965; Packer, 1979) or copulatory plug popping in macaques (e.g., Kurland, 1977) are good examples of such offensive mating tactics in nonhuman primates. The evolution of offensive mating interference will create strong selective pressure for the evolution of countertactics that can successfully defend the individual from the sexual interference of competitors (I B 2 in Table 11-1), for example reciprocal alliances (e.g., Packer, 1977) or protective consortships (e.g., Hausfater, 1975; Tutin, 1979) in primates.

Similarly, parental effort can be divided into two categories: *prezygotic* and *postzygotic* expenditures. Prezygotic effort includes the essential and minimal investment of *gametes* (II A 1 in Table 11-1) as well as "optional" *nongametic* forms of investment that increase the survival, and hence reproductive, success of as yet unformed offspring (II A 2 in Table 11-1). A good example of such nongametic, prezygotic investment is the courtship feeding common to many birds (e.g., Nisbet, 1973) and insects (e.g., Thornhill, 1976a). The postzygotic forms of parental investment include morphological, physiological, and behavioral mechanisms that nurture the developing offspring prior to and after the birth event. Thus, *preparturitional* investment (II B 1 in Table 11-1) includes such characteristics as egg shells, wombs, hives, and placentae, while the *postparturitional* (II B 2 in Table 11-1) expenditures include the kinds of traits that are in ordinary parlance referred to as "parental care," for example, suckling, carrying, or in other ways nurturing immature individuals.

Although some form of mating effort and gametic parental effort are absolutely necessary for any reproductive success, and thus are common to all taxa, *extragametic* investment, including nongametic, preparturitional, and postparturitional expenditures, are expected to vary as a function of other evolutionary and ecological variables. While Trivers (1972) and Low (1978) concentrate on the alternative allotments of parental investment that a parent can make between potential offspring, our *reproductive effort array* should make it apparent that natural selection will also operate on an individual's reproductive strategy to optimize allocations of reproductive effort among the various components of mating and parental effort.

It should be understood that the components within the array of reproductive effort (Table 11-1) are logically, and usually empirically, separable aspects of reproductive effort. But they should not be construed to be elements in the temporal ordering of the major functional stages of the reproductive process. For example, infanticide with reinsemination is an aspect of mating effort because it decreases a competitor's sexual, and hence reproductive, success, even though it occurs after a competitor's successful copulation and fertilization. Thus, despite the fact that infanticide occurs "postzygotically," it actually functions within the domain of mating, and not parental, effort (I B 1 in Table 11-1).

REPRODUCTIVE EFFORT AND REPRODUCTIVE TACTICS

The action of sexual selection, that is, the variation in mate number (Darwin, 1871; Bateman, 1948; Trivers, 1972), and the action of natural selection for other sexual and parental adaptations is dependent upon each sex's relative allocation of reproductive effort among the various mating and parenting components. Whenever allocations to the major subsets of reproductive effort are mutually exclusive, the degree to which an individual maximizes its reproductive success by emphasizing parental effort must of necessity lead to a deemphasis in mating effort. For example, the differences between the typical male-competitive and female-nurturant mammalian reproductive strategies are the result of the fact that males expend more on mating effort, which applies to a potentially large number of as yet unformed offspring, whereas females invest most of their reproductive effort in various forms of parental effort, which can be applied only to many fewer, particular, and easily identifiable offspring (Trivers, 1972; Kurland and Gaulin, 1978; Low, 1978; Wade, 1979).

In general, different reproductive effort arrays for the sexes predict different kinds of sexual and parental tactics and, thus, alternative strategies for reproduction. Whenever the reproductive effort array involves a larger apportionment of reproductive effort into parental expenditures for one sex, then that sex will be the reproductively limiting sex. That is, both male and female reproductive success will be limited by access to or allocations of such parental effort (Trivers, 1972). In mammals, and indeed in most vertebrates, females make a much greater outlay of parental effort than males, as made evident by large, nutrient-rich ova; gestation; lactation; feeding; and infant protection. By contrast, the male might only make the small, parental contribution of the single, necessary sperm cell. Given this reproductive effort array, female reproductive success is expected to be limited by the ability to efficiently convert resources into parental effort, whereas male reproductive success is expected to be limited by access to females and their parental effort, not by male parental effort (Trivers, 1972, 1976). Females evolve therefore into the "ecologic" sex, finely tracking the environment, garnering resources, and converting them into parental expenditures in the most cost-effective manner. But if female parental effort ultimately limits male reproductive success, then any male's increased access to fertile females represents a decrease in another male's reproductive success (exploitative competition). Consequently, males will evolve into the "competitive" sex, tracking females, and perhaps sacrificing feeding efficiency or other ecological adaptations in order to increase time for mating activity (Schoener, 1971; Trivers, 1972).

Any particular offspring represents for the female not only a greater portion of her lifetime reproductive effort, but also her lifetime potential reproductive success. An error in mate choice, leading to reproductive failure or poor-quality offspring, will therefore have more serious consequences for the average female than for the average male, because she will have greater difficulty replacing wasted reproductive effort (Williams, 1966a; Orians, 1969; Trivers, 1972). Females are expected to, and indeed do, select mates on the basis of a variety of morphological and behavioral traits that correlate with male genetic fitness or male parental contributions beyond the gametic stage (Fisher, 1958; O'Donald, 1962, 1967; Williams, 1966a; Orians, 1969; Trivers, 1972; Burley, 1977; Partridge, 1980).

Given a reproductive effort array in which males exhibit a mating effort bias and females a parenting effort bias, there exist two general classes of male reproductive strategy: aggressive competition (intrasexual selection) and display competition (epigamic selection) (Huxley, 1938; Fisher, 1958; Mayr, 1972). Under the aggressive competition regime, males are expected to translate their high mating effort into a variety of morphological traits and behavioral tactics designed to increase their competitive advantage against other males, who are also attempting to gain access to sexually receptive females, thus, the emergence of male interference and exploitative mate competition tactics (Table 11-1). For example, male mobility, and other aspects of male migratory patterns, may increase the number of available potential mates (Trivers, 1972). A large number of precopulatory and postcopulatory tactics may be brought into the male's armamentarium, including, for example, increased body size; increased aggressivity; weapons, such as horns and enlarged teeth; forced copulations; copulatory harassment; infanticide, with reinsemination; or isolation and guarding of a mate (see reviews in Trivers, 1972; Wilson, 1975; Ralls, 1976, 1977).

Because male reproductive success is limited by the mating effort of other males and because male reproductive success is potentially so large, given multiple matings, selection may favor males expending in any one breeding period a rather large amount of reproductive effort, in comparison to females breeding at the same time. Consequently, as males grow, develop male competitive traits, and attempt to reproduce, they will exhibit a higher age-specific mortality schedule than females due to the additional risks they incur from their reproductive attempts (Williams, 1966a, 1966b; Gadgil and Bossert, 1970; Trivers, 1972). Thus, male ontogeny also is expected to be modified by male reproductive effort. With body size or behavioral experience at a premium in male-male competition, males of less than optimal size or experience, that is, less than optimal age, may be severely handicapped. Selection is therefore expected to favor delaying maturation in males until such time as the reproductive payoff can be expected to offset the associated costs (Trivers, 1972; Wittenberger, 1979). Finally, because male reproductive success, but not female reproductive success, increases monotonically with mating success, males are expected to be more "indiscriminate" with respect to choice of sexual partner in the sense that they will copulate with as many different females as possible (Trivers, 1972).

Whenever historical circumstance, phylogenetic inertia, or ecological conditions preclude a male reproductive strategy of aggressive competition, males may instead increase their reproductive success by transforming mating effort into flamboyant colors, displays, and other adornments that can increase conspicuousness and, hence, attractiveness to females, and therefore increase the chances for successful copulation (Fisher, 1958; O'Donald

1962, 1967; Williams, 1966a; Trivers, 1972; and review in Halliday, 1978). These male display tactics, driven by powerful female choice, may also have associated high costs due to decreased time spent feeding, increased visibility to predators, or increased susceptibility to abiotic hazards. Thus pretty males may still incur greater mortality and delayed maturation, relative to highly discriminating conspecific females.

Although it is common to contrast male competitive (intrasexual) with female choice (epigamic) systems of sexual selection (e.g., Brown, 1975; Wilson, 1975; Alcock, 1979), the two selective processes are not mutually exclusive. For example, in species where male-male competition was initially assumed to be the sole driving force behind male and female reproductive strategies, females may effectively choose the best mate by inciting the males to competition and then copulating with the winner (Thornhill, 1976, 1979; Cox and LeBoeuf, 1977). Indeed, females may use the results of male-male competition to assay those genes that have demonstrated survival value or resource-accrual ability (Trivers, 1976). Male competition and female choice are thus two aspects of the same evolutionary process resulting from a male mating effort bias and a female parental effort bias in reproductive effort allocations.

Female traits, such as mate discrimination, or male traits, such as aggressive competition, are not intrinsic to females or males, but rather are properties of a particular reproductive effort array (Williams, 1966a). Whenever male reproductive effort, due to ecological, demographic, and historical factors, shifts toward parental expenditures, and female reproductive effort coevolves toward increased mate access, "sex-role reversal" in size, aggressivity, mortality, or copulatory behavior is expected and, indeed, often found (Williams, 1966a; Trivers, 1972; Emlen and Oring, 1977). Similarly, as the reproductive effort expenditures of the sexes converge on the same allocation pattern, members of both sexes are expected, and again found, to be equally choosy, competitive, or broody (Williams, 1966a; Trivers, 1972; Emlen and Oring, 1977; but see Ralls, 1976, 1977).

MATING SYSTEMS OR REPRODUCTIVE EFFORT ARRAYS?

Traditional "mating system" classifications, and their endless variants, appear often to be little more than semantic exercises in frustration as the author attempts to use the undefinable notion of the "pair-bond" to analyze reproductive patterns (e.g., Selander, 1972). Such mating-system taxonomies in effect avoid accurately describing the variability in reproductive patterns typically found within natural populations, and instead attempt to present a static, species-typical picture of reproduction in which variation becomes a source of embarrassment to the researchers. For example, given the observed variation in the number of mates male indigo buntings may have in the same habitat in different years (Carey and Nolan, 1975), to what extent is it possible to characterize the population, let alone the species, as "serially polygynous," "simultaneously polygynous," "monogamous," or "serially monogamous"? More importantly, such mating-system classifications obscure the understanding of the selective pressures that might have shaped male and female reproductive strategies because they concentrate on the emergent mating system as an irreducible result of "typical" male and female relations, thus failing to characterize differences between the reproductive tactics utilized by individual males and females in particular ecological contexts. By so focusing on the results of male and female interactions (e.g., Orians, 1969; Selander, 1972; Brown, 1975; Emlen and Oring, 1977), mating-system approaches overlook the detailed logic of the actual intra- and inter-sexual interactions and "conflicts of interest" that in fact produced the "mating system" (but see Trivers, 1972; Maynard Smith, 1977; Wittenberger and Tilson, 1980). Finally, mating-system taxonomies represent yet another example of the human propensity—and the occupational hazard of many behaviorists and ecologists—to project onto the animals human cultural, linguistic, and ideological practices; in this case, marriage systems, residence patterns, and terminologies find their zoological expression in, for example, happily married, pair-bonded, "monogamous" gibbons; harem-structured, male-dominated societies of "polygynous" baboons; and male-competitive, "promiscuous" macaques.

In contrast to the mating-system approach, the reproductive effort array discussed above allows the formulation of more precise functional hypotheses because it focuses the investigator's attention on the classes of reproductive tactics associated with each component of reproductive effort. This can be demonstrated by applying such a reproductive effort analysis to the hamadryas baboon (Kummer, 1968), American rhea (Brunning, 1974), and the threespine stickleback (Wooton, 1976), all of which could be classified as "polygynous" in that reproductive males "bond" to several different reproductive females. Male threespine sticklebacks compete with each other for nest sites in the beds of small streams. A successful male builds a tunnel-nest of aquatic vegetation and courts a series of females in an attempt to induce them to lay eggs in his nest. Rhea males also compete for sites where each builds a large scoop-nest, and for the possession of harems of females each of whom contributes one or more eggs to the nest over several days. After the male's clutch is complete, the female flock then moves on to mate with a different territorial male. In both the rheas and sticklebacks, females desert, leaving the male to care for the developing young. Male hamadryas obtain harems either by "kidnapping" immature females or by "inheriting" the females from the previous harem owner. Such harems are relatively stable over long periods. As in most mammals, female hamadryas contribute the preponderance of parental investment in the young.

Describing these three species as simply "polygynous" obscures major differences in reproductive patterns. The mating-system approach can be improved slightly by specifying whether the polygyny is sequential (i.e., series) or simultaneous (i.e., harem). Yet as soon as a female perspective is added, the utility of this mating-system taxonomy begins to break down: baboons are simultaneously polygynous with some promiscuity (polybrachygyny? kleptogamy?); rheas are simultaneously polygynous-sequentially polyandrous (promiscuous? serially monobrachygamous?); and sticklebacks exhibit sequential polygyny-polyandry (serial monogamy? serial polygamy?).

Clearly the reproductive effort arrays of these three "polygynous" species are markedly different (Table 11-2). Most apparent are the differences in male and female postzygotic investment. In the hamadryas baboon, the postzygotic investment is almost entirely a female task. In the rheas, female postzygotic investment ends with oviposition, while the male investment continues postparturitionally. And in the stickleback, where there is no preparturitional investment, postzygotic investment falls to the male. These alternate allocations of reproductive effort in turn define a unique set of possible sexual tactics and countertactics in each species. For example, in the threespine stickleback, some males capitalize on the absence of female preparturitional expenditures by parasitically mimicking female courtship displays. Such "pseudo-females" (Morris, 1952) may obtain some reproductive success if the nest-holding male mistakes them for potential mates, and allows them access to the eggs already in his nest. Clearly, such a tactic is not available to male rheas or baboons. Instead the complex interactions between male competition and female choice produce a system of highly ritualized agonistic contests and harem ownership in hamadryas baboons (Kummer et al., 1974; Bachman and Kummer, 1980), and nest-site defense and broodiness in male rheas, with group egg-laying in female rheas (Brunning, 1974).

Female preparturitional effort in the rhea allows the possibility of a different form of reproductive parasitism, namely, "cuckoldry," in which the female extracts postzygotic effort from a male other than the one who fertilized her eggs. Cuckoldry also reduces the reproductive success of a hamadryas baboon harem male, but not because he wastes his postzygotic investment. Instead, his reproductive success is decreased through exploitative mate competition due to the temporary loss of a potential mate's investment. In each of these species, the female's essential gametic investment exceeds that of the male. Therefore, the reproductive success of males is to this extent egg-limited; they should and do compete for access to such female gametic investment, regardless of their extragametic investment tactics.

Table 11-2. A Comparison of Some "Polygynous" Mating Systems in Terms of Their Reproductive Effort Arrays.

Reproductive Effort	Hamadryas Baboon *(Papio hamadryas)*	American Rhea *(Rhea americana)*	Threespine Stickleback *(Gasterosteus aculeatus)*
Mating Effort			
Exploitative competition	Male	Male**	Male**
Interference competition			
Offensive	Male*	? (No sex bias)	Male
Defensive	Male	? (No sex bias)	Male
Parental Effort			
Prezygotic			
Gametic	Female	Female	Female
Nongametic	? (Absent)	Male	Male
Postzygotic			
Preparturitional	Female	Female	Absent
Postparturitional	Female	Male	Male

For each component of reproductive effort, the entry indicates which sex expends greater reproductive effort.
A question mark refers to insufficient data.
*:Reported for females in gelada baboons (Dunbar and Dunbar, 1977).
**:Because the number of eggs a male incubates in his nest is limited, female exploitative competition occurs, but we would argue that males still show a greater commitment to this aspect of reproductive effort.

Considering the American rhea again (Table 11-2), there exists a rather distinctive and, in fact, taxonomically rare, pattern of male and female postzygotic parental effort. Although preparturitional expenditures of reproductive effort are made by the female rhea, males exhibit all the postparturitional investment. Consequently, females risk having eggs rejected by "coy" males whenever circumstances indicate that the male's relatedness to those zygotes is low and, hence, that the male would in effect squander limiting, and not easily replaced, postparturitional reproductive effort (see "Probability of Relatedness and Extragametic Effort" below). In fact, territorial male rheas do inspect the eggs of females and sometimes do not accept them (Brunning, 1974). Moreover, the rhea situation exemplifies a potential difference between the two classes of prezygotic parental effort. Whereas gametic investment is necessarily committed to a single potential offspring, nongametic investment is often uncommitted.Thus a female rhea wastes her gametic effort if her eggs are rejected by the male owner of the reproductively limiting scoop-nest; but the male's nongametic investment, territorial defense of optimal nest sites and nest building, can be applied with equal reproductive value to the subsequent young of the same or some other female. Here, as in other phases of the reproductive cycle, natural selection is expected to favor mechanisms that protect the organism from wasting reproductive effort. When, as in the rhea, gametic and nongametic effort are differentially vulnerable to loss, the individual may be selected to initiate such reproductive investment in response to different proximate cues in the social and physical environment. Indeed, we will argue that variability in the probability of genetic relatedness and the physiological and ecological consequences of body size will critically shape such facultative responses in reproductive effort.

PROBABILITY OF RELATEDNESS AND EXTRAGAMETIC EFFORT

Hamilton's (1964) kinship theory predicts that the evolution of social behavior critically depends upon the associated benefits *(B)* and costs *(C)* to the participants' fitness and the genetic relatedness *(r)* between them. From these considerations, it can be shown that "altruism" is adaptive only when $Br > C$, where B = the benefit gained by the recipient and C = the cost incurred by the altruistic actor (Hamilton, 1964). Ordinarily r refers to some average or expected degree of relatedness between members of a particular kin-class, for example, ½ for parents and offspring, ¼ for half-siblings, or ⅛ for full-cousins. But due to the inherent imperfections in proximate mechanisms for kin-discrimination and information about kin relationships, putative members of a given kin-class may or may not be so related. Thus, kin-discrimination within a particular kinship class may be expected also to evolve (Hamilton, 1964). If the *probability of relatedness (p)* itself varies with circumstances, then Hamilton's rule for adaptive altruism can be refined and reformulated such that $p(B)(r) > C$. Although p is obviously relevant to whether or not any social interaction is adaptive, it is frequently neglected in the literature. Here we are principally concerned with the effects of p on parent-offspring relationships, although it ought to be relevant also to siblings, cousins, and other kin relationships. While parents and offspring in diploid species are usually considered to be related by ½, this is, in fact, the maximum value of *parental relatedness,* i.e., $p(r)$, where $p \leq 1.0$ and $r = ½$. Because parents and offspring are ordinarily the most closely related members within a given population and because there is a fundamental "trophic" asymmetry in the ability of parents, as opposed to offspring, to provide aid, fitness benefits are expected and observed to flow from parent to offspring (Hamilton, 1964). There are, of course, some curious and critically important exceptions among the social hymenoptera and isoptera (Hamilton, 1964), and birds (Emlen, 1978), and other animals (reviewed in Kurland, 1980).

Although the benefits derived from such parental altruism will vary as a function of ecological circumstance (see "Extragametic Effort, Ecology, and Offspring Ontogeny" below), regardless of the magnitude of such potential benefits, whenever $p = 0$, parental effort beyond the minimal gametic investment is wasted. Since natural selection will act so as to minimize the inefficient expenditure of reproductive effort, mechanisms that guard organisms against the misapplication of parental effort are expected to evolve. Thus there is expected to be a coevolution of parental relatedness and extragametic parental investment. Therefore the *origin* of extragametic parental adaptations was dependent upon the ability of the individual to identify its own zygotes. Parenting in this sense could never have even begun to spread if the probability of relatedness were not greater than zero (cf. Werren, et al., 1980). That is, a nonzero probability of relatedness is necessary, although not sufficient, for such parental adaptations, but a critically low level of relatedness is sufficient for desertion of the zygotes (Kurland and Gaulin, 1978). On the other hand, once parenting is somewhat established in the population, the opportunity arises for some individuals to increase their reproductive success by parasitizing the extragametic parental expenditures of others (whether conspecific or not, see "Alternative Interpretations of Primate 'Parental Care'" below). Thus the *maintenance* of parental adaptations entails an evolutionary dynamic between cuckoldry and anticuckoldry tactics (cf. Trivers, 1972). Here again we see the interaction between sexual and parental tactics.

A BRIEF SURVEY OF PARENTAL ADAPTATIONS AND RELATEDNESS

It has been suggested that sexual dimorphism in parental investment across species is correlated with sex differences in the probability of relatedness (Williams, 1975; Graul, et al., 1977; Kurland and Gaulin, 1978; Maynard Smith, 1978a; Ridley, 1978; Alexander and Borgia, 1979; Blumer, 1979; Perrone and Zaret, 1979).

Species can be characterized by the presence of high parental relatedness for both, one, or neither sex (Fig. 11-1). In Table 11-3 all possible pairings of parental relatedness are illustrated with examples from diverse taxonomic groups. Reports of parental care and the locus of egg development are used to measure the relative extragametic parental investment of the sexes.

In those species where both sexes broadcast gametes widely, for example, oysters, and all teleosts with pelagic eggs (Budd, 1940; Breder and Rosen, 1966), neither sex can identify its own zygotes with any certainty. Parental expenditures beyond gametic production has not been reported in these groups. Although it may seem that oysters could not invest postzygotically, parental effort in the form of gastric brooding, for example, is possible. Note that among taxonomically diverse marine invertebrates, brooding occurs, but only in internally fertilizing hermaphroditic species (Ghiselin, 1974). The reciprocal exchange of gametes between hermaphroditic mates establishes the requisite parental relatedness.

In most arthropods, amphibians, reptiles, mammals, as well as certain families of fish and birds, maternal care is reported to be much more common than paternal care (Walker, 1964; Breder and Rosen, 1966; Lack, 1968; Wilson, 1971; Porter, 1972; Wells, 1977). Moreover, extragametic expenditures in these species occur both before and after parturition. Because a period of zygotic development within the female's body establishes maternity during the preparturitional stage, subsequent postparturitional investment may be adaptive so long as the mother's young are still identifiable (cf. Williams, 1975; Kurland and Gaulin, 1978; Maynard Smith, 1978a). On the other hand, this same developmental sequence reduces paternal relatedness such that male reproductive success may be maximized more readily by increasing the number of copulations with fertile females than by increasing the investment in unidentifiable offspring. Such a mating pattern may be self-perpetuating: if males bias their allocation of reproductive effort toward mating effort and away from parental effort, then "promiscuity" may increase such that paternity is further reduced. Thus, in such cases, Trivers's (1972) argument that previous investment by a female commits her to future investment is correct with respect to the relationship between pre-

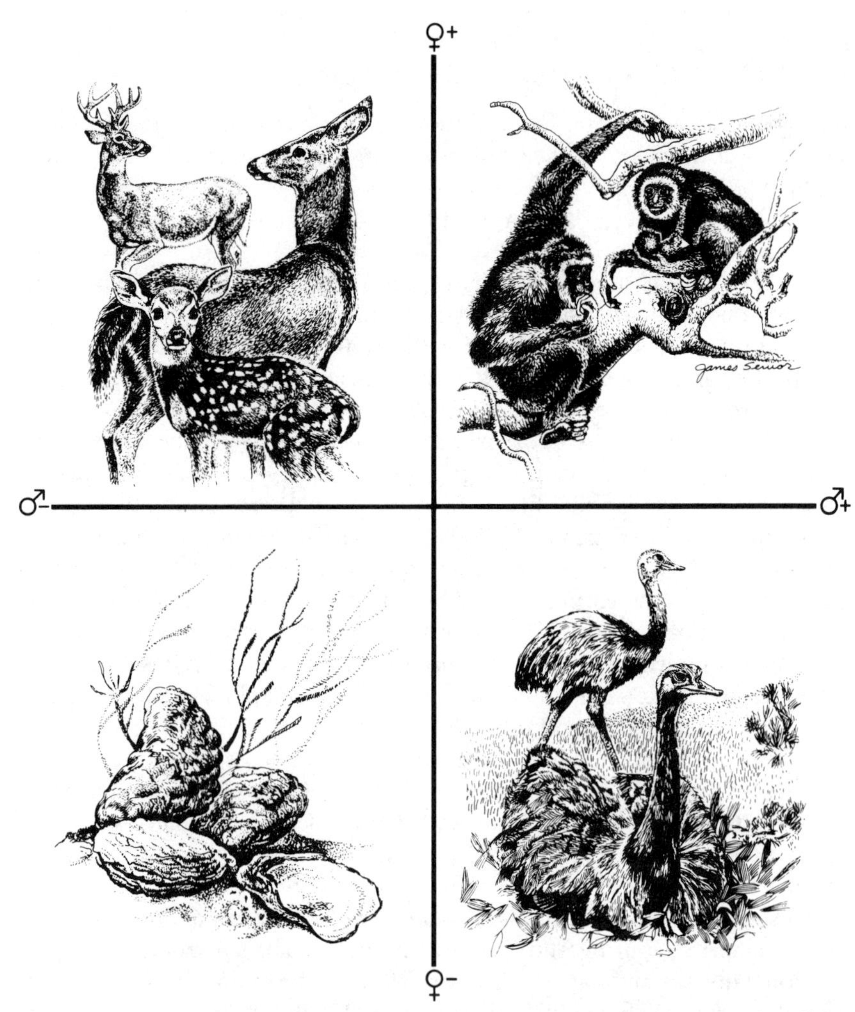

Fig. 11-1. Relationship between parental relatedness and extragametic effort. Association between relatedness and extragametic effort is schematically represented by means of an abscissa of high-to-low paternal relatedness and an ordinate of high-to-low maternal relatedness. The gibbon, red deer, oyster, and rhea exemplify the four possibilities and are situated in the appropriate quadrants in the "relatedness space."

and postparturitional investment, but for the wrong reason. It is neither the potential loss of past investment nor the prospective cost of bringing another offspring to an equivalent stage of development (Dawkins and Carlisle, 1978) that leads the female to continued investment. Rather over the course of preparturitional expenditures, male and female parental relatedness diverge,

Table 11-3. Some Examples Illustrating the Relationship between Parental Relatedness and Extragametic Parental Investment.

Parental Relatedness Male	Parental Relatedness Female	Extragametic Investment by	Selected Examples
–	–	Neither sex	Most marine invertebrates (bivalvia), Teleost fish with palagic eggs (clupeidae, alepocephalidae, plecoglossiade, congridae, halosauridae, anguillidae).
–	+	Females	Teleost fish with nonpelagic eggs (umbridae, solenostomidae, amblyopsidae, embioticidae, zoarcidae, poeciliidae) Amphibians (pipidae, phacophoridae, hylidae) Reptiles (all turtles and snakes) Birds (most anatidae, phasianidae, and many trochilidae) Mammals (many cervidae, bovidae, lemuridae, cebidae, and cercopithecidae and numerous rodent and chiropteran families)
+	+	Both sexes	Teleost fish with nonpelagic eggs (percidae, ictaluridae, doradidae) Birds (many passeriformes, and all psittaciformes and columbiformes) Mammals (castoridae and hylobatidae; some emballonuridae, hipposideridae, canidae, and bovidae)
+	–	Males	Teleost fish with nonpelagic eggs (gasterosteidae, syngnathidae, anabantidae), Birds (rheidae, tinamidae)

A "+" indicates the presence of high levels of parental relatedness, whereas a "–" indicates low levels.

thereby altering their optimal allocations of subsequent reproductive effort.

High degrees of extragametic investment for both sexes are found among some mammals, including, among others, hylobatid primates, castorid rodents, several canids, and some bovids; many passerine birds; some teleosts; and some coleopteran insects (e.g., *Necrophorus)* (Breder and Rosen, 1966; Lack, 1968; Milne and Milne, 1976; Kleiman, 1977). Prolonged courtship and continual isolation of a partner from other potential mates, for example by means of territiorial aggression, may lead to relatively high degrees of parental relatedness for both male and female members of a pair. For example, in *Hylobates lar,* the white-handed gibbon, the vast majority of aggression, and hence, territorial exclusion, occurs between adults of the same sex (Carpenter, 1940; Ellefson, 1967; Chivers, 1972). Among a large number of tropical birds, as well as hylobatid primates, complex and synchronized dueting occurs. This dueting is usually explained in terms of its environmental cause, namely, the necessity of vocal, as opposed to visual, contact for maintenance of a "pair-bond" within dense vegetation (Hooker and Hooker, 1969; Thorpe, 1973; Tembrock, 1976). Alternatively, both the long learning period and the complexity of the realized songs may serve to so monopolize a partner that cuckoldry and desertion become increasingly inefficient reproductive tactics. Experimental studies of pair formation in birds (Barash, 1976; Erickson and Zenone, 1976; Zenone, et al., 1979) demonstrate the existence of male behavior patterns that function to maintain high levels of paternal relatedness. All such pairing, courtship, and same-sex-exclusion behaviors may be interpreted as a means of sufficiently elevating paternal relatedness such that a male, as well as a female, can profitably invest.

Species in which a male's parental relatedness is higher than a female's, although rare, represent a critical test of our model. It is within such species that female extragamatic investment is expected to be significantly reduced. This situation is best exemplified by the rheid and tinamid birds; and gasterosteid, anabantid, and syngnathid fish (Breder and Rosen, 1966; Moodie, 1972; Brunning, 1974; Jenni, 1974; Wootton; 1976; Graul et al., 1977).

In rheas, from a female perspective, any given egg in the male's nest is of uncertain relatedness, and thus, postparturitional investment is less likely to increase her reproductive success. However, territorial exclusion of other males by the nest-holder, and the relatively short time lag between ovulation and egg-laying will tend to raise the male's parental relatedness such that male investment in the clutch can increase his reproductive success.

A very similar situation occurs in the threespine stickleback, *Gasterosteus aculeatus* (Moodie, 1972), and a labyrinth fish, *Macropodus cupanus* (Breder and Rosen, 1966), in both of which the male guards a nest containing eggs from several females. Many other fish exhibit male, but not female, parental care (e.g., syngnathids; see also Blumer, 1979). Because these species have only been studied in aquaria, where the potential mate pool is artificially restricted, it is not known whether under natural conditions males invest simultaneously in the clutches of several females. But whenever a male can collect and simultaneously invest in the zygotes of several females, reducing average maternal relatedness, the species is more likely to evolve male postparturitional investment. This may well explain the preponderance of paternal care in teleosts with external fertilization that so puzzled Ridley (1978). Moreover, the reproductive behavior of anabantid and gasterosteid fish makes clear that relative parental investment alone is not always an accurate predictor of sexual selection, because in such species males are both the competitive and the parental sex (e.g., Pressley, 1981). The breakdown in the application of Trivers' (1972) model to these fish results from a peculiarity of these species' ecology, namely, the territorial behavior of males. The territorial phenotype can be applied not only to mating, but also to prezygotic and postzygotic reproductive effort over the course of a reproductive cycle. Consequently, a given male can "buy" an area relatively free of members of the same sex, suitable for potential offspring, and easily defensible once eggs and frye are present, all with the same behavior.

The multiple adaptive functions possible for some phenotypes might also explain why human males typically compete for mates, yet also invest in offspring. In the human case, paternal investment often involves protection from violence and other dangerous aspects of the competition that is induced by the males themselves (e.g., Hrdy, 1979)! That is, some forms of human male reproductive effort operate simultaneously as mating and parenting expenditures. However, the major point should be clear: despite the higher male parental investment, the variance in male reproductive success is still higher than the variance in female reproductive success; and more importantly, because males are still egg-limited, they exhibit a higher association between mating and fertility than do females. In other words, Bateman's (1948) general principles of sexual selection still operate.

IS THE PROBABILITY OF RELATEDNESS REALLY RELEVANT TO PARENTS?

Despite the apparent association between parental relatedness and extragametic investment, several authors have argued that the probability of relatedness cannot explain sex-specific variation in parental behavior (Maynard Smith, 1978b; Grafen, 1980; Werren et al., 1980). Because individuals compete to maximize reproductive success only against members of their own sex, it may be inappropriate to compare male and female parental relatedness in order to explain male and female parental adaptations. Rather what should be compared is the net reproductive success resulting from alternative reproductive tactics available to members within each sex. And since parental relatedness is assumed to be constant across tactics, it drops out of consideration: parental relatedness is not only not a sufficient condition for parental behavior, it is not even necessary (Maynard Smith, 1978b; Grafen, 1980).

If we consider the American rhea case again, it can be shown that these criticisms are not necessarily valid. In fact, only by considering sex differences in parental relatedness can one demonstrate that male care was more likely to evolve in this species. The typical rhea male invests postparturitionally in the young of several females. The cost of his doing so is in part measured by the number of matings he foregoes as a result of his paternal duties (the "promiscuity cost," Werren et al., 1980). Consequently, his parental behavior is adaptive only when

$$p_m(r_o)(W_1 > p'_m(r_o)(W_2); \tag{2}$$

where r_o = the genetic relatedness between parent and offspring, W_1 = the fitness gained from expending a given quantity of reproductive effort as postparturitional investment, W_2 = the fitness gained from expending the same amount of reproductive effort in desertion and new mating effort, and where p_m and p'_m = the male's probability of relatedness to the eggs in situation 1 and 2, respectively. Here one might reasonably assume that p_m ~ p'_m, and thus the male ought to invest whenever $W_1 > W_2$. Whether $W_1 > W_2$ will depend upon several factors, including how great an effect postparturitional investment can have on the survival of the eggs and thus on the reproductive prospects of the resulting offspring, and also it will depend on the availability of females for new matings.

The parallel reproductive alternatives for a female rhea would be either to invest postparturitionally in the clutch of eggs in a given male's scoop-nest or desert and lay additional eggs in the nest of some other male. She is therefore expected to invest postparturitionally only if

$$p_f(r_o)(W_1) > p'_f(r_o)(W_3) \tag{3}$$

where r_o and W_1 are defined as in equation (2), but where W_3 = the fitness gained from expending an equivalent amount of reproductive effort in desertion, gametic, and preparturitional investment in additional eggs, and where p_f and p'_f = the female's probability of relatedness to the eggs in situation 1 and 3, respectively. Given internal fertilization, it must be the case that $p'_f = 1.0$. However, in situation 1, the female rhea's postparturitional investment would necessarily be applied to the eggs and young of other, presumably nonrelated females, so that $p_f = 1/n$; where n = the number of females laying eggs in the male's scoop-nest (anywhere from 2 to 15, Brunning, 1974).[3] Consequently, the female rhea's postparturitional investment would only be adaptive if $W_1 > n\ W_3$.

Thus, in analyzing the alternative reproductive tactics (parenting or deserting) available to male and female rheas, we find that although parental relatedness does drop out for the male, it does not for the female. The

[3]We assume for the sake of simplicity that each female lays the same number of eggs in a nest. If this is not the case, then it is, of course, necessary to measure the proportion of eggs a given female lays in order to estimate her probability of relatedness. In this case $p_f = x_f / \Sigma\ x_i$, where x_i = the number of eggs laid by the ith female, and x_f = the number of eggs laid by a particular female. Finally, if females rheas are related, the addition of an inclusive fitness effect will change our analysis only quantitatively, not qualitatively.

resultant male and female social and sexual interactions that define that rhea reproductive system can only be understood by comparing the results of the reproductive tactic analysis for males with the results of the reproductive tactic analysis for females (cf. Maynard Smith, 1977). Thus, a complete evaluation of whether natural selection will favor male or female post-parturitional effort will depend on the relationship between W_1, W_2, W_3. We have assumed that males and females are equally efficient investors and therefore we have used W_1 on the left side of both equations (2) and (3). For both sexes the relevant comparison is between this payoff (W_1) of investing further in already formed zygotes and the alternative of forming new zygotes (W_2 for males and W_3 for females). Thus, the issue remaining is whether or not W_3 is an nth as large as W_2 (i.e., $W_3 \geq W_2/n$). In other words, males would be less willing than females to invest postparturitionally only if male desertion resulted in n times more new zygotes than could female desertion. But in any case, given that $n = 1/p_f$, parental relatedness clearly is not irrelevant to the analysis of the evolution of parental tactics in the American rhea.

Although we allowed in the previous example that $p_m \sim p'_m$, this may be true only under special circumstances. Indeed, a male's alternative allocations of reproductive effort to mating or parenting may in fact affect these male probabilities of relatedness. Whenever males attempt to increase the number of zygotes they produce by increasing mating effort, the resultant "promiscuity" will lower a male's probability of relatedness to any given female's offspring. This is especially true in species that practice internal fertilization simply because monitoring a large number of females will be more difficult and thus more susceptible to error. If females are spatially clumped or can be induced by the male to aggregate, the effects of increased male mating effort on paternal relatedness may be reduced, but not obliterated. The overall consequence may be that attempts by males to maximize the number of mates may not maximize reproductive success. Increased expenditure by males on mating effort not only reduces the amount of reproductive effort available for parental investment, it also reduces its adaptiveness by increasing the probability that males would squander parental investment on unrelated offspring. We believe that this inverse relationship between the amount of mating effort and the payoff from parental effort holds generally for all organisms.

WHY MALE PIGEONS LACTATE, BUT MALE GIBBONS DON'T

More than 90% of all avian species exhibit high levels of male parental investment; whereas, this is the case for less than 3% of mammalian species (Lack, 1968; Kleiman, 1977).

A Proximate Analysis

Both Maynard Smith (1977) and Daly (1979) have asked "Why don't male mammals lactate?" Daly argues that physiological barriers are too great: because it is the female that gives birth, she has an appropriate cue for the initiation of lactation that the male lacks. This analysis, however, is a proximate and not an evolutionary one. Male birds do not lay eggs, but this has not precluded their evolving brood patches in many species and crop milk in the columbiformes (Wallace and Mahan, 1975; Welty, 1975). Obviously when male postparturitional effort is adaptive, selection can provide an appropriate proximate machinery to insure its emergence. Accounting for sex differences in terms of hormonal differences is description, not explanation. Hormones are simply the body's way of communicating with its various organs, but their effects can vary from organ to organ and species to species. Given the metabolic processes by which hormones are formed, act, and are degraded, even the notion of "male" and "female" hormones is somewhat questionable.

One example from avian physiology should serve to point up the incompleteness of hormonal explanations of behavior. In all known cases where both parents brood the young, brood patches are induced in both sexes by a surge of estrogen. This could be regarded as a hormone-induced parental adaptation, since across species females are more parental than males. The "feminine" parental tendencies of most male birds would then be regarded as due to their higher levels of "female" hormones. What, then, is the prediction regarding sex-role-reversed species with exclusive male postparturitional investment? Here one would have to argue that males must have more "female" hormones than females. In the two species where data exist, the Northern and Wilson's Phalaropes, only the male develops brood patches (since the female does not incubate or brood) and they are testosterone induced, as laboratory experiments with both sexes have shown (Johns and Pfeiffer, 1963). The body uses the hormones it has available, but the particular message read is shaped by selection in accordance with the range of possible adaptive strategies. Many sorts of proximate differences will be expected to be involved in orchestrating different reproductive effort allocations. No one would deny that differences in proximate mechanisms exist, for example between birds and mammals, but such differences should be regarded as much a result as a cause of evolution.

A Parental Relatedness Analysis

The absence of lactation among male mammals and, indeed, the characteristic absence of male parental care in this class follow necessarily from the male's uncertainty about the paternity of his mate's offspring. Among mammals, internal fertilization, estrous cycles, and gestation periods of 13 to 660 days (Vaughan, 1972) result in a significant time lag between ovulation and unambiguous signs of pregnancy. This interval represents the minimal period during which a male must sequester a female in order to insure a high level of paternal relatedness. It might be argued that estrus does not occur during pregnancy, so there is no possibility of cuckoldry between the end of estrus and parturition. It would appear that a male mammal need therefore only guard a sexually receptive female till the end of estrus, not till she becomes unambiguously pregnant. However, estrous behavior does occur during pregnancy, for example, in several primate species (i.e., Loy, 1970, 1971; Hrdy, 1977, 1979). The possible adaptive function of such "sham" estrus can be understood if one realizes that males have no independent corroboration of fertilization other than the absence of subsequent estrus. Consequently, males are vulnerable to manipulation by "deceptive" female reproductive tactics. Indeed, such tactics are possible because polyestrous female mammals may not be fertilized during their first estrus. Thus if males responded to mere copulation as a proximate cue for the initiation of extragametic parental effort, they would waste a significant proportion of this reproductive expenditure (cf. Kurland, 1977). Therefore, if males are to increase their fitness by means of parental investment, they must have effective anticuckoldry adaptations. The cost of such anticuckoldry tactics are proportional to the number of additional matings that must be foregone in order to insure high paternal relatedness to the recipients of male extragametic investment (Werren et al., 1980). Although phylogenetic, demographic, and ecological factors may intervene, the most general statement that can be made is that this "opportunity" (Grafen, 1980) or "promiscuity" (Werren et al., 1980) cost will be proportional to the length of the guarding interval.

In birds, the analogous interval, ovulation to oviposition, is only 2 to 41 hours (Wing, 1956; Welty, 1975). Indeed, ovulation and various parental behaviors are so highly correlated that male birds are susceptible to cuckoldry for an extremely short period (Barash, 1976; Erickson and Zenone, 1976; Zenone et al., 1979). Consequently, the cost in missed mating opportunities is reduced and thus guarding-investing, as opposed to mating-deserting, tactics are more likely to yield a superior fitness payoff in birds than in mammals.

It is true that sperm storage in some avian species allows fertile eggs to be laid as much as 70 days after a single copulation (Welty, 1975). But in most cases, fertility drops off rapidly. For example, in mallards and domestic fowl, eggs are typically infertile six days after copulation (Marshall, 1961; Wallace and Mahan, 1975). In addition, the aggressive components of courtship in some birds seem to be nicely adjusted to delay ovulation to the point where the courting male's sperm will have an advantage over the aging sperm already in the female's reproductive tract (Warren and Kilpatrick, 1929; Erickson and Zenone, 1978). Moreover, sperm storage seems to be most highly developed in those species where paternal care is conspicuously absent; for example, most anatidae, phasianidae, and meleagrididae.

Where mammalian male postparturitional investment occurs, it is characteristically behavioral (e.g., territorial defense), whereas female postparturitional investment entails the development and maintenance of particular physiological and morphological substrates. Such a greater commitment to postparturitional parental effort is a feasible reproductive tactic for a female mammal, because maternal relatedness is high and invariant. In contrast, male parental relatedness is not only lower than the female's, it is also more variable. Given this variability, natural selection is expected to favor highly facultative male investment traits that can be turned on and off rapidly in response to variation in paternal relatedness during the male's reproductive career. On the other hand, male birds exhibit a variety of physiological and morphological adaptations for parental care such as crop milk, brood patches, vascularized pouches, and pedal capillary webs (Welty, 1975). In contrast to mammals, the characteristically higher male parental relatedness in avian species may have favored the evolution of such commitments to parental investment.

In summary, then, the significantly shorter avian ovulation-to-parturition interval means that birds, in comparison to mammals, are that much closer to a condition of external fertilization in which males can more easily establish paternity by means of highly cost-effective anticuckoldry tactics.[4]

Mixed Reproductive Strategies in Birds and Mammals

Differential mate-guarding intervals in birds and mammals have further effects on male reproductive tactics in these two groups. Short mate-guarding intervals in birds adequately raise paternal relatedness to a level where male extragametic parental expenditures may be adaptive without a high "promiscuity cost," so that some reproductive effort is still available for the male to spend on additional mate access. The larger mate-guarding interval of mammals makes it much less likely that such "mixed reproductive strategies" (Beecher and Beecher, 1979) would be successful. Males who attempted to mix their tactics would probably suffer reduced fitness relative to males who played pure strategies because time spent on mate access would seriously compromise paternal relatedness and thus the efficacy of paternal care. Thus in most mammals, males attempt to maximize mate access and show little or no extragametic investment (Wilson, 1975). In cases where mam-

[4]Whenever parental care occurs, female uniparental care predominates among mammals, reptiles, and amphibians, but biparental care characterizes most birds. Although there are some birds that exhibit male uniparental care, this has not been reported for any mammal. However, male uniparental care, particularly with external fertilization, is a rather common pattern in fish. Since aquatic environments and external fertilization preceded terrestriality and internal fertilization, we are led to make what would initially appear to be a counterintuitive suggestion, namely, that uniparental care by males may actually be phylogenetically "primitive."

malian males do invest extragametically in offspring, one finds both extensive and prolonged mate-guarding and a drastic reduction in mate number, usually to one (e.g., hylobatid primates, Chivers, 1972). This dichotomy in tactics is, as predicted not found among birds where males of many species have more than one mate and exhibit extragametic parental investment (e.g., Nice, 1937; Bent, 1953; Beecher and Beecher, 1979).

It should be apparent from this analysis that male parental investment in no way necessarily implies "monogamy" (cf. Ralls, 1976, 1977; Kleiman, 1977). Without specification of other evolutionary and ecological factors, the emergence of male-female "bonding," whether seasonal or perennial, and the appearance of a system of mating that the investigator could comfortably classify as "monogamous" is simply moot (cf. Wittenberger and Tilson, 1980). Instead, the actual dynamics of the reproductive system are much more precisely described and more fully understood in terms of the allocation of male and female reproductive effort. It should be noted that this evolutionary-functional perspective on male and female reproductive effort, by eliminating preoccupation with "pair-bonds" or "mono-/poly-gamy" demonstrates that there is nothing particularly anomalous or unique about human sexual and parental patterns (cf. Campbell, 1966; Alexander and Noonan, 1979). Indeed, human males, like other endotherms (e.g., song sparrows, Nice, 1937; lions, Bertram, 1975) can, under the right environmental conditions, manifest rather high levels of extragametic investment while at the same time effectively monopolizing the reproductive effort of several females. Relatedness also has been shown to be a key factor for human male extragametic investment (Kurland, 1979; Gaulin and Schlegel, 1980).

Sexual dimorphism is thought to be at a minimum when the sexes commit equal effort to parenting (e.g., Trivers, 1972; Alexander et al., 1979). But this is imprecise. Sexual selection will favor dimorphism to the extent that members of one sex are in a position to sum the reproductive effort of members of the opposite sex, that is, obtain multiple mates. All things being equal, these two different perspectives are not that dissimilar, because the mating effort required to obtain multiple mates will detract from the resource pool available for parental effort. However, we have argued that the shorter mate-guarding interval in birds, compared to mammals, will allow avian males to increase fitness by contributing extragametic parental investment and simultaneously seeking additional matings (e.g., Beecher and Beecher, 1979). In such cases, where mating and parenting are less in conflict, considerable sexual dimorphism may be present despite the parental investment made by males, because such males are still involved in intense mating competition to which females do not take part. This may explain why many passerine birds (e.g., parulidae and fringilidae) exhibit considerable plumage and behavioral dimorphism—only males sing, for example—but at the same time maintain rather extensive male parental care.

Given the basic gametic dimorphism, anisogamy, males have a higher potential reproductive success than females and thus will compete for the reproductively limiting female gametic effort (Bateman, 1948; Williams, 1966a). But whenever, for whatever reasons, a male shifts reproductive effort from mating to parenting, he necessarily foregoes some of his potential reproductive success, thus creating what would seem to be an insurmountable evolutionary barrier to male parenting. However, if paternal relatedness is sufficiently high, and if a territorial male's prezygotic investment is a valuable commodity to more than one female (e.g., "resource defense polygyny" in icterids, Orians, 1969; Altmann, et. al., 1977; Emlen and Oring, 1977), then the male, in effect, sums the reproductive effort of several females without significant increase in mating effort.

ALTERNATIVE INTERPRETATIONS OF PRIMATE "PATERNAL CARE"

The primate literature is rife with examples of supposed male parental investment in species characterized by multi-male groups. Many of these putative cases of primate paternal investment are mistaken either because (presumably beneficial) effects are confused with function (Williams, 1966a) or because the various components of reproductive effort are not properly separated and analyzed. For example, in discussion of troop progression, DeVore and Washburn (1963) and Rhine (Rhine and Owens, 1972; Rhine et al., 1980) posit a sociospatial organization of moving baboon troops in which dominant adult males, who are assumed to be the prime group-protectors, are nearest infants and their mothers (but see Altmann, 1979, for an opposing view). Moreover, Bernstein (1976), Hrdy (1976), and Kurland (1977) argue that if dominant males sire the majority of young in multi-male groups of primates, they could increase their inclusive fitness more by defending resident immatures from either predators or conspecifics than could subordinate males. Indeed, such preferential defense by resident, dominant, adult males does seem to occur (Hrdy, 1976, 1977; Kurland, 1977).

If we compare the relative fitness of dominant and subordinate males before defense with their relative fitness after a successful (i.e., undefended) predatory attack on the group, it can be shown that their relative fitness remains unchanged.[5] Indiscriminate defense of troop members may very well alter the *absolute* fitness, but it has no effect on the *relative* fitness, of the resident males—it is relative fitness that is maximized by natural selection

[5]Because adult males in most multi-male socially organized groups of nonhuman primates spend their reproductive career in nonnatal groups, their inclusive fitness ("kinship") effect is zero within that group, whereas their immediate fitness ("personal") effect may or may not be greater than zero, depending on success in mate competition. Consequently, each male's total Hamiltonian inclusive fitness in our example reduces to the classical Darwinian fitness (reproductive success) measured by numbers of surviving offspring rather than surviving kin of any genealogic relationship.

(Williams, 1966a). This can be demonstrated by comparing the fitness payoffs such that

$$\frac{p_d(r_o)(\overline{W}_d) - m(p_d(r_o)(\overline{W}_d)}{p_s(r_o)(\overline{W}_s) - m(P_s)(r_o)(\overline{W}_s)} < \frac{p_d(r_o)(\overline{W}_d)}{p_s(r_o)(W_s)}; \qquad (4)$$

where r_o = the parent-offspring degree of relatedness, p = the male's probability of relatedness to putative offspring, W = the achieved fitness of a male, m = the chance that a randomly chosen immature in the group will be killed if undefended, and where the subscripts d = dominant male and s = subordinate male. For troop defense to be adaptive, the left-hand side of (4) must be less than the right-hand side, that is, dominant males lose fitness by failing to defend. But a little bit of algebraic manipulation is sufficient to demonstrate that the left-hand side, the relative fitness of nondefending dominant males or nondefending subordinate males, is exactly equal to the right-hand side of the equation, the relative fitness of dominant to subordinate males prior to the successful predatory event. This analysis does assume indiscriminate defense of group members, particularly infants, by the resident males; that is, the effects of troop defense are generalized and not directed at any particular individual. There is only anecdotal evidence (e.g. Kurland, 1977) that this may ever not be the case.

Not only does indiscriminate troop defense not change the relative fitness of dominant, resident males, it also allows transient low-ranking males who father one or few offspring in each of several troops to effectively parasitize the parental efforts of the resident males. The young of such "floating," "cheater" males would be defended and the males themselves would incur none of the attendant risks. Thus, transient, nonparental males might reproductively do as well as, if not better than, the resident males (cf. Bernstein, 1976). Such parasitism of indiscriminant troop defense is effectively successful cuckoldry and therefore renders the resident male's parental behavior "evolutionarily unstable" (sensum Maynard Smith, 1976). On the other hand, if resident males adjusted their defensive tendencies, not in response to their dominance, but rather to their reproductive value (Fisher, 1958), indiscriminate defense might well be evolutionarily stable for older males, because they risk losing a smaller reproductive future if killed while defending. Nevertheless, the preceding discussion should serve to emphasize that, in general, male extragametic investment is often diluted by the inability of males to finely target it on their own offspring.

Much of what often passes for male "parental care" may, in primates and in mammals in general, in fact be more parsimoniously interpreted as male mating effort rather than male parental investment. For example, considering baboons again, a male defending the troop as a whole against a predatory attack is defending the pool of potential mates most directly and easily under his control and, therefore, is protecting the reproductively limiting resource (e.g., Ransom and Ransom, 1971; or in macaques, Kurland, 1977). Sexual dimorphism in body size, dentition, musculature, maturation, mobility, and coalitional behavior are all aspects of the baboon sexual dimorphism in reproductive effort allocation, predicated on the higher mating effort of males and the higher parental effort of females (Hausfater, 1975). Because of this male mating effort bias, males may be likely to investigate novel situations where potential mates need protection or where new mates can be found (Buskirk et al., 1974; Hamilton et al., 1975; Lindburg, 1969; Cheney and Seyfarth, 1977). Consequently, their reproductive strategy may bias males toward a greater involvement in situations in which predatory or conspecific aggression is increasingly likely. Viewed this way, "troop defense" might sometimes increase a male's attractiveness to potential mates, but it certainly increases the price paid for being "masculine"—it is the cost of male mating effort, not male parental investment (Williams, 1966a).

In a very few primate species, there regularly occur one-male/one-female "family" groups in which male parental care is particularly well developed and in which (as expected) anticuckoldry tactics promote high levels of paternal relatedness (e.g., Chivers, 1972; Kleiman, 1977; Vogt, chapter 14; and especially, Cubbicciotti and Mason, 1978). Given a uni-male social organization and the resultant intended male-male competition for group tenure, a resident male may be important for protecting immatures from the infanticidal attacks of other males (e.g., Hrdy, 1977). With no other reproductive males in residence, male "troop defense" in uni-male groups may very well be male parental effort in that the resident protects immatures that he presumably sired. However, adult male-infant interactions in multi-male groups of primates, and indeed in most mammalian social groups, seem to be characterized more by "exploitation" than "care" (e.g., Hrdy, 1976, 1979; Packer, 1980; see chapter 15).[6]

EXTRAGAMETIC EFFORT, ECOLOGY, AND OFFSPRING ONTOGENY

The mere fact of high parental relatedness will not necessarily make parental effort beyond the minimal gametic investment adaptive. It may be that given the prevailing ecological circumstances and juvenile and adult mortality schedules, parents of one or both sexes can obtain a higher fitness payoff by expending their reproductive effort as mating effort and gametic parental

[6]Most recently—that is, after completion of this paper—Busse and Hamilton (1981) reported that adult males in the multi-male groups of chacma baboons may very well be parental. They find that resident males carry infants, probably offspring, in order to protect them from infanticidal, immigrant males. These findings, of course, further support our argument that adaptive primate parental care is not expected to be indiscriminate, but rather precisely targeted on the immature most likely to be the male's own offspring (see chapter 8).

effort than by expending it as extragametic parental effort. For example, it is frequently argued that the need for parental care (postparturitional parental effort in our taxonomy) is reduced in species with precocial young, and that as a result, such species tend to show uniparental care (e.g., Orians, 1969; Emlen and Oring, 1977; Wittenberger and Tilson, 1980). This is probably basically correct, but it is far from a complete evolutionary analysis of the problem. For example, it is conceivable that, given the conjunction of precociality, readily harvestable food for immatures, and low effective predation rates (or the inability of parents to deter predators), postparturitional parental investment does not sufficiently increase the reproductive prospects of offspring to justify its associated costs. Such circumstances, rather common among various exothermic vertebrates (e.g., reptiles, Porter, 1972; amphibians, Wells, 1977; and fish, Breder and Rosen, 1966) but apparently absent among endotherms, would lead to the evolution of reproductive systems characterized by the total absence of parental care—by either sex.

If one parent can effectively contribute to offspring survival but two parents are only marginally more effective, desertion tactics and uniparental care will evolve (Trivers, 1972; Maynard Smith, 1977), but the issue of which parent will desert and which will invest has not been addressed. It is precisely this question that has been explored in the preceding section.

Because it will be helpful in exploring some of the ecological factors that shape parental tactics, we will continue to examine the preceding argument concerning precociality and uniparental care. The argument as stated ignores the fact that if one compares similarly sized adults, one of which is producing a precocial and the other an altricial young, the former makes a larger preparturitional investment to bring the young to an advanced stage of development but the difference in parental effort between the two may simply be the shifting back or forward of preparturitional investment in relation to the time of birth, with total postzygotic investment in fact equal. Moreover, the argument takes the developmental program as given and therefore constraining the rest of the reproductive strategy. This is surely not the case. Whenever postparturitional effort is more cost-effective than preparturitional effort, for example if postparturitional investment can be more efficiently distributed among several potential recipients (Low, 1978), a reduction in preparturitional effort and the birth of altricial young would be favored by natural selection. Conversely, whenever a given expenditure of reproductive effort is more productively (in terms of fitness) applied preparturitionally than postparturitionally—as when offspring mortality can be sufficiently reduced (Williams, 1966a)—increased preparturitional effort and the birth of precocial young would be favored.

If we regard the perinatal developmental stage as evolutionarily plastic, it now becomes possible to explore its ecological determinants. Assuming that the degree of offspring precociality does determine the relative payoff of additional parental investment, what we are in fact doing in exploring the ecological causes of precociality and altriciality is unpacking at least some of the ecological causes of parental effort tactics.

Among the most fundamental problems faced by all animals is the need to find food and, with the exception of certain parasites, the need to avoid becoming food for other animals (Bertram, 1978). Due to the trophic structure of animal communities, predation may be greater on smaller and less developed young. Thus, the degree of precociality might be expected to be positively correlated with the level of predator pressure on immatures. Alternative antipredator tactics are available when other factors favor altriciality. For example small, altricial young can be cached as they are in some rodents (e.g., Eisenberg, 1968); ungulates (e.g., Geist, 1974); and prosimians (e.g., Jolly, 1972). Alternatively, when paternal relatedness is high enough to make it adaptive for males to invest postparturitionally, early delivery of the young may unburden the mother, thereby reducing the chance of predation on her and the brood, and permit more effective biparental predator defense.

Certain kinds of feeding ecologies will tend to favor precociality. When food is widely available and simply harvested, young could feed themselves and parents could begin their next reproductive venture sooner by providing offspring with enough equipment at birth to feed alone. Moreover, if such food is of low nutrient density, it will be inefficient for the parent to attempt to provision the young because of the compensatorily large quantities that would be required. Thus, high food abundance, low food quality, and uncomplicated foraging strategies singly and, even more so, in combination will favor the production of precocial young. Low food abundance, high food quality, and complex foraging strategies favor altriciality.

BODY SIZE AND DIET: THE JARMAN-BELL PRINCIPLE

Just as predator pressure may vary with body size, so may feeding ecology. Hemmingsen (1960), Kleiber (1961), Munro (1969) and others show that, across species, metabolic requirements scale to about the ¾ power of body weight. For example if an animal weighing x requires a total of M calories per day, a larger animal weighing ax requires only bM calories, where $a > b > 1$. Thus the larger animal requires more total nutrient intake than the smaller animal ($bM > M$), but per unit of body weight, the larger animal actually needs less food:

$$\frac{M}{x} > \frac{bM}{ax} \tag{5}$$

because $a > b$. This negative allometric relationship between metabolic requirement and body weight has clear consequences for the coevolution of diet and body size (Jarman, 1968, 1974; Bell, 1971; Geist, 1974; Gaulin and Konner, 1977; Gaulin, 1979). Because larger animals have higher total nutrient requirements, it would be difficult for them to derive a major portion of these requirements from uncommon foods. They would simply be unable to harvest sufficient quantities of such foods to fill their metabolic needs and would therefore have to concentrate on common foods. Although the total requirements of larger animals are higher, their per-unit-weight requirements are lower than those of smaller-bodied forms. Thus, larger animals do not have to maintain a very high rate of nutrient flow to their tissues and may therefore base their diets on relatively low-quality food, that is, foods in which nutrients are not very concentrated.

For smaller animals the problems are reversed. Since their total requirements are lower, uncommon foods could supply a significant fraction of their needs. But their higher per-unit-weight requirements will necessitate higher rates of nutrient extraction and therefore favor the selection for high-quality (nutrient-dense) foods. These patterns—larger animals eating common, low-quality foods, and smaller ones eating uncommon, high-quality foods—are further reinforced by the trophic structure of natural communities. The bulk of the biomass in any habitat tends to be made up of material that contains little nutrient per unit weight. However nutrients may be quite concentrated in temporally and spatially uncommon items (Odum, 1970); Gaulin, 1979). The ecological consequence of body size on diet is referred to as "The Jarman-Bell principle" (Geist, 1974; Gaulin, 1979; and summarized in Fig. 11-2). In general the predicted diet-body weight associations are confirmed by the available evidence (Geist, 1974; Jarman, 1974; Gaulin and Konner, 1977; Gaulin, 1979).

Body size, like any adaptation, is a compromise between the selection pressures acting on it. Natural selection may therefore sometimes produce

	total nutrient requirement	$\frac{\text{nutrient requirement}}{\text{body weight}}$
large animal	large (abundant foods)	small (poor quality foods)
small animal	small (rare foods)	large (high quality foods)

Fig. 11-2. The Jarman-Bell principle. Expected characteristics of food choice as a function of the total and relative nutrient requirements of large and small animals.

deviations from the predicted dietary patterns, but such deviations will always be accompanied by certain compensatory specializations. A large animal occupying a small-animal feeding niche (i.e., feeding on uncommon high-quality foods) would have no problem with nutrient-extraction rates but might be unable to obtain enough of such foods to satisfy its higher total needs. Thus, harvest rates are problematic for dietarily deviant large animals, and they would require locomotor, food-finding, and food-handling specializations that effectively increase the supply of uncommon foods. A small animal feeding like a large one could certainly obtain sufficient supplies of common foods but might well starve to death even if it fed continuously, because it would be unable to extract nutrients rapidly enough for body maintenance. Deviant small animals would have to possess digestive specializations to effectively increase nutrient-extraction rates (Gaulin, 1979). Although primates conform rather well to the dietary patterns predicted from the Jarman-Bell principle, certain species also illustrate deviant feeding strategies. For example, among the prosimians, the aye-aye *(Daubentonia)* feeds extensively on rare, high-quality insects (Petter and Peyrieras, 1970), despite its relatively large size; but virtually all the specialization that set it apart from the rest of the suborder are food-harvesting adaptations. Conversely, *Lepilemur* is quite a small folivore but coprophagy (Charles-Dominique and Hladik, 1971) allows it to effectively increase the amount of nutrient it can extract from such low-quality food (Gaulin, 1979).

BODY SIZE AND EXTRAGAMETIC EFFORT

To the extent that animals conform to the predicted diet-body weight patterns, body size can be shown to be important in determining the payoff of alternative reproductive effort allocations. Considering larger animals first, their common, but low-quality, foods could be harvested by sufficiently developed young but would be hard for the parent to supply in adequate quantities. That is, they conform to the set of ecological attributes that favor precociality and hence uniparental postparturitional investment. Conversely, the smaller animal's uncommon, high-quality foods may require special skills to locate and harvest but could, because of their high nutrient density, be supplied to the offspring in sufficient quantities. Such conditions will favor altricial young with biparental postparturitional investment (see "Body Size, Ecology, and Offspring Otogeny" above).

Relatedness Effects

We have argued in previous sections that the adaptiveness of extragametic effort is importantly conditioned by the probability of relatedness. We have also focused on some of the ecological factors favoring precociality or altriciality and the relative benefits of investing in precocial as compared to altricial young. But there is a parallel line of logic that directly links body size to postparturitional investment strategies via its effects on the probability of relatedness, reinforcing our arguments on the ecological sources of reproductive effort.

Due to higher total nutrient requirements, larger animals will often have to forage over considerable areas to harvest sufficient low-quality food. Large home ranges render territorial strategies inefficient due to high defense costs (Brown, 1964; Davies, 1978). Similarly, the necessity to forage widely would make mate-guarding difficult—such mate-guarding being typically a male task. This in turn will lower paternal, as compared to maternal, relatedness, making extragametic investment a less viable male reproductive tactic. The reverse analysis can be applied to small animals. Low total nutrient requirements and high-quality food may combine in such a way that small and therefore defendable areas contain sufficient food resources to permit adaptive territoriality. Whenever food resource defense can also function (perhaps at some slight additional cost) as mating-resource defense, paternal, but not maternal, relatedness is increased, particularly with internal fertilization. This should be reflected in male reproductive tactics for two reasons: such territoriality will increase the payoff associated with male extragametic, and especially post-parturitional, investment because it reduces the chances that such investment is wasted on unrelated young *and* it simultaneously increases the cost of his seeking additional matings due to necessary lapses in territory defense resulting from such mating effort. In this case, male prezygotic investment, in the form of a feeding territory, conditions or preadapts the male for subsequent postparturitional investment both by constraining his reproductive tactics and by altering the profitability of alternative reproductive effort allocations through raising his paternal relatedness.

Trivers (1972) predicts that low levels of parental investment by one sex should favor increased mobility as a means of maximizing access to the limiting parental investment of the opposite sex. In reviewing the literature on home range utilization in terrestrial endotherms and ectotherms, Gaulin and Crandlemire (in prep.) find that when one sex emphasizes mating effort at the expense of parental investment, that sex usually exhibits a significantly larger home range. Such larger ranges are obviously more difficult to patrol and thus, on average, the greater number of potential mates contained therein, the lower the parental relatedness of each of these mate's offspring. The predicted association between body size, on the one hand, and both resource-exploitation patterns and male parental care, on the other hand, is expected to be most evident in a comparison of phylogenetically close species, because this minimizes the confounding influence of independently evolved specializations. For example, among hominoid primates, the largest species, *Gorilla,* shows large, undefended home ranges; low-quality diet; multi-male groups; and exceedingly low levels of male parental care (Fossey, 1979). In marked contrast, the smallest species, *Hylobates,* shows small, defended territories; high-quality diet; uni-male groups; and rather high levels of male parental care (Chivers, 1972; see chapter 15).

The effects of body size on the efficacy of territorial behavior and on feeding ecology combine to form a coherent pattern in which small size tends to produce conditions where male extragametic, and especially postparturitional, investment is valuable and where such investments can be adaptively expended on altricial offspring. Large body size tends to remove the need for biparental expenditures on precocial neonates and simultaneously increases the likelihood that such investment can be parasitized (summarized in Fig. 11-3).

Other Reproductive Effects of Body Scaling

Williams (1966a) argues that, due to higher mortality rates, small animals should exhibit a greater reproductive effort than larger ones. Data on egg-maternal weight ratios in fish appear to support Williams's contention

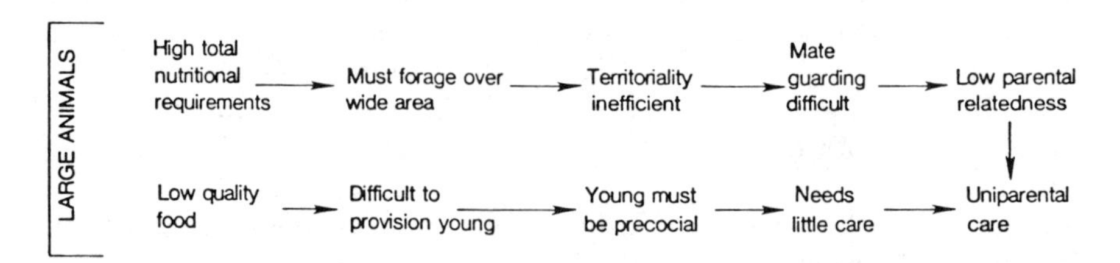

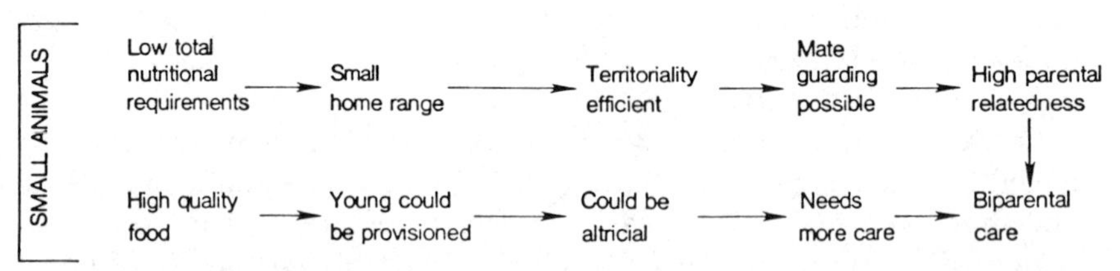

Fig. 11-3. Effects of body size on parental adaptations. Total nutrient requirements shape ranging patterns and thus the efficacy of mate guarding. Food quality conditions the relative effectiveness of parental provisioning of the young as opposed to independent foraging by the young.

(Williams 1966a). However, Williams makes no prediction as to how this greater reproductive effort should be allocated. Leutenegger's demonstration that the ratio of litter weight to maternal weight in primates (Leutenegger, 1973) and in eutherian mammals (Leutenegger, 1976) is negatively correlated with maternal weight is consistent with Williams's expectations. If litter size were held constant, obstetric problems would increase with increasing litter-maternal weight ratios, consequently, smaller animals are expected to produce larger litters. That is, within any particular taxonomic group (e.g., ceboid primates), females of the smallest species (e.g., callitrichids) need to produce a larger number of more easily deliverable neonates, whereas in the larger species (e.g. cebids), females can give birth to one or only a few rather larger offspring—the very pattern that Leutenegger finds (but see chapter 13).

But Leutenegger (1980) extends the argument to parental care. On analogy with Maynard Smith's (1978a) suggestion that more exhaustive female preparturitional investment would select for increased male postparturitional investment, Leutenegger suggests that higher female gestational costs in smaller animals will favor the evolution of male parental care. Larger litters may also increase the value of investment by two, as opposed to one, parent. Leutenegger (1979, 1980) further suggests that these scaling effects may explain why the smallest ceboid primates, the callitrichids, exhibit twin births and male parental care. These facts are also consistent with our ecological and relatedness arguments (summarized in Fig. 11-3). But there is, of course, no a priori reason to assume that only one of the two explanations is correct. It is quite possible that the various consequences of body size act in consort to shape reproductive strategies.

Empirical Corroboration

The question of whether these theoretical predictions concerning body size and reproductive effort allocation are fully supported will have to await a very extensive literature review, although they are confirmed by Jarman's (1974) work on ungulates, where the smallest species tend to exhibit selective, high-quality diets; small, defended ranges; fewer cases of males with multiple mates; and increased male parental effort (see also Alexander et al., 1979). Although small-bodied mammalian species, primarily rodents, are not well-known because they are nocturnal, and hence not easily observed, it may be that "monogamy," stable family groups, and high paternal relatedness predominate (e.g., the oldfield mouse, *Peromyscus polionotus,* Foltz, 1981). On the other hand, the best-known mammals are in fact the large, diurnal forms in which male parental care and "pair-bonds" are quite rare (Foltz, 1981; Kleiman, 1977; see above "Why Male Pigeons Lactate, But Male Gibbons Don't"). The primates, another mammalian group for which reasonably good data exist, also conform fairly well to the predicted pattern (Clutton-Brock and Harvey, 1977; Clutton-Brock et al., 1977; Leutenegger, 1978; Alexander et. al., 1979), as do the birds where, although biparental care is the rule, the larger and more terrestrial species tend toward uniparental postparturitional investment (Lack, 1968; Wiley, 1974).

The well-known positive correlation between body size and degree of sexual dimorphism (Rensch, 1950; Maynard Smith, 1977; Leutenegger, 1978) provides a more general, but indirect, support for the proposed model. Although a variety of hypotheses have been advanced to explain why larger-bodied species show relatively more sexual dimorphism in body size than closely related smaller-bodied forms (Clutton-Brock et al., 1978), none has received any independent empirical support. In terms of current sexual selection theory (e.g., Williams, 1966a; Trivers, 1972), high degrees of sexual dimorphism are the result of divergent reproductive effort allocations by the sexes (see "The Components of Reproductive Effort" above). Sexual dimorphism in a variety of traits is expected to be at a minimum whenever the sexes allocate reproductive effort most similarly. Thus only when males contribute some parental effort to equalize the female's initially greater parental expenditures could sexual dimorphism be reduced. The fact that smaller-bodied species in a number of phylogenetically distinct groups show relatively less sexual dimorphism than larger-bodied ones is entirely consistent with our expectation that smaller individuals should exhibit increased male parental effort due to the effects of body size on parental relatedness and feeding ecology (Fig. 11-3).

Some species undoubtedly represent exceptions to the predicted patterns. One reason may simply be that they deviate from their expected feeding ecologies and such cases should, of course, be examined on the basis of dietary adaptations and food distribution rather than merely body size (cf. Ralls, 1976, 1977). Alternatively, while the influence of body size on feeding ecology is expected to favor altriciality and hence biparental care in small animals, but precociality and uniparental care in large animals, predator pressure may sometimes counteract this pattern. Thus the greater predator pressure on smaller-bodied animals may favor increased precociality whenever offspring concealment or biparental predator defense are ineffective antipredator tactics.

An interesting exception to the predicted relationship between body size and extragametic effort is the indris, *Indri indri.* With a body weight of 10 to 15 kg., it is not only the largest living prosimian, but also in fact a rather large primate. Given our arguments, the indris appears to be simply too big for biparental care, "family" social groups, male tolerance of infants, and monomorphic morphology (e.g., Pollock, 1977, 1979). However, the indris also possesses a number of harvesting and alimentary specializations: "dental comb" ingestion, vertical-clinging and leaping, large salivary glands, enlarged stomach, and enlarged caecum (Hill, 1953; Pollock, 1977). These adaptations may allow the indris to live in a "small-bodied," defensible, nutritionally sufficient territory that includes easily assimilable and evenly distributed prey items, but that cannot include more than one mate.

Among birds, and at the low end of the body-size spectrum, hummingbirds provide the converse exception: a small-bodied form that is decidedly not biparental (see review and citations in Emlen and Oring, 1977). Because the hummingbird's primary limiting resource, flowers, is temporally transient, there is little opportunity for a male to guard both his mate and resource-base effectively. Thus, there would be a low return on extragametic reproductive effort expended on the male's mate's offspring. The peculiarly evanescent food resource and the extremely high bioenergetics of the trochiliformes necessitates rather brief "promiscuous" sexual encounters ("resource defense polygyny," Emlen and Oring, 1977, but what really may be an "exploded" lek). The same may hold for the convergent sunbirds (Nectariniidae) and Hawaiian honeycreepers (Drepanididae) (e.g., Carpenter and MacMillen, 1976). In general, body size may correlate rather predictably with various aspects of reproductive effort (Fig. 11-3); however, the primate and avian exceptions serve to emphasize that the key evolutionary determinant of reproductive strategy is the *interaction* between an organism's physiology and ecology rather than simply its body size. The idiosyncratic ecological physiology of indris and hummingbirds appears to falsify predictions from the proposed model, but upon deeper inspection neither species actually violates the model's basic principles and logic. The hummingbird's feeding ecology and energetics preclude mate guarding and thus high paternal relatedness, whereas the indris' physiology and morphology provides for cost-effective mate guarding, and hence high paternal relatedness, but at the same time makes the inclusion of multiple mates and their offspring within the territory unaffordable.

RELATEDNESS AND INVESTMENT: THE PROBLEM OF CAUSALITY

We have concentrated on those key evolutionary and ecological factors that importantly condition the allocation of reproductive effort. The close association between parental relatedness and reproductive effort allocation is indubitable, but the direction of causation is as yet unclear (Trivers, pers. comm.). Thus selection may have favored some particular reproductive physiology that *indirectly* altered parental relatedness and the concomitant reproductive effort patterns of males and females *or* selection may have *directly* shaped parental relatedness patterns due to the adaptive response of males and females to this change in the reproductive environment. Depending on choice of causal direction, there exists an "indirect-relatedness effect" and a "direct-relatedness effect" hypothesis.

Williams's (1966a) discussion of the evolution of viviparity provides a good example of the indirect-relatedness effect. Because egg-laying species always suffer some loss of reproductive effort due to egg predation before that expenditure can even begin to be compounded by means of offspring growth, viviparity will be expected to evolve whenever the preadaptation of internal fertilization exists. Now if viviparity does evolve, for this or any other ecological reason, it clearly will tend to reduce the parental relatedness of the nonincubating sex, leading to the emergence of uniparental care. Such an analysis therefore suggests that reproductive effort allocation patterns are the byproduct of other more fundamental reproductive adaptations.

Another concrete example of the indirect-relatedness effect can be advanced as a possible explanation for the markedly different reproductive effort patterns found among birds and mammals. The characteristic avian adaptation of flight may make the carrrying of developing fetuses too costly, either due to increased energetic expenditure or predation risk, such that females who unburdened themselves as early as possible after fertilization had higher reproductive success than those that did not (cf. Neill, pers. comm., cited in Stearns, 1976). This physiological response to an ecological problem would have increased paternal relatedness levels and thus might have favored further modifications in reproductive effort allocations, in particular, increased male postparturitional effort.

Under the direct-relatedness effect hypothesis, reproductive physiology is designed by natural selection to produce the parental relatedness levels that will exact optimal amounts of investment from one's mate. For example, where offspring mortality could be reduced by the intervention of two parents, selection may favor alteration of parental relatedness patterns such that increased postzygotic investment from a mate can be adaptively expended on the young.

It is also possible to exemplify the direct-relatedness effect in avian and mammalian investment patterns. Williams (1966a) argues that selection will favor accelerated development during periods of high mortality. Additional parental care could importantly facilitate such ontogenetic acceleration. All things being equal, the increased protection and resources that could be provided by biparental, as opposed to just uniparental, investment should permit more rapid growth, particularly during critical stages of development. Most passerine birds are thought to be subject to quite high mortality during the nestling phase but to markedly lower mortality after fledging (Ricklefs, 1969a, 1969b). No such mortality threshold has been reported for mammals. If biparental care could sufficiently accelerate attainment of fledgling status, natural selection would favor parental relatedness mechanisms permitting adaptive male postparturitional investment. That is, females whose behavior and physiology led to high levels of paternal relatedness produced more fledglings than females exhibiting alternative reproductive patterns, because their mates were more likely to invest postparturitionally. This kind of selective pressure in mammals would have been essentially inoperative given the more uniform mortality schedule of early mammalian postnatal development. Williams (1966a) comments that the absence of viviparity among birds is puzzling because they possess the preadaptation of internal fertilization. If the direct-relatedness effect hypothesis is correct, the absence of live births is easily explained: mutant females who retained developing zygotes reduced paternal relatedness and thereby lost essential male postparturitional investment.

We provide two specific hypotheses regarding the evolution of divergent reproductive effort patterns in birds and mammals principally to exemplify alternative causal relationships among key variables. It may well be that neither is correct although it seems, in principle, possible to evaluate them.

For example, the second particular example of the indirect-relatedness effect—flighted forms must deliver young early for energetic reasons, incidentally altering paternal relatedness—could be tested by examining whether biparental care is more prevalent among flighted than flightless birds. Our own initial survey of the avian literature does indeed support the prediction that flightless birds are more likely to exhibit uniparental care and flighted birds, biparental care (Gaulin and Kurland, in prep.). In the light of this argument, it would also be informative to consider the parental adaptations of volant and nonvolant mammals so that the hypothesis can be tested independently. In contrast, our review of the literature on both Megachiroptera and Microchiroptera (e.g., Leen and Novick, 1969; Bradbury, 1977) suggests that uniparental care is overwhelming predominant in these flighted mammals and in fact biparental care is conspicuously absent when compared with nonvolant rodents of about the same body size and habitat. However, such a simple rejection of the indirect-relatedness effect is complicated by the fact that other aspects of chiropteran physiology may very well modify reproductive physiology, setting the stage for divergent male and female reproductive effort allocations. It is generally the case that temperate species practice delayed implantation or delayed fertilization, presumably as a result of the winter hibernation (Allen, 1937; Davis, 1970; Orr, 1970), a situation not present among the many tropical species (e.g., Lyman, 1970; Bradbury and Emmons, 1974; Bradbury and Vehrencamp, 1977a). In addition, many temperate, and a few tropical, Microchiropterans migrate seasonally (Griffin, 1970). Thus compared to most tropical species, a male in a temperate zone species would have a lower probability of relatedness to any mate's offspring born the spring following the winter hibernation or migration. Thus overall male bats seem to bias their allocations of reproductive effort to mating effort, while at the same time, females expend reproductive effort primarily as parental investment. However, in some tropical species (Bradbury and Vehrencamp, 1977b), males show considerable extragametic investment, although not any obvious "parental care," in that males "inherit" ecologically and reproductively critical territories from their presumed fathers. Moreover, some tropical species exhibit moderate levels of parental relatedness (e.g., McCracken and Bradbury, 1977). To date, neither male parental care nor any other indication of male extragametic parental investment has been reported for any temperate Microchiropteran. This is at least consistent with the idea that temperate, as opposed to tropical, bats face an ecological and concomitant physiological barrier to male extragametic investment due to effects on male, but not female, parental relatedness. Flight itself, therefore, may not be a sufficient condition biasing parental relatedness and thus parental extragametic expenditures. But it would seem that other aspects of mammalian physiology and ecology (thermoregulation in highly variable habitats) can alter the reproductive system sufficiently to dissimilate male and female parental relatedness. Clearly, a much more detailed analysis of both volant and nonvolant birds and mammals must be undertaken before the more general form of this hypothesis can be rejected.[7]

The direct-relatedness effect—young are delivered early in order to allow adaptive male investment in cases where such additional postparturitional

---

[7]Penguins are an interesting exception to the specific prediction that flightless birds will tend to uniparental care, and flighted birds, biparental care, since these flightless species are biparental (e.g., Richdale, 1957). However, because the sphenisciformes are also subject to rather high predation from other birds and pinnipeds, particularly during the breeding season, there may have been a selective premium on the evolution of biparental care of eggs and chicks.

Finally, we wish to call attention to a bizarre case of uniparental male postparturitional effort in an ostensibly flighted bird, the American finfoot *(Heliornis fulica),* one of three species comprising the Heliornithidae, a group morphologically similar to rails. Courtship is intense and each member of the courting pair vigorously excludes same-sex conspecifics (Alvarez del Toro, 1971). Both sexes contribute to nest construction and incubation of two eggs which are hatched after only 11 days (Alvarez del Toro, 1971). Upon hatching, the rather altricial nestlings are carried in a skinpleat which operates as a brood pouch under the male's wing (Alvarez del Toro, 1971)! This seems to be a counter-example to our flight-indirect-relatedness effect hypothesis. However, the American finfoot rarely flies to its perch and instead typically swims, scrambles, and climbs through its lowland, riverine habitat. Indeed, as Alvarez del Toro (1971, p. 80) notes, "When escaping from danger, real or potential, the finfoot almost always swims." Thus this peculiarly uniparental species is effectively flightless! The intense courtship and prezygotic and preparturitional biparental effort may lead to high levels of paternal relatedness and facilitate the emergence of uniparental postparturitional investment. The small size of the altricial nestlings presumably minimizes the extra load that the male must bear.

investment would be most beneficial—could be evaluated by comparing parental care patterns across species with known life-tables. Those that show high mortality followed by an abrupt increase in survivorship should show biparental care during the perilous period of high mortality. Because such demographic/life-historical data are difficult and time-consuming to gather, a reasonable approximation to testing the second hypothesis might be to compare the frequency of biparental care among nidicolous and nidifugous birds, since the former are expected to exhibit a sharper mortality-threshold.

A PREDICTABLE CONCLUSION

Throughout this paper, we have analyzed the natural history of reproduction in a variety of fish, birds, and mammals, such reproductive systems being ordinarily lumped into categories of "polygamy" or "monogamy." But these diverse patterns of reproductive adaptation can be better described and summarized by means of the components of reproductive effort (Table 11-1) than by means of the categories of mating-system taxonomy. And indeed, we have so redescribed and differentiated some familiar avian and mammalian reproductive systems. More importantly, we hope to have demonstrated that the formulation of functional hypotheses about the evolution of reproductive tactics can be facilitated by specifying precisely patterns of male and female reproductive effort in a given population.

Trivers (1972) argues that the action of sexual selection is determined by each sex's relative allocation of reproductive effort to mate competition and parental effort—mating and parental effort in our system. Furthermore, he suggests that an evolutionary instability in isogamy leads to the emergence of anisogamy, and given this gametic dimorphism, natural selection could only favor increased parental investment by the sex producing the large gamete—by definition, females. On economic analogy, Dawkins and Carlisle (1976) and Boucher (1977) argue that previous investment does not always make future investment adaptive. Consequently, the relationship between gametic investment and all forms of extragametic parental effort requires further elucidation. Were anisogamy the sole explanation for the evolution of relative parental investment patterns, sex-role reversal, as in the avian phalaropodidae, could not occur. However, by considering in detail the entire pattern and interaction between male and female reproductive effort allocations, it becomes possible to specify what other evolutionary and ecological factors may have shaped the evolution of male and female reproductive tactics. Indeed, we argue that parental relatedness and body size are expected to have profound repercussions on male and female reproductive strategy. For example, in the case of a moderately "sex-role reversed" species, the American rhea, we demonstrate the important effects that differences in male and female parental relatedness have on male and female extragametic parental investment.

In an attempt to develop some broadly applicable conclusions about the evolution and ecology of reproductive effort allocation, we have developed a number of sometimes rather rarefied theoretical arguments. However, we believe that many of the ideas presented in this paper generate predictions that are eminently testable. Although we are in the process of testing some of these predictions, we present below a number of testable propositions that are easily derived from our particular analysis. Refinement of the concepts presented here and detailed hypothesis testing will, of course, ultimately decide the significance of our presentation. It is in this spirit that we offer the following predictions that should and could be tested in any number of phylogentically-restricted group of organisms.

Predictions About Parental Relatedness Effects

1. Due to anisogamy (Parker, et al., 1972), females are more likely to be internally fertilized and therefore more likely to exhibit extragametic effort.
2. External fertilization should remove the sex bias in parental care patterns *except* where males can assemble the eggs of several females, in which case there will be a male uniparental care bias.
3. Long "gestation" periods, that is, long periods of preparturitional investment, will be associated with uniparental care by the gestating sex.
4. Biparental care, in particular postparturitional investment, will be associated with shorter "gestation" periods.
5. The likelihood of male parental care, particularly postparturitional investment, is inversely proportional to the duration of sperm storage. Where females store sperm, anticuckoldry tactics by investing males are expected to be particularly well developed and highly effective (e.g., Smith, 1979).

Predictions About Ecological Effects

6. High food abundance, low food quality, and uncomplicated foraging strategies singly, and especially in combination, favor precociality and uniparental care.
7. Low food abundance, high food quality, and complex foraging strategies favor altriciality and biparental care.
8. Biparental care will be associated with territorial resource-defense tactics.
9. Uniparental care will occur predominantly among larger animals.

10. Biparental care will occur predominantly among smaller animals.
11. Indirect-relatedness effect (see text):
    a. Flighted birds are more likely to exhibit biparental care and flightless birds, uniparental care.
    b. Volant mammals (bats) are more likely to exhibit uniparental care and nonvolant mammals, biparental care, especially when comparing animals of about the same body size.
12. Direct-relatedness effect (see text):
    a. Nidicolous birds are more likely to exhibit biparental care and nidifugous birds, uniparental care.
    b. In general, within any phylogenetically close group, those species that exhibit a more pronounced mortality threshold during ontogeny are more likely to exhibit biparental care.

Finally, our arguments suggest that sexual dimorphism in behavior, physiology, and morphology is not due to how parental a male or female is, but rather to how much males and females vary in mating success (cf. prediction 2). Consequently, we predict that:

13. The degree of sexual dimorphism should be proportional to sex differences in mate number, not to sex differences in parental investment.

ACKNOWLEDGMENTS

We thank Robert L. Trivers for original insights that ultimately made this paper possible and for helpful criticisms of earlier versions that we hope have improved the final product. We gratefully acknowledge expert bibliographic assistance by Joy L. Lightcap, faultless manuscript preparation by Shelvia Hummel, and the organization of the subject index entries for this chapter by John Driscoll. During the development of this paper, J.A.K. was supported by the Harry Frank Guggenheim Foundation and S.J.C.G. was supported by NIH Grant 5SO7RR07081-14. Authorship was determined by a modest improvement in the first author's squash game.

POSTSCRIPT

Our chapter was completed in December, 1980. Due to unavoidable delays in the final development of this volume, a wealth of new hypotheses, arguments, and data relevant to some of the topics originally addressed here have appeared. Rather than undertaking a revision, we decided to leave the chapter as it was first conceived. However, extensions and tests of the ideas developed in this chapter can be found in new papers by Gaulin and Sailer, and Gaulin and Kurland.

REFERENCES

Alcock, J. 1979. *Animal Behavior: An Evolutionary Approach.* Sunderland, MA: Sinauer.

Alexander, R.D., and Borgia, G. 1979. On the origin and basis of the male-female phenomenon. In M.S. Blum and N.A. Blum (eds.), *Sexual Selection and Reproductive Competition in Insects,* pp. 35-77. New York: Academic Press.

Alexander, R. D., and Noonan, K. M. 1979. Concealment of ovulation, parental care, and human social evolution. In N. A. Chagnon and W. Irons (eds.), *Evolutionary Biology and Human Social Behavior: An Anthropological Perspective,* pp. 436-53. North Scituate, MA: Duxbury.

Alexander, R. D.; Hoogland, J. L.; Howard, R. D.; Noonan, K. M.; and Sherman, P. W., 1979. Sexual dimorphism and breeding systems in pinnipeds, ungulates, primates, and humans. In N. A. Chagnon and W. Irons (eds.), *Evolutionary Biology and Human Social Behavior: An Anthropological Perspective,* pp. 402-35. North Scituate, MA: Duxbury.

Allen, G. M. 1962. *Bats.* Reprint of 1937 edition. New York: Dover Publications.

Altmann, S. A. 1979. Baboon progressions: order or chaos? A study of one-dimensional group geometry. *Animal Behaviour* 27: 46-80.

Altmann, S. A.; Wagner, S. S.; and Lenington, S. 1977. Two models for the evolution of polygyny. *Behavioral Ecology and Sociobiology* 2: 397-410.

Alvarez del Toro, M. 1971. On the biology of the American Finfoot in Southern Mexico. *The Living Bird* 10: 79-88.

Arnold, S. J. 1976. Sexual behavior, sexual interference, and sexual defense in the salamanders *Ambystoma maculatum, Ambystoma tigrinum* and *Plethodon jordani. Zeitschrift für Tierpsychologie,* 42: 247-300.

Bachmann, C., and Kummer, H. 1980. Male assessment of female choice in hamadryas baboons. *Behavioral Ecology and Sociobiology,* 6: 315-21.

Barash, D. P. 1976. Male response to apparent female adultery in the mountain bluebird *(Sialia currucoides):* An evolutionary interpretation. *American Naturalist,* 110: 1097-1101.

Bateman, A. J. 1948. Intra-sexual selection in Drosophila. *Heredity* 2: 349-68.

Beecher, M. D., and Beecher, I. M. 1979. Sociobiology of bank swallows: Reproductive strategy of the male. *Science* 205: 1282-85.

Bell, R. H. V., 1971. A grazing ecosystem in the Serengeti. *Scientific American* 225(1): 86-93.

Bent, A. C. 1953. *Life Histories of North American Wood Warblers (Order Passeriformes).* Washington D.C.: U.S. Government Printing Office (U.S. National Museum, Bulletin 203).

Bernstein, I. S. 1976. Dominance, aggression, and reproduction in primate societies. *Journal of Theoretical Biology* 60: 304-472.

Bertram, B. C. R. 1975. The social system of lions. *Scientific American* 223(5): 54-65.

———. 1978. Living in groups: Predators and prey. In J. R. Krebs and N. B. Davies (eds.), *Behavioural Ecology,* pp. 64-96. Oxford: Blackwell.

Blumer, L. S. 1979. Male parental care in the bony fishes. *Quarterly Review of Biology* 54: 149-61.

Boggs, C. L., and Gilbert, L. E. 1979. Male contribution to egg production in butterflies: Evidence for transfer of nutrients at mating. *Science* 206: 83-84.

Boucher, D. H. 1977. On wasting parental investment. *American Naturalist* 111: 786-88.

Bradbury, J. W. 1977. Social organization and communication. In W. A. Wimsatt (ed.), *Biology of Bats* vol. 3, pp. 1-72. New York: Academic Press.

Bradbury, J. W. and Emmons, L. H. 1974. Social organization of some Trinidad bats. I. Emballonuridae. *Zeitschrift für Tierpsychologie* 36: 137-83.

Bradbury, J. W., and Vehrencamp, S. L. 1977a. Social organization and foraging in emballonurid bats. III. Mating systems. *Behavioral Ecology and Sociobiology* 2: 1-17.

———. 1977b. Social organization and foraging in emballonurid bats. IV. Parental investment patterns. *Behavioral Ecology and Sociobiology* 2: 19-29.

Breder, C. M., and Rosen, D. E. 1966. *Modes of Reproduction in Fishes.* New York: Natural History Press.

Brown, J. L. 1964. The evolution of diversity in avian territorial systems. *Wilson Bulletin* 76: 160-69.

———. 1975. *The Evolution of Behavior.* New York: Norton.

Brunning, D. F. 1974. Social structure and reproductive behavior in the Greater Rhea. *The Living Bird* 13: 2512-94.

Budd, P. L. 1940. Development of eggs and early larvae of six California fishes. *California Division of Fish and Game, Fish Bulletin* 56: 1-53.

Burley, N. 1977. Parental investment, mate choice, and mate quality. *Proceedings of the National Academy of Science, U.S.A.* 74: 3476-79.

Buskirk, W. H.; Buskirk, R. E.; and Hamilton III, W. J. 1974. Troop-mobilizing behavior of adult male chacma baboons. *Folia Primatologica* 22: 9-18.

Busse, C., and Hamilton III, W. J. 1981. Infant carrying by male chacma baboons. *Science* 212: 1281-83.

Campbell, B. G. 1966. *Human Evolution: An Introduction to Man's Adaptations.* Chicago: Aldine.

Carey, M., and Nolan, Jr., V. 1975. Polygyny in indigo buntings: A hypothesis tested. *Science* 190: 1296-97.

Carpenter, C. R. 1940. A field study in Siam of the behavior and social relations of the gibbon. *Comparative Psychology Monograph* 16(5): 1-168.

Carpenter, F. .L., and MacMillen, R. E. 1976. Threshold model of feeding territoriality and test with a Hawaiian honeycreeper. *Science* 194: 639-42.

Charles-Dominique, P., and Hladik, C. M. 1971. Le Lepilemur du sud de Madagascar: Ecologie, alimentation et vie socilae. *Terre et la Vie* 1: 3-66.

Cheney, D. L., and Seyfarth, R. M. 1977. Behavior of adult and immature male baboons during inter-group encounters. *Nature* 289: 404-06.

Chivers, D. J. 1972. The siamang and the gibbon in the Malay peninsula. *Gibbon and Siamang* 3: 103-35.

Clutton-Brock, T. H., and Harvey, P. H. 1977. Primate ecology and social organization. *Journal of Zoology* 183: 1-39.

Clutton-Brock, T. H.; Harvey, P. H; and Rudder, B. 1977. Sexual dimorphism, socionomic sex ratio and body weight in primates. *Nature* 269: 797-800.

Cox, C. R., and LeBoeuf, B. J. 1977. Female incitation of male competition: A mechanism in sexual selection. *American Naturalist* 111: 317-35.

Crook, J. H. 1972. Sexual selection, dimorphism, and social organization in the primates. In B. Campbell (ed.), *Sexual Selection and the Descent of Man 1871-1971,* 231-81. Chicago: Aldine.

Cubbicciotti, D. D., and Mason, W. A. 1978. Comparative studies of social behavior in *Callicebus* and *Saimiri:* Heterosexual jealousy behavior. *Behavioral Ecology and Sociobiology* 3: 311-22.

Daly, M. 1978. The cost of mating. *American Naturalist* 112: 771-74.

———. 1979. Why don't male mammals lactate? *Journal of Theoretical Biology* 78: 325-45.

Darwin, C. 1871. *The Descent of Man and Selection in Relation to Sex.* London: John Murray.

Davies, N. B. 1978. Ecological questions about territorial behaviour. In J. R. Krebs, and N. B. Davies (eds.), *Behavioral Ecology,* pp. 317-50. Oxford: Blackwell.

Davis, W. H. 1970. Hibernation: ecology and physiological ecology. In W. A. Wimsatt (ed.), *The Biology of Bats,* vol. 1, pp. 266-300. New York: Academic Press.

Dawkins, R., and Carlisle, T. R. 1976. Parental investment, mate desertion and a fallacy *Nature* 262: 131-32.

DeVore, I. 1965. Male dominance and mating behavior in baboons. In F. A. Beach (ed.), *Sex and Behavior,* pp. 261-89. New York: Wiley.
DeVore, I., and Washburn, S. L. 1963. Baboon ecology and human evolution. In F. C. Howell (ed.), *African Ecology and Human Evolution,* pp. 335-67. New York: Aldine.
Dunbar, R. I. M., and Dunbar, E. P. 1977. Dominance and reproductive success among female gelada baboons. *Nature* 266: 351-52.
Eisenberg, J. F. 1968. Behavior patterns. In J. A. King (ed.), *Biology of Peromyscus (Rodentia).* American Society of Mammalogists Special Publication 2: 451-95.
Eisenberg, J. G., Muckenhirn, N. A. and Rudran, R. 1972. The relation between ecology and social structure in primates. *Science* 176: 863-74.
Ellefson, J. O. 1967. A natural history of gibbons in the malay peninsula. Ph.D. dissertation, University of California, Berkeley.
Emlen, S. T. 1978. The evolution of cooperative breeding in birds. In J. R. Krebs and N. B. Davies (eds.), *Behavioural Ecology, An Evolutionary Approach,* pp. 245-81. Oxford: Blackwell.
Emlen, S. T., and Oring, L. W. 1977. Ecology, sexual selection, and the evolution of mating systems. *Science* 197: 215-23.
Erickson, C. J., and Zenone, P. G. 1976. Courtship differences in male ring doves: Avoidance of cuckoldry? *Science* 192: 1353-54.
———. 1978. Aggressive courtship as a means of avoiding cuckoldry. *Animal Behaviour* 26: 307-08.
Fisher, R. A. 1958. *The Genetical Theory of Natural Selection,* 2nd ed. New York: Dover.
Foltz, D. W. 1981. Genetic evidence for long-term monogamy in a small rodent *Peromyscus polionotus. American Naturalist* 117: 665-75.
Fossey, D. 1979. Development of the mountain gorilla *(Gorilla gorilla beringei):* The first thirty-six months. In D. A. Hamburg and E. R. McCown (eds.), *The Great Apes. Perspectives on Human Evolution* vol. V, pp. 139-84. Menlo Park, CA: Benjamin/Cummings.
Gadgil, M., and Bossert, W. H. 1970. Life historical consequences of natural selection. *American Naturalist* 104: 1-24.
Gaulin, S. J. C. 1979. A Jarman/Bell model of primate feeding niches. *Human Ecology* 7:1-20.
Gaulin, S. J. C., and Crandlemire, J. In preparation. Sexual Selection, Home Range and Spatial Ability.
Gaulin, S. J. C., and Konner, M. J. 1977. On the natural diet of primates, including humans. In R. Wurtman, and J. Wurtman (eds.), *Nutrition and the Brain,* vol. I, pp. 1-86. New York: Raven Press.
Gaulin, S. J. C., and Kurland, J. A. In preparation. Flight, precociality, and the evolution of biparental care.
Gaulin, S. J. C., and Schlegel, A. 1980. Paternal confidence and paternal investment: A cross-cultural test of a sociobiological hypothesis. *Ethology and Sociobiology* 1: 301-09.
Geist, V. 1974. On the relationship of social evolution and ecology in ungulates. *American Zoologist* 14(1): 205-20.
Ghiselin, M. T. 1974. *The Economy of Nature and the Evolution of Sex.* Berkeley: The University of California Press.
Goodman, D. 1974. Natural selection and a cost ceiling on reproductive effort. *American Naturalist* 108: 247-68.
Grafen, A. 1980. Opportunity cost, benefit and degree of relatedness. *Animal Behaviour* 28: 966-67.
Graul, W. D.; Derrickson, S. R.; and Mock, D. W. 1977. The evolution of avian polyandry. *American Naturalist* 111: 812-16.
Griffin, D. R. 1970. Migration in Bats. In W. A. Wimsatt (ed.), *Biology of Bats,* vol. I, pp. 233-64. New York: Academic Press.

Halliday, T. R. 1978. Sexual selection and mate choice. In J. R. Krebs, and N. B. Davies (eds.), *Behavioural Ecology,* pp. 180-213.Oxford: Blackwell.

Hamilton, W. D. 1964. The genetical evolution of social behavior. *Journal of Theoretical Biology* 7: 1-51.

Hamilton III, W. J.; Buskirk, R. E.; and Buskirk; W. H. 1975. Chacma baboon tactics during intertroop encounters. *Journal of Mammalogy* 56: 857-70.

Hausfater, G. 1975. Dominance and reproduction in baboons *(Papio cynocephalus):* A quantitative analysis. *Contributions to Primatology* 7: 1-150.

Hemmingsen, A. M. 1960. Energy metabolism as related to body size and respiratory surfaces, and its evolution. *Reports of the Steno Memorial Hospital and the Nordisk Insulin-laboratorium,* Copenhagen, 9: 1-110.

Hill, W. C. O. 1953. *Primates: Comparative Anatomy and Morphology. I. Strepsirhini.* Edinburgh: Edinburgh University Press.

Hooker, T., and Hooker, B. I. 1969. Duetting. In R. A. Hinde (ed.), *Bird Vocalizations,* pp. 185-205. Cambridge: Cambridge University Press.

Horn, H. S. 1978. Optimal tactics of reproduction and life-history. In J. R. Krebs, and N. B. Davies (eds.), *Behavioural Ecology,* pp. 411-29, Oxford: Blackwell.

Hrdy, S. B. 1976. Care and exploitation of nonhuman primate infants by conspecifics other than the mother. *Advances in the Study of Behavior* 6: 101-58.

———. 1977. *The Langurs of Abu: Male and Female Strategies of Reproduction.* Cambridge: Harvard University Press.

———. 1979. Infanticide among animals: A review, classification, and examination of the implications for the reproductive strategies of females. *Ethology and Sociobiology* 1: 13-40.

Huxley, J. S. 1938. The present standing of the theory of sexual selection. In G. R. DeBeer (ed.), *Evolution: Essays on Aspects of Evolutionary Biology,* pp. 11-42. Oxford: Clarendon.

Jarman, P. J. 1968. The effect of the creation of Lake Kariba upon the terrestrial ecology of the middle Zambezi Valley, with particular references to the large mammals. Ph.D. dissertation, Manchester University.

———. 1974. The social organization of antelope in relation to their ecology. *Behaviour,* 58(3,4): 215-67.

Jenni, D. A. 1974. Evolution of polyandry in birds. *American Zoologist* 14:129-44.

Johns, J. E., and Pfeifer, E. W. 1963. Testosterone-induced incubation patches of phalarope birds. *Science* 140: 1225-26.

Jolly, A. 1972. *The Evolution of Primate Behavior.* New York: MacMillan.

Kleiber, M. A. 1961. *The Fire of Life: An Introduction to Animal Energetics.* New York: Wiley.

Kleiman, D. G. 1977. Monogamy in mammals. *Quarterly Review of Biology* 52: 39-69.

Kummer, H. 1968. *Social Organization of Hamadryas Baboons.* Chicago: The University of Chicago Press.

Kummer, H.; Götz, W.; and Angst, W. 1974. Triadic differentiation: An inhibitory process protecting pair bonds in baboons. *Behaviour* 49: 62-87.

Kurland, J. A. 1977. Kin selection in the Japanese monkey, *Contributions to Primatology* 12: 1-145.

———. 1979. Paternity, mother's brother, and humans sociality. In N. A. Chagnon, and W. Irons (eds.), *Evolutionary Biology and Human Social Behavior: An Anthropological Perspective,* pp. 145-80. North Scituate, MA: Duxbury.

———. 1980. Kin selection theory: A review and selective bibliography. *Ethology and Sociobiology* 1: 255-74.

Kurland, J. A., and Gaulin, S. J. C. 1978. Aspects of sexual selection I. Parental certainty and reproductive effort. University Park, circulated manuscript.

Lack, D. 1968. *Ecological Adaptations for Breeding in Birds.* London: Methuen.

Leen, N., and Novick, A. 1969. *The World of Bats.* New York: Holt, Rinehart and Winston.

Leutenegger, W. 1973. Maternal-fetal weight relationships in primates. *Folia Primatologica* 20: 280-93.
———. 1976. Allometry of neonatal size in eutherian mammals. *Nature* 263: 229–30.
———. 1978. Scaling of sexual dimorphism in body size and breeding system in primates. *Nature* 272: 610-11.
———. 1979. Evolution of litter size in primates. *American Naturalist* 114: 525-31.
———. 1980. Monogamy in callitrichids: A consequence of phyletic dwarfism? *International Journal of Primatology* 1: 95-98.
Lindburg, D. G. 1969. Rhesus monkeys: Mating season mobility of adult males. *Science* 166: 1176-78.
Low, B. S. 1978. Environmental uncertainty and the parental strategies of marsupials and placentals. *American Naturalist* 112: 197-213.
Loy, J. 1970. Perimenstrual sexual behavior among rhesus monkeys. *Folia Primatologica* 13: 286-97.
———. 1971. Estrous behavior of free-ranging rhesus monkeys *(Macaca mulatta). Primates* 12: 1-31.
Lyman, C.P. 1970. Thermoregulation and metabolism in bats. In W.A. Wimsatt (ed.), *The Biology of Bats,* vol. 1, pp. 301-30. New York: Academic Press.
Marshall, A. J. 1961. Reproduction. In A. J. Marshall (ed.), *Biology and Comparative Physiology of Birds,* pp. 180-220. New York: Academic Press.
Maynard Smith, J. 1976. Evolution and the theory of games. *American Scientist* 64:41-46.
———. 1977. Parental investment: A prospective analysis. *Animal Behaviour* 25: 1-9.
———. 1978a.The ecology of sex. In J. R. Krebs, and N. B. Davies (eds.), *Behavioural Ecology,* pp. 159-79. Oxford: Blackwell.
———. 1978b. *The Evolution of Sex.* Cambridge: Cambridge University Press.
Mayr, E. 1972. Sexual selection and natural selection. In B. G. Campbell (ed.), *Sexual Selection and the Descent of Man, 1871-1971,* pp. 87-104. Chicago: Aldine.
McCracken, G. F., and Bradbury, J. W. 1977. Paternity and genetic heterogeneity in the polygynous bat, *Phyllostomus hastatus. Science* 198: 303-06.
Milne, L. J., and Milne, M. 1976. The social behavior of burying beetles. *Scientific American* 235(2): 84-89.
Moodie, G. E. E. 1972. Morphology, life history, and ecology of an unusual stickleback *(Gasterosteus aculeatus)* in the Queen Charlotte Islands, Canada. *Canadian Journal of Zoology* 50: 721-32.
Morris, D. 1952. Homosexuality in the ten-spined stickleback. *Behaviour* 4: 233-61.
Mullins, D. E., and Keil, C. B. 1980. Paternal investment of urates in cockroaches. *Nature* 283: 567-69.
Munro, H. N. 1969. Evolution of protein metabolism in mammals. In H. N. Munro (ed.), *Mammalian Protein Metabolism* vol. 3, pp. 87-101, New York: Academic Press.
Nice, M. 1937. *Studies in the Life History of the Song Sparrow.* Reprinted in 1964. New York: Dover.
Nisbet, I. C. T. 1973. Courtship-feeding, egg-size and breeding success in common terns. *Nature* 241: 141 42.
O'Donald, P. 1962. The theory of sexual selection. *Heredity* 17: 541-52.
———. 1967. A general model of sexual and natural selection. *Heredity* 22: 499-518.
Odum, H. T. 1970. Summary: An emerging view of the ecological system at El Verde. In H. T. Odum (ed.), *A Tropical Rainforest,* pp. 191-289. Washington, D.C.: Office of Information Services.
Orians, G. H. 1969. On the evolution of mating systems in birds and mammals. *American Naturalist* 103: 589-603.
Orr, R. T. 1970. Development: Prenatal and postnatal. In W. A. .Wimsatt (ed.), *Biology of Bats* vol. I, pp. 217-31. New York: Academic Press.
Packer, C. 1977. Reciprocal altruism in *Papio anubis. Nature* 265: 441-43.

———. 1979. Male dominance and reproductive activity in *Papio anubis. Animal Behaviour* 27: 37-45.

———. 1980. Male care and exploitation of infants in *Papio anubis. Animal Behaviour* 28: 512-20.

Parker, G. A.; Baker, R. R.; and Smith, V. G. F. 1972. The origin and evolution of gamete dimorphism and the male-female phenomenon. *Journal of Theoretical Biology* 36: 529-33.

Partridge, L. 1980. Mate choice increases a component of offspring fitness in fruit flies. *Nature* 283: 290-91.

Perrone, M., and Zaret, T.M. 1979. Parental care patterns of fishes. *American Naturalist* 113: 351-61.

Petter, J. J, and Peyrieras, A. 1970. Nouvelle contribution a l'etude d'un lemurien malgache, le aye-aye (*Daubentonia madagascariensis* E. Geoffroy). *Mammalia* 34: 167-93.

Pollock, J. I. 1977. The ecology and sociology of feeding in *Indri indri.* In T. H. Clutton-Brock (ed.), *Primate Ecology,* pp. 37-69. New York: Academic Press.

———.1979. Female dominance in *Indri indri. Folia Primatologica* 31: 143-64.

Porter, K. R. 1972. *Herpetology.* Philadelphia: W. B. Saunders.

Pressley, P. H. 1981. Parental effort and the evolution of nest-guarding tactics in the threespine stickleback, *Gasterosteus aculeatus L. Evolution* 35: 282-95.

Ralls, K. 1976. Mammals in which females are larger than males. *Quarterly Review of Biology* 51: 245-76.

———. 1977. Sexual dimorphism in mammals: Avian models and unanswered questions. *American Naturalist* 111: 917-38.

Ransom, T. W., and Ransom, B. S. 1971. Adult male-infant relations among baboons *(Papio anubis). Folia Primatologica,* 16: 179-95.

Rensch, B. 1950. Die abhängigkeit der relativen sexualdifferenz von der körpergrosse. *Bonner Zoologische Beiträge* 1: 58-69.

Rhine, R. J., Hendy, H. M., Stillwell-Barnes, R., Westlund, B. J., and Westlune, H. D. 1980. Movement patterns of yellow baboons *(Papio cynocephalus):* Central positioning of walking infants. *American Journal of Physical Anthropology* 53: 159-67.

Rhine, R. J., and Owens, N. W. 1972. The order of movement of adult male and black infant baboons *(Papio anubis)* entering and leaving a potentially dangerous clearing. *Folia Primatologica* 18: 276-83.

Richdale, L. E. 1957. *A Population Study of Penguins.* Oxford: Clarendon.

Ricklefs, R. E. 1969a. An analysis of nesting mortality in birds. *Smithsonian Contributions to Zoology* 9: 1-48.

———. 1969b. Natural selection and the development of mortality rates in young birds. *Nature* 233: 922-25.

Ridley, M. 1978. Paternal care. *Animal Behaviour* 26: 904-32.

Schoener, T. W. 1971. Theory of feeding strategies. *Annual Review of Ecology and Systematics* 2: 369-404.

Selander, R. K. 1972. Sexual selection and dimorphism in birds. In B. Campbell (ed.), *Sexual Selection and the Descent of Man, 1871-1971,* pp. 180-230. Chicago: Aldine.

Smith, R. L. 1979. Repeated copulation and sperm precedence: Paternity assurance for a male brooding water bug. *Science* 205: 1029-31.

Stearns, S. C. 1976. Life-history tactics: A review of the ideas. *Quarterly Review of Biology* 51: 3-47.

———. 1977. The evolution of life-history tactics. *Annual Review of Ecological Systems* 8: 145-72.

Tembrock, G. 1976. Sound production of *Hylobates* and *Symphalangus. Gibbon and Siamang* 3: 176-205.

Thornhill, R. 1976a. Sexual selection and nuptial feeding behavior in *Bittacus apicalis (Insecta: Mecoptera). American Naturalist* 110: 529-48.

———. 1976b. Sexual selection and paternal investment in insects. *American Naturalist* 110: 153-63.
———. 1979. Adaptive female-mimicking behavior in a scorpionfly. *Science* 205: 412-14.
Thorpe, W. H. 1973. Duet-singing birds. *Scientific American* 229(3): 70-79.
Trivers, R. L. 1972. Parental investment and sexual selection. In B. G. Campbell (ed.), *Sexual Selection and the Descent of Man, 1871-1971,* pp. 136-79. Chicago: Aldine.
———. 1974. Parent-offspring conflict. *American Zoologist* 14: 249-64.
———. 1976. Sexual selection and resource-accruing abilities in *Anolis garmani. Evolution* 30: 253-69.
Tutin, C. E. G. 1979. Mating patterns and reproductive strategies in a community of wild chimpanzees *(Pan troglodytes schweinfurthii). Behavioral Ecology and Sociobiology* 6: 29-38.
Vaughan, T. A. 1972. *Mammalogy.* Philadelphia: W. B. Saunders.
Wade, M. J. 1979, Sexual selection and variance in reproductive success. *American Naturalist* 114: 742-46.
Walker, E. P. 1964. *Mammals of the World.* Baltimore: Johns Hopkins University Press.
Wallace, G. D., and Mahan, H. D. 1975. *An Introduction to Ornithology,* 3rd ed. New York: MacMillan.
Warren, D. C., and Kilpatrick, L. 1929. Fertilization in the domestic fowl. *Poultry Science* 8: 237-56.
Wells, K. D. 1977. The social behaviour of anuran amphibians. *Animal Behaviour* 25: 666-93.
Welty, J. C. 1975. *The Life of Biurds,* 2nd ed. Philadelphia: W. B. Saunders.
Werren, J. H.; Gross, M. R.; and Shine, R. 1980. Paternity and the evolution of male parental care. *Journal of Theoretical Biology* 28: 619-31.
Wiley, R. H. 1974. Evolution of social organization and life-history patterns among grouse. *Quarterly Review of Biology* 49: 201-27.
Williams, G. C. 1966a. *Adaptation and Natural Selection.* Princeton: Princeton University Press.
———. 1966b. Natural selection, the cost of reproduction, and a refinement of Lack's principle. *American Naturalist* 100: 687-90.
———. 1975. *Sex and Evolution.* Princeton: Princeton University Press.
Wilson, E. O. 1971. *The Insect Societies.* Cambridge: The Belknap Press of Harvard University Press.
———. 1975. *Sociobiology: The New Synthesis.* Cambridge: The Belknap Press of Harvard University Press.
Wing, L. W. 1956. *Natural History of Birds.* New York: Ronald Press.
Wittenberger, J. F. 1979. A model for delayed reproduction in iteroparous animals. *American Naturalist* 114: 439-46.
Wittenberger, J. F., and Tilson, R. L. 1980. The evolution of monogamy: Hypotheses and evidence. *Annual Review of Ecology and Systematics* 11: 197-232.
Wootton, R. J. 1976. *The Biology of the Sticklebacks.* New York: Academic Press.
Zenone, P. G.; Sims, E. M.; and Erickson, C. J. 1979. Male ring dove behavior and the defense of genetic paternity. *American Naturalist* 114: 615-26.

12

Significance of Paternal Investment by Primates to the Evolution of Adult Male-Female Associations

William J. Hamilton III
University of California
Davis, California

INTRODUCTION

The purpose of this chapter is to examine some scenarios for the evolution of mating patterns and male-female associations during human evolution. The term *bond* is used here to identify persistent male-female associations that include some kind of support by one or both individuals for the other. Degrees of strength of male-female association under conditions of captivity may be measured by monitoring maintenance of proximity, grooming, and the performance of various other kinds of associated behavior. In the field, and for purposes of this analysis, I use no such measures; instead I attempt to evaluate qualitatively and, where possible, quantitatively the relative costs and benefits to males and females of maintaining proximity, of mating, and of supporting one another and progeny. My focus is upon specific biological characteristics of primates, savanna baboons (*Papio* spp.) in particular. It has been suggested that the polygynous mating system and multi-male/female social organization of savanna baboons approximates hypothetical early hominid archetypes that, with certain changes, could converge upon the contemporary human sociotype. Savanna baboons have not been chosen as the focus here however, on this basis; rather this analysis is meant to identify problems with which other current evaluations of hominid social evolution must contend.

Analyses dealing with specific adaptations based upon knowledge of living primates and interpretations relating these observations to human characteristics often suffer from failure to consider how specific evolutionary steps could have taken place and, in particular, how constraints establishing contemporary equilibria for hypothetical archetypal social analogues could have been bypassed. For example, recent explanations of continuous sexual receptivity and lack of estrous swellings in humans consider the characteristics of estrous cycles, parental, and especially paternal, care of infants, and male-female bonds in nonhuman primates (e.g., Alexander and Noonan, 1979, Strassmann, 1981). These scenarios assume specific changes from a generalized hypothetical ancestor with specific estrous swelling characteristics, assumptions that must at best be considered speculative.

A general synthesis relating primate biology to the evolution of human social organization and reproductive patterns is not yet possible given available evidence and concepts. Many parts of the picture are missing, justifying both speculation and generalized data collection. The problems in analysis, in increasing order of difficulty, are as follows: (1) to identify the general characteristics of contemporary human male-female associations; (2) to identify a range of possible characteristics of archetypal societies, based at least in part upon the comparative analysis of contemporary primate societies; and (3) to suggest and evaluate conditions that led from ancestor to human.

PRIMATE CAREGIVING

Paternal care is the focus of this analysis. Trivers (1972) emphasized that sexual rivalry is greatest for the sex making the greatest investment in progeny, which is the female for all nonhuman primates, with the possible exception of some of the monogamous species. Evolutionary transitions from primate to human or from known primate sociotypes to a prehuman ancestor possessing some humanlike paternal care characteristics may thus have involved changes in patterns of male caregiving, resulting in prolonged association of individual males and females. Or, archetypes may already have been socially organized in a manner similar to contemporary humans (Lovejoy, 1981).

This analysis of caregiving considers the value of this behavior to progeny and its costs to both parents, including the reciprocal relationship of caregiving by one parent to the costs and benefits of caregiving by the other. Parents of most primate species, especially savanna baboons, are seldom closely related to one another. Males may allocate care primarily to their probable progeny, or in the case of adolescent males, to their relatives, and they do not aid females beyond the extent to which the investment also benefits their own infants. Exceptions include those situations where male support of females may enhance the probability that the male will gain mating access to females. The extent of male support of females may depend upon the extent to which such caregiving enhances a mother's, or prospective mother's, care of that male's offspring. Evaluation of such care can, in an evolutionary analysis, be substituted for an evaluation of male-female bonds, the strength of which are measured here as supportive behavior allocated to one's mate or offspring rather than in terms of attachment.Thus, certain forms of support may be directed to the other parent, while other care is allocated to progeny or, as in the case of predation defense, may be more diffusely oriented to group members if it results in enhanced inclusive fitness of self, mate, or progeny.

Because this analysis is limited to savanna baboons, the social and reproductive biology of these primates is reviewed briefly here. Female savanna baboons reside continuously in their troops of birth and begin to reproduce at approximately six years of age (S.A. Altmann et al., 1977); they then give birth approximately every two years. In contrast, males attain high social rank and sire most of their offspring during a relatively short interval. Males leave their natal troops at approximately 8 to 10 years of age (J. Altmann et al., 1977; Packer, 1979a, b), when they are nearing adult size and their canines reach maximum length and sharpness (Popp, 1978). Evidence from our study area suggests that these young males have a good chance of becoming alpha males in the troops to which they transfer (Busse, 1981). Eight of nine males observed when they first attained alpha rank were relatively young adult males who had recently immigrated. The exception was a male who attained alpha rank in what was probably his natal troop. Like the other alpha males, this individual was a young adult. These nine males, plus two males who held alpha rank when the study began, held alpha rank for a median interval of five months. After losing alpha rank, some males dropped rapidly in rank (Busse and Hamilton, 1981). One male regained alpha rank after a tenure of four months as the second-ranking male. These results, combined with evidence that alpha males sire a majority of infants (K. S. Smith, in prep.), indicate that males in these troops sire most of their offspring during a short interval of their lifetimes.

PATTERNS OF CAREGIVING

How do male primates allocate care to progeny? What is the benefit of such caregiving to mothers and their infants? What are the costs and benefits of caregiving to males? Here I limit potential answers to these questions to savanna baboons, focusing in particular on one population of chacma baboons, *Papio ursinus*.

The fitness value to a male of giving care to other individuals in the group at any point in time is the summed value of caregiving to recipients, times the probability of paternity or relatedness to those individuals, divided by respective value of that caregiving to these individuals at that time, further reduced by the extent to which caregiving is not shared, i.e., is depreciable (S. A. Altmann et al., 1977; see also Maynard Smith, 1977, and Ridley, 1978). Because male caregiving may be redundant with maternal caregiving, especially when mothers are present, the benefit to fathers of substituting their care for that of mothers will accrue in proportion to the extent that by doing so, the future reproductive effort of those females and the care to their infants is thereby enhanced. This relationship was recognized in its general form by Trivers (1972), who called male ability to contribute to offspring, relative to that of a female, complementarity.

Infant Transport

All primate mothers transport their infants in the immediate postpartum interval. But the extent to which others, including the infant, share this cost varies dramatically between species. For chacma baboons, infant transport by nonmothers is expressed especially in the case of preadult males and females who adopt orphaned infants (Hamilton et al., 1982). Six of 11 observed adoptions were by four- to five-year-old males. There was little or no opportunity for them to adopt more than one infant at a time. The limited evidence (Hamilton et al., 1982) suggests that males are, on the average, adopting their mothers' infants by a subsequent father ($R = \frac{1}{4}$). Evaluation of relationships in this case is simplified because the mother is gone and caregiving is not depreciable.

The advantage to the infant is energy saved, since foster parents do not provide food to infants but travel with them and may carry or walk with them to foraging places. Being transported reduces an infant's energy consumption and, while the benefit cannot be determined in terms of inclusive fitness, it can be roughly measured as a percent of daily energy expenditure.

For the closely related and morphologically similar hamadryas baboons, *P. hamadryas,* the actual energy savings to infants by being transported can be calculated since the energetic cost of walking at 1.3 km. per hour is about 2.5 times greater than resting (Taylor et al., 1982). Thus, for a 5.2 km. per day route taken during a 12-hour day, a reasonable average for chacma baboon troops studied by us, it will cost an infant 50% more energy to walk than to be transported during a 12-hour daytime of activity. If an infant is transported during half the daily movement, a figure approximating what we have observed for foster parents, its savings thus will be 25% of its daytime energy output.

For carrying mammals, increases in energy costs of transporting a mass are directly proportional to the relative mass of the transported object and the lean body mass of the carrier (Taylor et al., 1980). The energy cost of transporting an infant will thus increase with infant mass and decrease as males increase in size. Transporting an adopted infant weighing 1.5 kg. when six months old will increase the energy cost of locomotion to a 15 kg. male by 10% for the interval of carrying while eliminating the added cost of locomotion to the infant. Energy cost to such a male carrier will involve an increased metabolic cost of 5.6% if an infant is carried over all a day route and 2.8% for half a day route of his energy expenditure during a 12-hour day.

While these values are estimates, they are not so far removed from reality as to drastically misrepresent relative costs and benefits to infant and transporter. We can see from these figures that in terms of net energy budgets, the carrying individual saves the infant a considerable cost at a much lower personal cost. Elsewhere (Hamilton et. al., 1982) we conclude that costs other than energy to four- to five-year-old foster parents are low because they have limited alternative opportunities to express fitness-enhancing behavior, especially mating opportunities. The asymmetry of costs and benefits to recipient and donor measured in terms of energy expenditures is increased by the probable costs to the infant of not being carried. Since the costs of travel to a smaller individual are somewhat greater than to a larger one (Taylor et al., 1980), and since the metabolic rate of the infant is considerably higher, the asymmetry of costs and benefits to transporter and transported will actually be greater than calculated above. In the particular case of adoption described here, the asymmetry of costs to foster parent and orphan is acute because adopted infants were nursing when their mothers disappeared.

The extent to which energy is limiting to baboons is not known but can be assumed to be significant, especially for prematurely weaned orphans. These calculations suggest that infant carrying may be an important feature of primate paternal care. For larger paternal males, whose lean body mass is about 30 kg., the value of carrying an infant will be greater than that for a four- to five-year-old male in terms of relatedness ($R = \frac{1}{2}$ rather than $\frac{1}{4}$) and energetics. While no absolute value can be adopted now in quantitative terms because relatedness and energetics are not equivalent fitness units, it does suggest that paternal males would carry infants unless other conditions prevailed. Yet paternal chacma baboons seldom transport infants, and when they do it is usually for short distances and not more than a few minutes per day. Males are often fathers of several infants at the same time, and carrying for them becomes depreciable. Nevertheless, division of carrying care, unless it reduces the value of such effort to individual infants, should in sum still be equally valuable to infants and thus to fathers. This would not be the case for orphans.

This complementarity of infant carrying by males will be relatively large because, unlike conditions for orphans, the mother is available to transport infants, and the value of being transported by male or female is greater for orphans because maternal milk is not being provided. It is thus worthwhile for males to reduce maternal infant transport costs only to the extent that such males may be fathers of additional offspring by the same females or that such relief may enhance the mothers' subsequent ability to care for an infant. Thus the value of male investment increases with complementarity of paternal care.

Close proximity of foster parents to adopted infants and their tendency to carry them may also reduce infant vulnerability to predation and convey thermal advantages. Adopted infants are harassed by other troop members and may be attacked less frequently than unprotected orphans. Also, protected infants may gain a thermal advantage through being held, huddling, or being carried across water. One orphan, LnrJ, was not carried at water crossings and, for several evenings before his disappearance, shivered and squealed for long intervals before dusk and after swimming alone at water crossings (Busse, pers. comm.).

Countercarrying

Temporary carrying of infant baboons, usually those less than one year old, in confrontations with other usually higher-ranking males has been discussed by Ransom and Ransom (1971), Popp (1978), Packer (1980), Busse and Hamilton (1981), Busse (chapter 8), Strum (chapter 7), and many others. Countercarrying should be strongly distinguished from infant transport during locomotion. Countercarrying tends to be brief, does not result in extensive progressive movement along the day route, and is usually oriented to other higher-ranking males.

For chacma baboons, Busse and Hamilton (1981) conclude that one function of this behavior is protection of infants from the infanticidal tendencies of immigrant males. Evaluation of the actual benefits to infants would require elimination or removal of carriers: probable fathers ($R = ½$). While such destructive experiments are not likely to be done, one accidental removal of infant protection resulted in immediate death from infanticide by a high-ranking immigrant male (Busse and Hamilton 1981). If infanticide is an adaptive strategy, as suggested by Hrdy (1979), its probability should be near one for unprotected infants, and the male contribution should be the difference between one and the extent that females or other relatives can prevent it.

The potential costs of this countercarrying behavior to males include injuries sustained during encounters with higher-ranking males. These costs seem to be low based upon observed rates of wounding to countercarrying males and their infants. Thus infant carrying is another low-cost, probably high-benefit male chacma baboon caregiving behavior. In general, protection of young infants is a high-value form of caregiving by males, especially in dimorphic species where male intergroup transfer is the rule (see Hausfater 1975; Packer 1979a; Busse and Hamilton 1981; Busse, chapter 8) and where the probability that the countercarrying male is a close relative is high.

While the value of infanticide to nonpaternal males may decline with time (Hrdy 1979; Hausfater et al., 1981), the duration of its effectiveness will increase with an increase in the duration of nonsharable paternal investment in offspring, and it cannot be treated solely in the context of advances in the schedule of fertile female reproductive cycles. A more general statement is possible.

The value of infanticide is a function of the probability that a male is or will be the father of a female's next infant times the extent to which mothers give synchronous care to infants not sired by him, depreciating the value of care given to the subsequent (infanticidal) male's offspring by her. For baboons this additional value of maternal care to older infants may be low.

Because males are the infanticidal sex among baboons and are nearly twice as large as females and invariably outrank them, female defense against infanticide is not as effective as male defense. Thus the complementarity of male infanticide protection behavior is high, and in species possessing the transfer and dimorphism characteristics of baboons, evolution of male caregiving in the form of infant protection from infanticide is a likely behavior.

The significance of male countercarrying as infanticide protection has been stressed here and elsewhere (Busse and Hamilton, 1981). It is clear from our reported observation and those of others that this is not a complete explanation of the behavior, which may vary within and between populations and species. The previous discussion emphasizes the analysis of the potential costs and benefits related to that part of the behavior associated with infant protection.

Protection of Infants Against Social Interference

Intense interest in infants by females other than the mother has been viewed as a relatively harmless expression of maternalism by other potential mothers (e.g., Lancaster, 1971). Another explanation is also possible. Irrespective of motivation, attention by other baboon females may cut sharply into the time budget of new mothers when energy demands upon them are relatively great. High infant mortality during the postparturition interval (J. Altmann, 1980; Busse, 1981) is correlated with a strongly reduced time budget for foraging because of the need to carry the infant against the breast with one hand, as a result of the "overattentiveness" of other females, and because of the energy demands of late pregnancy and lactation.

Male associates of new mothers threaten and displace other females, preventing some of their interference with the mother and infant (pers. obs.). Because adult males outrank adult females, and because this status is seldom if ever seriously challenged, prevention of social interference is a potentially low-cost, high-benefit paternal behavior.

Food Sharing

Baboons seldom share specific food items with other individuals, including infants. When concentrations of fruits are available, dominant males often occupy optimum foraging spaces.

Juveniles and infants approach feeding males processing foods such as the massive fruits of *Kigelia pinnata.* Juveniles are unable to open or tear apart these fruits but remain nearby to sort through the chips of parts discarded by adult males and females, obtaining overlooked or ignored seed fragments. Rather than being active sharing, this aegis seems to be an incidental consequence of the greater tolerance for approach by much-lower-ranking infant and especially juvenile individuals, possibly relatives, and of the negligible value of such food to males. Regardless of intent, these are low-cost caregiving events representing an unmeasured advantage to infants.

One category of food that is sometimes directly shared by male baboons is meat (discussed by Strum, 1975, 1981; Harding and Strum, 1976; Hausfater, 1975). We have not observed meat sharing with infants and juveniles by chacma baboon males. Instead, for our study population, the general rule is that the captor eats the prey it captures or the prey is taken from it by a higher-ranking individual (Hamilton and Busse, 1982). However, Strum (1981) has seen juveniles and females feeding at impala fawns killed by adult males. The relatedness of males to juveniles they share with has not been reported. Prey capture is an uncommon event for baboons and meat sharing is even less frequent. Hence, any costs and benefits to males and juveniles they share meat with cannot in the aggregate represent a particularly important form of male caregiving. Nevertheless, even the infrequent observation of the sharing of meat, a resource unique for primates in terms of its high concentration of usable energy, is of particular interest because of the importance of meat in human diets.

Van den Berghe's (1980) suggestion that food sharing and family formation are inseparable events may be incorrect. Emergence of an ability by prehumans to kill large prey would have made possible the allocation of food to associated infants if prehumans lived in small groups, and the potential to kill large prey could have enhanced the ability of males to allocate major caregiving to infants and older progeny at low cost to themselves. I emphasize large prey and small groups because, for large primate groups (e.g., baboons), expression of hierarchical behavior limits access of most subordinate individuals to prey killed by other group members (Hamilton and Busse, 1978). For meat to play a prominent role in paternal care, prey items would have to be large, or groups small, or both.

Status Allocation

Juvenile females take on the status of their mothers from an early age (J. Altmann, 1980). Young males gain status (rank) as they mature. In conflicts with other individuals they are often supported by other males. While this may represent an important form of paternal and alloparental care, little data are available concerning the relatedness of individuals participating in such coalitions. But the possible importance of such relationships should not be underestimated, especially because they emerge at a time when the potential for expression of other forms of paternal care is declining. However, if such support is allocated to independent offspring, it may have little or no bearing upon male-female associations. Unless the infant receiving support by a male is not also being supported by the mother, the benefits derived by that infant are independent of all females who are not mothers of any half-siblings of that infant (i.e., half-siblings who have been sired by the supporting male).

Grooming

Males, especially fathers, may intensively groom infants, sometimes for extended intervals, especially following countercarrying. But this behavior probably has little functional value to infants in terms of ectoparasite removal because these same infants are persistently groomed by their mothers and other females. That is, complementarity is low and the extent of male grooming of infants is by comparison trivial. Male grooming probably serves primarily to reinforce male-infant associations, enhancing the effectiveness especially of countercarrying and its benefits.

Predation Protection

Predation protection for infant and juvenile baboons is an almost completely sharable action in the case of alarm signals and active defense against predators. Numerous explanations for protective behavior patterns are available but are beyond the scope of this paper. They are certainly relevant to further evaluation of paternal care by baboons.

Subadult males sometimes carry orphans during predator alarms; these males may even run back into a stream of fleeing baboons to pick up and carry their orphans (pers. obs.). Adult males, including fathers, have not been observed exhibiting this behavior. In November 1981, I observed an adult male, MD, attending a nine-month-old infant who had received a serious injury from an unknown source the previous day. MD trailed the group and carried the infant upon my close approach. The mother did not closely associate with the injured infant, who died later in the day. MD was the probable father of this infant, having consorted with the mother at the time of her estrous cycle, which probably led to conception. Thus, while limited, this observation suggests that probable fathers may occasionally extend to infants protection other than countercarrying beyond the interval of close infant association with their mothers.

INDIRECT SUPPORT TO PROGENY

In addition to the forms of caregiving detailed above, all male behaviors that support the troop as a whole may also benefit progeny and may, in fact, be an adaptive basis of such behavior patterns. Such behavior includes, but is not necessarily limited to, (1) predation defense, (2) participation in intergroup encounters, and (3) coordination of group movements. Evaluation of these behavior patterns vis-a-vis male investment in relatives is, again, beyond the scope of this paper. But such behaviors may have a significant aggregate value to progeny and may influence the characteristics of these male behavior patterns.

In the multi-male/multi-female baboon society studied by us, the emerging picture of male caregiving is as follows:

1. There is a tendency for males to care for infants they probably sired.
2. Male caregiving may also directly benefit mothers, but the orientation of the behavior is in most cases directed more toward the infant than the mother.
3. Fathers do not feed infants but may enhance their foraging efficiency by providing a place to feed that is relatively free from social disturbance.
4. Male caregiving shows a high degree of complementarity to female caregiving, and when complementarity is low for a particular behavior maternal caregiving is the rule.
5. The multiple mateships of polygynous primates preclude effective caregiving of depreciable paternal behavior (S. A. Altmann et al., 1977).

The mating characteristics of savanna baboons, emphasizing semisynchronous mating by individual males and sequential progeny production by several males, establish limits to and opportunities for certain patterns of paternal caregiving. Predation protection oriented to the group as a whole may not be adaptive if the risks taken by males do not differentially benefit those males relative to males not taking such risks. However, since benefits to males that result from group-supportive behavior may be considerably greater during their individual reproductive pulses, it may be advantageous for males to express troop-supportive behavior at those times.

Data derived from the study of chacma baboons (Table 12-1) support the generalizable prediction for savanna baboons that high complementarity of male caregiving and low (or asymmetrical) cost-benefit ratios to progeny and parent favoring both are conditions that promote allocation of specific, direct caregiving to infants. Most observed paternal behavior has low depreciability, in part because of the asynchronous birth schedule in savanna baboon groups. Benefits an infant can obtain from a male probably vary greatly during ontogeny, and possibly other factors could contribute to the relatively asynchronous birth season of this species.

WHAT IS TO BE DERIVED?

What were the characteristics of early hominid social organization and reproductive behavior? What are the characteristics of contemporary human social organization and reproductive behavior? What changes resulted in the evolutionary transformation from archetype to human?

Each of these questions has a series of alternative answers, some more possible than others, none particularly certain. A preliminary identification of the end product, human social structure, is first attempted, then some beginning points are noted and, finally, alternative hypotheses concerning transitions are considered.

Human Social Organization

Contemporary human social organization, regardless of whether it is identified as monogamous (Lovejoy, 1981) or polygynous (Alexander et al., 1979), is characterized by strong and enduring male-female associations and enduring care of offspring by both parents. The extended nuclear family is, on a comparative basis, a demonstrably general characteristic of diverse recent human populations (Murdock, 1967; Alexander, 1979). Hence, unless a monogamous prehuman ancestry is assumed, analysis of the transition from hypothetical prehuman precursors to the contemporary human social condition must consider how male-female associations were strengthened and how extended paternal care developed during hominid or prehominid evolution. Derivation of the extended family from this beginning is, at least conceptually, a less difficult problem.

Table 12-1. Kinds of Caregiving to Infants by Male Savanna Baboons, *P. ursinus*.

Type of Care	Caregivers	Cost to Caregiver	Depreciability	Complementarity
Direct Caregiving				
Orphan infant carrying, protection	Foster fathers, 4-5 years, rarely adults	Energy burden	High	High for adult males, subadults
Other infant carrying	Mothers	Energy burden	Low	Low
Countercarrying (infanticide protection, other)	Fathers, other related males, other males	Risk of intraspecific conflict	Moderately high, low if mothers associate closely with fathers	High for adult males
Social interference, protection from nonmaternal females	Fathers	Low	Moderate to low, as above	High for adult males
Food sharing	Males	Limited to extent food is shared by unsatiated males	Low	High for adult males
Status allocation	Males	Low	Low	High for adult and juvenile males
Grooming	Fathers, foster fathers, juvenile males	Low	Low	Low
Indirect Caregiving				
Aegis at food	Adult males	Low	Low	High
Predation alarm	Adult males, juvenile males	Low?	Low	Intermediate

The character of male-female and other social associations has received much attention (summarized by Reynolds, 1976), but the possible nature of specific transitions from one social organization to another has not. Alexander and Noonan (1979) and Strassmann (1981) offer one scenario for this transition, emphasizing selection for reduced estrous swellings and signs of ovulation, development of extended individual male association with particular females, and extended paternal care.

Lovejoy (1981) concludes that Miocene hominids were monogamous, based primarily on the small degree of sexual dimorphism in canine size of fossil hominids compared with contemporary polygynous primates. Other authors emphasize the multi-male/multi-female group as a probable human social archetype (Lee and DeVore, 1968; Brace, 1979; Campbell, 1979), in part because most savanna-dwelling primates live in multi-male, multi-female groups, in agreement with current thought concerning the probable habitat of hominid evolution. Alexander et al., (1979) extend the argument, concluding that most contemporary humans and their precursors were polygynous. Their evidence is that polygyny is widespread among contemporary human populations and that the degree of sexual dimorphism in height for contemporary humans, ranging from 5 to 12%, agrees with the degree of body size dimorphism in other mildly polygynous mammals. Here I consider both alternatives.

Polygyny is widespread among vertebrates, especially birds (Lack, 1968), where the term territorial polygyny (Wittenberger, 1981a) is often used. Since primates considered here, including humans, are not necessarily territorial, but do possess some of the mateship characteristics of territorial polygynous birds (i.e., one male may synchronously attract a second and occasionally more mates), the term facultative or threshold polygyny is used to incorporate primates and other mammals into the comparison. Facultatively polygynous, as opposed to harem polygynous, birds and mammals are not highly sexually dimorphic. Most polygyny among contemporary human societies is facultative, i.e., a relatively small proportion of males have more than one wife (Murdock, 1967). The degree of sexual dimorphism characteristic of the fossil hominid record (Leutenneger, 1977) and contemporary human societies is thus also compatible with an interpretation of monogamy and facultative polygyny, including serial monogamy.

Archetypal Human Societies

Four types of organization exhibited by contemporary primates are considered here as human social archetypes:

Strict monogamy: the social group and social unit are the same and consist of one male, one persistently associated and sexually mature female, and their dependent offspring, if any. This social situation characterizes many contemporary human populations and most individuals of most human populations.

Facultative polygyny: groups consist of one male associated with one or several females; i.e., for some but not all individuals, the polygyny threshold is crossed (Orians, 1969). Populations of such species vary in their ratios of polygynous to monogamous groups.

Harem polygyny (single male, multi-female): one male is associated with several sexually mature females at one time.

Multi-male/multi-female polygyny: most contemporary human populations include several adult males and several adult females, while the degree of polygyny varies. The term specifically used here is applied to non-human primate groups.

Some species may exhibit more than one of these patterns of social organization within the same or different populations.

Monogamous Primates. Numerous monogamous primate species, New World callitrichids, cebids, Old World gibbons and siamangs, and two Mentawai Island langurs (Wittenberger and Tilson, 1980), as well as lemurs and tarsiers, would be suitable human archetypes if several of the monogamous social units of these species could be assembled spatially into social groups. But since none of the extant monogamous species ever are, neither their social units nor their social organization seem to offer a suitable human social archetype. For gibbons and siamangs, social organization is essentially the same for all nine living species; one adult male, one adult female, and their dependent, sexually immature offspring live together in a territory defended by both adults but more extensively by males. Females also defend the territory, especially against other females. Although several populations of these species have been intensively studied, the only deviation from the one-male/one-female monogamous group was observed by Srikosamatara (in press)—three adult groups, including at least one interspecific hybrid or a member of a second gibbon species. The one significant variation in paternal behavior within this species group is that male siamangs but not gibbons persistently carry one- to two-year-old infants during daily movements (Chivers, 1974, 1977). This suggests a possibly higher degree of complementarity for siamangs than for gibbons, possibly greater longevity for siamangs, or other factors enhancing the value of caregiving by males.

The strictness of monogamy as expressed by monogamous primates and their limited development of extended familial groups suggest that they are not useful archetypes for models deriving societies accomodating facultative polygyny. To the extent that such facultative polygyny is a necessary step, as

I assume here and as is assumed by Alexander and Noonan (1979), these species do not seem to be relevant hypothetical sociotypes for derivation of human familial social groups. Thus, regardless of what the social character of the evolutionary precursors of strictly monogamous primates may have been, evolution of monogamy and its correlation with monomorphism could lead to prolonged evolutionary stasis. Further speciation, if any, would result in the production of allopatric species that are organizational replicates of one another with significant changes in social behavior. I predict, therefore, that if the phyletic history of contemporary monogamous primates other than humans can be determined, it will reveal persistent sexual monomorphism, probably for extended evolutionary periods.

Possible limits to social evolution, suggested here for monogamous primates, may also characterize other primate species with different social units and social organizations. For the analysis of human social organization, a comparison of monogamous primate societies emphasizes the differences between strict monogamy and facultative polygyny. Mateship characteristics of both kinds of societies may be essentially the same, i.e., one female may persistently associate with one male and vice versa, but the potential for social change probably differs greatly for societies represented by these alternatives.

Facultative Polygyny. As a social archetype, facultative polygyny as considered here has the same general characteristics and limits as monogamy. That is, group membership can be regulated by group members. In the case of birds, this involves admission or rejection of additional breeding adults, usually by same-sex individuals. While many contemporary human families operate on this basis and possibly according to constraints applicable to birds (acceptance by males of all females to the group and choice by females according to the quality of the holdings of the adult male), the problem remains of how to incorporate several such units into a larger, integrated society. The same problems for making this transition apply to humans. How could such units be incorporated into *groups* of such social units? It is possible that such a transition was made, i.e., that several facultatively polygynous social units combined to form larger social units, but since facultatively polygynous primates do not exhibit such variation in their range of social expression, this route to sociality is not considered further here.

Harem Polygyny. Harems, i. e., one or two males and mutiple females, may also closely control group membership. Within the tiered social organization of the hamadryas baboons, the social units do assemble into larger social groupings. Kummer (1968) draws attention to this degree of flexibility of the hamadryas social system and its potential to accommodate coordinated family units.

Hamadryas baboons are of special interest relative to problems of baboon paternal care because males sequester females and may form male-female associations for prolonged intervals (Kummer, 1968). Unfortunately, there appear to be no data identifying the actual duration of male harems. The interval of male-female association for some females begins as early as 1½ years before sexual activity (Kummer, 1968). While females begin estrous cycling as early as two years of age, first pregnancy does not occur earlier than for savanna baboons. Early association is not, however, the only way in which females come to be associated with adult males. As males age, their harems are taken over in part or entirely by younger and more vigorous males. Since the mean interval of male tenure as harem holder is unreported, the degree to which the hamadryas male-female association differs in duration from that of other savanna baboons is not known. Probably the interval is significantly longer than for other baboons, and, if this is so, the duration of tenure-paternal care relationship suggested here would predict more extensive paternal care by hamadryas than by savanna baboons. This tentative prediction is fulfilled, since hamadryas males regularly transport infants and juveniles (Kummer, 1968), a form of paternal caregiving that is not a regular feature of male savanna baboon caregiving. But the boundaries of the one or two male multi-female groups are sharply defined and, for multi-male/multi-female baboon groups, tenure of males as harem masters is based upon the continued ability of males to hold females by force.

Multi-male/Multi-female Groups. While numerous archetypal forms of social organization could be legitimately advanced as candidates for the human archetype, I now focus on the multi-male/multi-female social organization and, in particular, evaluate the possible significance of our observations of chacma baboons to the problem. This approach does not imply that I think a baboonlike society gave rise to humans, nor that limits to further social evolution identified for savanna baboons are specifically applicable to hypothetical attempts to derive humans from their ancestors. Rather, I suggest that this approach has value in its identification of specific relationships which may be evaluated in hypothetical attempts to derive one form of society from another within a particular phyletic line.

Because of their variability, multi-male/multi-female primate groups offer a particularly appealing example for deriving social evolutionary scenarios. Sizes of these groups range from only a few individuals to nearly 200 in the case of some macaque groups. There are numerous correlations between group size and environmental conditions. These correlations can be considered relative to alternative hypotheses deriving contemporary human social organization from multi-male/multi-female groups.

Transition Hypotheses

Comparison of a wide range of mammalian species emphasizes a shift to increasing female investment in individual offspring with the shift from polygyny to monogamy (Zereloff and Boyce, 1980). This transition, whether partial or complete, probably characterizes prehominid to human evolution. Here I concentrate on the analysis of changes in expression of paternal care by human social archetypes, focusing on the multi-male/multi-female group.

What patterns and processes of evolutionary change could have transformed a multi-male/multi-female group—especially one organized on a priority of access to females model (Altmann, 1962)—to human social organization? Two questions arise concerning possible evolutionary increments in male caregiving:

1. Can benefits be obtained by males by increments in caregiving within multi-male/multi-female groups or other forms of social organization?
2. How would such changes influence the reproductive characteristics of females?

A review of the specific forms of caregiving identified above suggests that the most beneficial, lowest cost forms of male caregiving known are carrying infants, protecting infants from infanticide, and protecting mothers from harassment. Other nondepreciable or less depreciable forms of caregiving (e.g., predation alarms, counterattacks against predators) include all group members, are not necessarily oriented to particular infants, or are incidental. The one major exception may be the as yet ill-defined support given to older progeny by males in encounters with other group members. Additional forms of caregiving may have become available, for example from successful hunting of large prey.

Reduced Estrous Swelling Hypothesis

Numerous alternative explanations of reduced ovulation sign, not involving relationships to paternal care, are offered elsewhere (e.g., Reynolds, 1976; Geist, 1978; Benshoof and Thornhill, 1979; Burley, 1979; Dickeman, 1979; Symons, 1979, 1980; Barash, 1980; van den Berghe, 1980). Only Alexander and Noonan (1979) consider the relationship of paternal care and concealed ovulation. Here I limit my discussion to that aspect of the problem.

Alexander and Noonan (1979) assume that in addition to a multi-male/multi-female social organization, human precursors had conspicuous ovulation signs, probably perineal estrous swellings, that were eliminated during hominid evolution. Selection led to a contemporary human condition called "concealed ovulation," characterized by absence of any conspicuous sign identifying the time of ovulation. The selection pressure proposed to accommodate this change is that loss of ovulation sign enabled females to entice males into enduring relationships long enough to reduce their success in additional matings, while simultaneously raising associated male confidence of paternity by failing to inform other males of the time of ovulation (Alexander and Noonan, 1979).

According to this scenario, females with reduced estrous signals would gain an advantage by prolonging the interval of association with individual males (Alexander and Noonan, 1979), who would be, in most cases, subordinate individuals (Strassmann, 1981), inferior in direct agonistic competition for mates with dominant males. This hypothetical transition was suggested primarily to allow for the allocation of extended paternal care, identified as a necessity for derivation of the human familial social unit. The kinds of paternal care that may have been allocated to infants and juveniles are not specified.

The Alexander-Noonan (1979) and Strassmann (1981) scenario is evaluated here especially from the perspective of contemporary baboon biology. Females of this and other savanna baboon species have conspicuous estrous swellings, reaching their maximum size near the time of ovulation (Hendrickx and Kraemer, 1969), a presumed characteristic of the hypothetical human ancestor (Alexander and Noonan, 1979). How could selection potentially modify the reproductive characteristics of chacma and other savanna baboons toward a less polygynous condition?

Alexander and Noonan (1979) make two potentially erroneous assumptions concerning characteristics of multi-male/multi-female primate groups. First, they assume that female estrous swellings are necessarily associated with low confidence of paternity, and second, that males in multi-male/multi-female groups provide little care to infants. Observations of chacma baboons in Botswana lead to the opposite conclusion on both counts. Estrous swellings by females of this species are associated with high paternity confidence and are correlated with several forms of paternal care (Busse and Hamilton, 1981; Busse, 1981 and Chpt. 8).

By signaling optimal mating time, estrous swellings and copulatory vocalizations (Hamilton and Arrowood, 1978) may incite male-male competition, increasing the probability that receptive females will mate with the dominant male at the time of ovulation (Clutton-Brock and Harvey, 1977; Popp, 1978). The competition incitation hypothesis, an extension of sexual selection theory (Fisher, 1930), suggests that females benefit from advertising sexual receptivity and the time of ovulation because it is advantageous for their sons to inherit traits of superior males.

Analysis of the competition incitation hypothesis is confounded by the availability of an alternative hypothesis, that females may signal ovulation to enhance paternity confidence and the probability of paternal care allocation by high-ranking males (Busse and Hamilton, 1981; Busse, 1981). Dominant male baboons sire most infants (Hausfater, 1975; Packer, 1979a, b; K. S. Smith, in prep.) and subsequently care for the infants they sire (Packer, 1980; Busse and Hamilton, 1981). These findings run counter to interpretations by Alexander and Noonan (1979) and Wrangham (in press), namely that estrous swellings are necessarily associated with low confidence of paternity. Proximate mechanisms that could identify to males the infants they have probably fathered could include (1) the timing of mating relative to the degree of estrous swelling, and (2) the development of a red perineal coloration shortly after conception (S. A. Altmann, 1970). The sequence of some events in the male and female reproductive cycles is summarized in Table 12-2. The female cycle and male response to it may not have been characteristic of the estrous swelling system of prehominid or early hominid females, if indeed they had one. Nevertheless, it suggests that alternative assumptions (Alexander and Noonan, 1979) are also suspect; and given this ambiguity, other starting points may represent equally viable alternatives.

Strassmann's (1981) extension of the Alexander-Noonan (1979) hypothesis suggests that subordinate males gain access to females with reduced estrous sign because dominant males have fewer clues identifying optimal mating time. Would subordinate males mating with these females provide more paternal care than dominant males mating with females who still have conspicuous estrous signals? Evidence, at least for savanna baboons living in multi-male/multi-female groups, indicates that estrous signs are already associated with parental care.

The proposed estrous sign reduction-paternal care enhancement scenario presents a transition problem. Implementation of the proposed changes would reduce or eliminate advantages conveyed by a form of paternal care

Table 12-2. Possible Bases for Paternity Identification in Chacma Baboons, Arranged in Sequential Order.

Possible Signal	Male Response
Early estrous swelling	Relatively unselective association of troop males with female
Full estrous swelling	Consortship by dominant available male
Red perineal pregnancy sign	May identify to male his success in impregnating female, male may protect this female
Birth of infant	Father guards infant from interference by other troop members, including other females and infanticidal transfer males
Resumption of female estrous cycling	Same as above; previous male excluded if not high ranking

already available to all group females. Loss of estrous sign would reduce paternity confidence and paternal care based upon it. Hence, characteristics of the social organization of multi-male/multi-female groups may limit further evolution extending male-female parentally based bonds within such groups. Enhancement of an advantage, in this case parental care, cannot be mediated by a change that would at the same time reduce the same advantage. Thus, I view the scenario provided by Alexander and Noonan (1979) as an unestablished and perhaps unlikely evolutionary avenue for development of extended male-female bonds and extended parental care, especially within the context of female estrous swellings in multi-male/multi-female groups. The Alexander-Noonan (1979) and Strassmann (1981) scenario requires either (1) a different function of estrous swellings than is proposed here for savanna baboons, or (2) a stepwise evolutionary change from conditions as they prevail for contemporary savanna baboons.

The estrous swelling reduction hypothesis and other discussions of the lack of estrous sign by humans offer ideas that may explain the development of this condition from a nonhuman primate analogue. But the degree of confidence in these arguments remains low because (1) the biology of the primate species referred to is either poorly known or inadequately considered; (2) related biological processes that may modify or change several interpretations are not considered; (3) specific evolutionary transitions and evolutionary processes associated with them are not considered relative to alternatives, but are instead evaluated in isolation; and (4) the approach is one of hypothesis development rather than falsification.

Troop Fragmentation Hypothesis

One additional alternative is to consider transitions from the existing pattern of paternal care that will (1) extend its duration, and (2) isolate its allocation to a single female. One way that this change may have taken place is through troop fragmentation.

A simple weapon such as a sharpened stick, perhaps developed from a digging stick as is variously used by at least three chimpanzee populations (Goodall, 1964, 1965; Suzuki, 1966; Jones and Pi, 1969), could have been developed to provide effective individual defense against predators. Such a development could have led to social group fragmentation. Comparative evidence for such fragmentation during primate evolution comes from observations of social groups of baboons living in relatively predator-free environments. Anderson (1980) found that chacma baboon troops in a relatively predator-free environment near Johannesburg, South Africa, fragment to form smaller social groups wth persistent membership. Troops living nearby in the same kind of habitat but subject to more severe leopard predation remain coalesced (Stoltz and Saayman, 1970).

Troop fragmentation has also been noted in Ethiopia, where *P. anubis* scatter into smaller parties, but remain within a few hundred meters of one another (Aldrich-Blake et al., 1971). While individual identity of members, and, therefore, subgroup members was not known because of the short duration of this study, particular effort was made to identify age and sex composition of subgroups. These observations showed nonrandom association, and it is especially noteworthy to the troop fragmentation hypothesis that males tended to associate with females and to avoid other males. Thus, on a temporary basis, daily troop fragmentation, in this case presumably a response to particularly low predation pressure and scattered resources, leads to subgroup formation sometimes resembling familial groups in composition. However, these subgroups reassemble in the evening and individuals sleep together in large trees. Use of communal sleeping trees and cliffs is generally regarded as a predation evasion strategy (Hall and DeVore, 1965; S. A. Altmann and J. Altmann, 1970). Daily reassembly of subgroups makes male choice of female associate possible on a daily basis and thus does not fully satisfy requirements for extension of individual male-female association over prolonged intervals. It does, however, identify the possibility that if subgroups develop, male-female association may form on a nonrandom basis into groups that, if further contacts were more limited, would increase the duration of male-female association. If one accepts the paradigm that predation is the major factor promoting primate aggregation (Alexander, 1974; but see Wrangham, 1981, for a dissenting view) and that there are social and competitive costs to living in social groups, it follows that there may be strong selection for dispersion once pressures from predation are reduced (e.g., Myers et al, 1981, for shorebirds).

Additional, limited evidence of flexibility in savanna baboon social structure comes from Hamilton and Tilson's (1982) observations of males living a persistent, solitary existence under special conditions in a Namib Desert environment. This evidence suggests that certain environments may not support troops but can sustain isolated individuals. If such conditions were more general for any reason, it might be to the advantage of females to associate with such males, contributing to the origin of small bands from multi-male/multi-female groups.

Comparative evidence supports the conclusion that dispersed, evenly distributed food resources lead to dispersed spatial distribution of individuals (Magnuson, 1962). Dispersion can lead to territoriality if resources are defensible, and resources become more defensible as group size and the space occupied by the group diminish (Hamilton and Watt, 1970). Territoriality may lead to monogamy because the integrity of the social unit can be maintained by the territorial occupants, as is the general case for monogamous primates and many monogamous bird species.

In the absence of intense predation, primate troop fragmentation could result from resource competition. Subordinate females, in particular, might benefit from reduced competition because of their low competitve ability relative to males. In the chacma baboon troops we have observed, subordinate females are most frequently supplanted at foods, and the likelihood that their dependent offspring will survive is reduced (Busse, 1981). Similar correlations between infant mortality and low rank of mothers have been reported for other primates. Thus, other factors held constant, less competitive conditions resulting from group dispersal may enhance female fitness. I view extension of parental care as an unrealized potential benefit to females and their progeny in some extant primate groups. The dynamics of competition within social groups preclude realization of these potential advantages given contemporary social organizations.

Thus, development of defensive and/or offensive weaponry by prehumans could have tipped the adaptive balance in favor of troop fragmentation. Regardless of how smaller subgroups assorted themselves, there would be fewer males and fewer females per group, and the interval of individual male association with females could increase. There would, therefore, be an enhanced potential for formation of persistent social units. Within smaller social groups, selection pressure for male-male competition would be reduced or absent. As duration of adult male-female associations increases, the opportunity for extended allocation of paternal care and extended duration of infant dependence upon both parents could also increase.

The troop fragmentation hypothesis is distinguished from the alternative identified above because it does not depend upon a gradual social transition within a prototypic society resembling some extant society (Fig. 12-1). The proposed scenario for social change is developed from known primate social organization and depends upon a known unique characteristic of early hominid development; extensive use of tools.

Social groups as defined here are viewed most commonly as adaptations to predation avoidance (Hamilton, 1971; Alexander, 1974). Life in a social group imposes costs to the social unit, and when constraints imposed by predation are reduced or removed, social groups may fragment spatially to form smaller groups with an enhanced probability of forming closer social units uninterfered with by other group members competing sexually and for other resources. The potential for a transition to an alternative form of social organization emphasizing social units and their subsequent modification is probably greatest for complex societies responding facultatively to changing environmental conditions. Hence, one alternative to gradual change from the multi-male/multi-female group to a smaller, facultatively polygynous group is a basic change in social organization. The possible change suggested here should not be confused with invocation of sudden evolutionary change, as emphasized recently by several authors challenging evolutionary gradualism (e.g., Gould, 1977, 1980; Stanley, 1979). Anderson's (1980) obser-

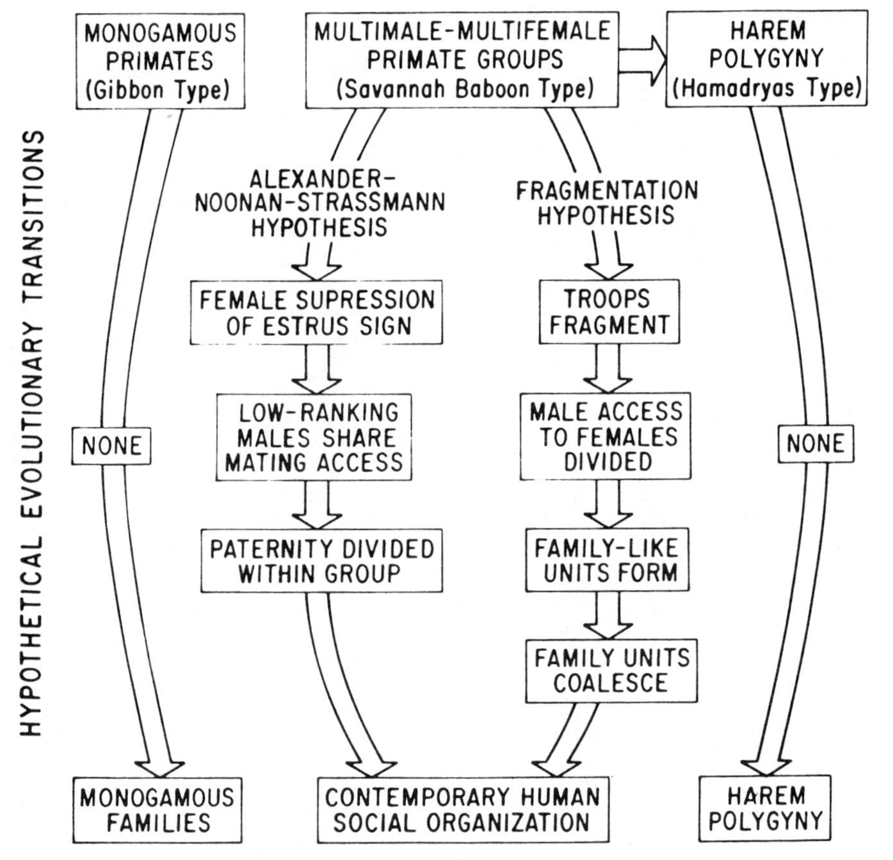

Fig. 12-1. Hypothetical evolutionary transitions.

vations show that part of the range of variation proposed by the troop fragmentation hypothesis is already included within the social potential of savanna baboons. The troop fragmentation hypothesis proposed here suggests that the evolutionary development of tools by hominids facilitated further evolution of an existing, flexible social organization.

I view the last hypothesis offered here, that of troop fragmentation following development of defensive weaponry, as one of numerous potential scenarios to be explored and evaluated. Close evaluation of other scenarios is in order. We should not be deceived to conclude that we are approaching a solution to a problem that is difficult and may yield no firm resolution.

Alternative models and possible evolutionary transitions associated with them are summarized in Figure 12-1. This is by no means an exhaustive or correct survey of evolutionary events. It is provided to more precisely define some available scenarios and the discussion presented in this chapter.

REFERENCES

Aldrich-Blake, F. P. G.; Bunn, T. K.; Dunbar, R. I. M.; and Headley, P. M. 1971. Observations on baboons, *Papio anubis,* in an arid region in Ethiopia. *Folia Primatol.* 15: 1-35.

Alexander, R. D. 1974. The evolution of social behavior. *Ann. Rev. Ecol. Syst.* 5: 325-83.

Alexander, R. D. 1979. *Darwinism and Human Affairs.* Seattle: University of Washington Press.

Alexander, R. D.; Hoogland, J. L.; Howard, R. D.; Hoogland, K. M.; and Sherman, P. W. 1979. Sexual dimorphism and breeding systems in pinnipeds, ungulates, primates and humans. In N. A. Chagnon and W. G. Irons (eds.), *Evolutionary Biology and Human Social Behavior: An Anthropological Perspective,* pp. 436-53. North Scituate, MA: Duxbury Press.

Alexander, R. D., and Noonan, K. N. 1979. Concealment of ovulation, parental care, and human social evolution. In N. A. Chagnon and W. G. Irons (eds.), *Evolutionary Biology and Human Social Behavior: An Anthropological Perspective,* pp. 436-53. North Scituate, MA: Duxbury Press.

Altmann, J. 1980. *Baboon Mothers and Infants.* Chicago: University of Chicago Press.

Altmann, J.; Altmann, S. A.; Hausfater, G.; and McCuskey, S. A. 1977. Life history of yellow baboons: Physical development, reproductive parameters, and infant mortality. *Primates* 18: 315-30.

Altmann, S. A. 1962. Social Behavior of Anthropoid Primates: Analysis of Recent Concepts. In E. R. Bliss (ed.) Roots of Behavior. New York: Harper and Brothers.

Altmann, S. A. 1970. The pregnancy sign in savannah baboons. *Lab. Anim. Digest* 6: 7-10.

Altman, S. J., and Altmann, J. 1970. *Baboon Ecology.* Chicago: University of Chicago Press.

Altmann, S. A.; Wagner, S. S.; and Lenington, S. 1977 Two models for the evolution of polygyny. *Behav. Ecol. Sociobiol.* 2: 397-410.

Anderson, C. 1980. Chacma baboon *(Papio ursinus)* social groups and their interrelationships in the Suikerbosvand Reserve, South Africa. Ph.D. dissertation, University of California, Riverside.

Barash, D. 1980. Human reproductive strategies: A sociobiological overview. In J. S. Lockard (ed.), *The Evolution of Human Social Behavior,* pp. 143-64, New York: Elsevier.

Benshoof, V., and Thornhill, R. 1979. The evolution of monogamy and concealed ovulation in humans. *J. Sociol. Biol. Struct.* 2: 95-106.

Brace, C. L. 1979. Biological parameters and Pleistocene hominid lifeways. In I. S. Bernstein and E. O. Smith (eds.), *Primate Ecology and Human Origins: Ecological Influences on Social Organization* New York: STPM Press.

Brockelman, W. Y.; Ross, B. A.; and Pantuwatana, S. 1974. Social interactions of adult gibbons *(Hylobates lar)* in an experimental colony. In D. M. Rumbaugh (ed.), *Gibbon and Siamang* vol. 3, pp. 137-56. Basel: S. Karger.

Brockelman, W. Y., and Srikosamatara, S., in press. Sex, territoriality and monogamy in gibbons. In H. Preuschoft; D. J. Chivers; W. Y. Brockelman; and N. Creel (eds.), The Lesser Apes. Edinburgh: University of Edinburgh Press.

Burley, N. 1979. The evolution of concealed ovulation. *Am. Nat.* 114: 830-58.

Busse, C. 1981. Infanticide and parental care by male chacma baboons, *Papio ursinus.* Ph.D. dissertation, University of California, Davis.

Busse, C. D., and Hamilton III, W. J. 1981. Infant carrying and paternal care by male chacma baboons. *Science* 212: 1281-83.

Campbell, B. 1979. Ecological factors and social organization in human evolution. In I. S. Bernstein, and E. O. Smith (eds.) *Primate Ecology and Human Origins: Ecological Influences on Social Organization.* New York: STPM Press.

Chivers, D. J. 1974. The siamang in Malaya: A field study of a primate in tropical rain forest. *Contr. Primatol.,* vol. 4. Basel: S. Karger.

———. 1977. The lesser apes. In Prince Rainier III and G. H. Bourne (eds.), *Primate Conservation,* pp. 539-98. Academic Press, New York.

Clutton-Brock, T. H., and Harvey, P. H. 1977. Primate ecology and social organization. *J. Zool., London* 183: 1-39.

Dickeman, M. 1979. Female infanticide, reproductive strategies, and social stratification: A preliminary model. In N. A. Chagnon and W. Irons (eds.), *Evolutionary Biology and Human Social Behavior,* pp. 321-67. North Scituate, MA:, Duxbury Press.

Fisher, R. A. 1930. *The Genetical Theory of Natural Selection.* Oxford: Clarendon.

Geist, V. 1978. *Life Strategies, Human Evolution, and Environmental Design: Towards a Biological Theory of Health.* New York: Springer-Verlag.

Goodall, J. 1964. Tool-using and aimed throwing in a community of free-living chimpanzees. *Nature* 201: 1264.

———. 1965. Chimpanzees of the Gombe Stream Reserve. In I. DeVore (ed.), *Primate Behavior: Field Studies of Monkeys and Apes.* New York: Holt, Rinehart and Winston.

Gould, S. J. 1977. *Ontogeny and Phylogeny.* Cambridge: Harvard University Press.

———. 1980. Is a new and general theory of evolution emerging? *Paleobiology* 6:119-30.

Gulik: R. H. van. 1967. *The Gibbon in China: An Essay on Chinese Animal Lore.* Leiden: E. J. Brill.

Hall, K. R. L., and DeVore, I. 1965. Baboon social behavior. In I. DeVore (ed.), *Primate Behavior,* pp. 53-110. New York: Holt, Rinehart and Winston.

Hamilton, W. D. 1971. Geometry of the selfish herd. *J. Theoret. Biol.* 31: 295-311.

Hamilton, W. J. III and Arrowood, P. C. 1978. Copulatory vocalizations of chacma baboons *(Papio ursinus),* gibbons *(Hylobates hoolock),* and humans. *Science* 200: 1405-09.

Hamilton, W. J. III, and Busse, C. 1978. Primate carnivory and its significance to human diets. *BioScience* 28: 761-66.

———. 1982. Social dominance and predatory behavior of chacma baboons. *J. Human Evol.* 11:567-73.

Hamilton, W. J. III; Busse, C.; and Smith, K. S. 1982. Adoption of infant orphan chacma baboons. *Anim. Behav.* 30: 29-34.

Hamilton, W. J. III, and Tilson, R. L. 1982. Solitary male chacma baboons in a desert canyon. *Am. J. Primatol.*

Hamilton, W. J. III, and Watt, K. E. F. 1970. Refuging. *Ann. Rev. Ecol. Syst.* 1: 263-86.

Harding, R. O., and Strum, S. C. 1976. The predatory baboons of Kekopey. *Nat. Hist.* 85: 46-53.

Hausfater, G. 1975. Dominance and reproduction in baboons. *(Papio cynocephalus). Contrib. Primatol.* Basel: Karger.

Hausfater, G.; Saunders, C. D.; and Chapman, M. 1981. Some applications of computer models to the study of primate mating and social systems. In R. D. Alexander and D. W. Trinkle (eds.), *Natural Selection and Social Behavior: Recent Research and New Theory.* New York: Chiron Press.

Heisler, I. L. 1981. Offspring quality and the polygyny threshold: a new model for the "sexy son" hypothesis. *Am. Nat.* 117: 316-28.

Hendrickx, A. G., and Kraemer, D. C. 1969. Observations on the menstrual cycle, optimal mating time and pre-implantation embryos of the baboon, *Papio anubis* and *Papio cynocephalus. J. Reprod. Fertil.,* suppl. 6: 119-28.

Hrdy, S. 1979. Infanticide among animals. *Ethol. Sociobiol.* 1: 13-40.

Jones, C. and Pi, J. S. 1969. Sticks used by chimpanzees in Rio Muni, West Africa. *Nature* 223: 100-01.

Kummer, H. 1968. *Social Organization of Hamadryas Baboons: A Field Study.* Chicago: University of Chicago Press.

Lack, D. 1968. *Ecological Adaptations for Breeding in Birds.* London: Methuen.

Lancaster, J. B. 1971. Play-mothering: The relations between juvenile females and young infants among free-ranging vervet monkeys *(Cercopithecus aethiops). Folia Primatol.* 15: 161-82.

Lee, R. B., and Devore, I. 1968. *Man the Hunter.* Chicago: Aldine.

Leutenegger, W. 1977. Sociobiological correlates of sexual dimorphism in body weight in South African Australopiths. *S. Afr. J. Sci.* 73: 143-44.

Lovejoy, C. O. 1981. The origin of man. *Science* 211: 341-50.

Magnuson, J. J. 1962. An analysis of aggressive behavior, growth, and competition for food and space in medaka. *Canadian J. Zool.* 40: 313-63.

Maynard Smith, J. 1977. Parental investment: A prospective analysis. *Anim. Behav.* 25: 1-9.

Murdock, G. P. 1967. *Ethnographic Atlas.* Pittsburg: University of Pittsburg Press.

Myers, P.; Connors, P. G.; and Pitelka, F. A. 1981. Optimum territory size and the sanderling: Compromises in a variable environment. In A. C. Kamilag and T. D. Sargent (eds.), *Foraging Behavior: Ecological, Ethological and Psychological Approaches,* pp. 135-58. New York: Garland Press.

Orians, G. 1969. On the evolution of mating systems in birds and mammals. *Am. Nat.* 103: 589-603.

Packer, C. 1979a. Inter-troop transfer and inbreeding avoidance in *Papio anubis. Anim. Behav.* 27: 1-36.

———. 1979b. Male dominance and reproductive activity in *Papio anubis. Anim. Behav.* 27: 37-45.

———. 1980. Male care and exploitation of infants in *Papio anubis. Anim. Behav.* 28: 512-20.

Popp, J. L. 1978. Male baboons and evolutionary principles. Ph.D. dissertation, Harvard University.

Ransom, T. W. and Ransom, B. S. 1971. Adult male-infant relations among baboons *(Papio anubis). Folia Primatol.* 16: 179-95.

Reynolds, P. C. 1976. The emergence of early hominid social organization: 1. The attachment systems. *Yrbk. Phys. Anthropol.* 20: 73-95.

Ridley, M. W. 1978. Paternal care. *Anim. Behav.* 26: 904-32.

Slatkin, M. and Hausfater, G. 1976. A note on the activities and behavior of a solitary male baboon. *Primates* 17: 311-22.

Srikosamatara, S. in press. Ecology of the pileated gibbon *(Hylobates pileatus)* in southeast Thailand. In H. Preuschoft; D. J. Chivers; W. Y. Brockelman; and N. Creel (eds.), *The Lesser Apes.* Edinburgh: University of Edinburgh Press.

Stanley, S. M. 1979. *Macroevolution: Pattern and Process.* San Francisco: Freeman and Co.

Stoltz, L. P., and Saayman, G. S. 1970. Ecology and behaviour of baboons in the northern Transvaal. *Ann. Transvaal Mus.* 26: 99-143.

Strassmann, B. I. 1981. Sexual selection, paternal care, and concealed ovulation in humans. *Ethol. Sociobiol.* 2: 31-40.

Strum, S. C. 1975. Primate predation: Interim report on the development of a tradition in a troop of olive baboons. *Science* 187: 755-57.

———. 1981. Processes and products of change: Baboon predatory behavior at Gilgil, Kenya. In R. S. O. Harding, and G. Telcki (cds.), *Primate Dietary Patterns,* New York: Columbia University Press.

Suzuki, A. 1966. On the insect-eating habits among wild chimpanzees living in the savanna woodland of Western Tanzania. *Primates* 7: 481.

Symons, D. 1979. *The Evolution of Human Sexuality.* New York: Oxford University Press.

———. 1980. Precis of the evolution of human sexuality. *Behav. Brain Sci.* 3: 171-214.

Taylor, C. R.; Heglund, N. C.; McMahon, T. A.; and Looney, T. R. 1980. Energetic cost of generating muscular force during running. *J. Exp. Biol.* 86: 9-18.

Taylor, C. R.; Heglund, N. C.; and Maloiy, G. M. O. 1982. Energetics and mechanics of terrestrial locomotion. I. Metabolic energy consumption as a function of speed and body size in birds and mammals. *J. Exp. Biol.*

Tilson, R. L. 1983. Family formation strategies of Kloss' gibbons. *Folia Primatol.*, in press.

Trivers, R. I. 1972. Parental investment and sexual selection. In B. Campbell (ed.), *Sexual Selection and the Descent of Man*, Chicago: Aldine.

van den Berghe, P. L. 1980. The human family: A sociobiological look. In J. S. Lockard (ed.), *The Evolution of Human Social Behavior*, New York: Elsevier, pp. 67-86.

Weatherhead, P. J. and Robertson, R. J. 1979. Offspring quality and the polygyny threshold: The "sexy son" hypothesis. *Am. Nat.* 113:201-08.

———. 1981. In defense of the "sexy son" hypothesis. *Am. Nat.* 117: 349-56.

Wittenberger, J. F. 1981a. *Animal Social Behavior.* North Scituate, MA: Duxbury Press.

———. 1981b. Male quality and polygyny: The "sexy son" hypothesis revisited. *Am. Nat.* 17: 329-42.

Wittenberger, J. F., and Tilson, R. L. 1980. The evolution of monogamy: Hypotheses and evidence. *Ann. Rev. Ecol. Syst.* 11: 197-232.

Wrangham, R. W. 1983. An ecological model of female-bonded primate groups. *Behaviour* in press.

Zereloff, S. I.; and Boyce, M. S. 1980. Parental investment and mating systems in mammals. *Evolution* 34: 973-82.

13

The Evolutionary Role of Socio-ecological Factors in the Development of Paternal Care in the New World Family Callitrichidae

A. G. Pook
Jersey Wildlife Preservation Trust
Channel Island, England

In recent years a great deal of research has been carried out on the reproduction of marmosets and tamarins (Callitrichidae), almost entirely on captive animals (Bridgewater, 1972; Kleiman, 1977a). A fairly consistent pattern of reproductive behavior has emerged. All species studies so far have been found to reproduce most satisfactorily in captivity in monogamous family groups and to generally give birth to twins; in all species the male parent also has been found to be actively involved in caring for the infants from very shortly after birth (see chapter 14).

This pattern differs markedly from that of the other major family of New World primates (Cebidae)—and indeed of primates generally—who usually live in large polygynous or polygamous groups, have single offspring, and do not involve the males directly in rearing newborn infants. There has been considerable speculation as to the reasons for this difference.

It is generally agreed that the small body size of the callitrichids is associated with their system of reproduction, but beyond that, two distinct points of view have emerged. Hershkovitz (1977) concluded that both small body size and multiple births were primitive characteristics that had been retained by the Callitrichidae. However, other authors (Leutenegger, 1973;

Hampton, 1975; Eisenberg, 1977) have regarded these features as secondary adaptations. This was largely because certain features in the callitrichid reproductive physiology are the same as those found in larger primates normally bearing single infants (e.g., Leutenegger, 1980).

If we adopt the latter, more widely accepted view that the Callitrichidae evolved from a slightly larger form that gave birth to single offspring, it is necessary to propose reasons why these "regressive" steps were taken. The general evolutionary trend is for body size to increase and for larger animals to produce smaller numbers of offspring (Cope's law). Why, therefore, should body size have diminished in this group? In fact, little attention has been paid to this question. However, Eisenberg (1977) has suggested that after primates colonized South America, probably during the Eocene period, the Callitrichidae evolved to occupy ecological niches as diurnal frugivore-insectivores, which in the Old World were already occupied by small rodents of the squirrel family. Such rodents probably did not appear in South America until the Pliocene.

The development of producing multiple offspring has been discussed by Leutenegger (1973), given that there was a reason for evolving smaller body size ("phyletic dwarfism"). Leutenegger noted a strong inverse relationship in primates between the body weight of the adult female and the relative total weight of offspring at birth. In other words, the infants of smaller primates were of a larger proportion of their mother's weight than were larger primates. Leutenegger (1980) has claimed that this reaches its limit in *Saimiri sciureus,* in which the single infant is about 15% of its mother's weight, and that, in captivity at least, this leads to frequent obstetric difficulties. This suggests a threshold value for the weight of a single infant as a percentage of the weight of its mother. In order to comply with the trend just described, primates smaller than *Saimiri,* such as the Callitrichidae, are thus "obstetrically obligated to produce twins rather than single offspring" (Leutenegger, 1980, p. 97).

From this point, explanations of the monogamous family group structure and the high level of paternal investment in infant care follow. For the smaller callitrichids, the offspring represent such a large proportion of the mother's weight that help is necessary in caring for them from a very early age. In addition, two infants require a greater amount of feeding than one infant twice as large does since basal metabolism varies with the 0.75 power of body weight. Multiple births thus represent an even greater drain of the adult female's resources. The most reliable source of aid for the mother in such a situation is the male parent, and thus a monogamous group structure becomes necessary. In this way the father of the offspring can be assured of his paternity and it is therefore in his genetic interest to help with their care from a very early age.

The basis of this model is therefore very simple. Phyletic dwarfism in the Callitrichidae has led to the need for multiple births, which in turn has led to the need for close paternal involvement in infant rearing and thus for monogamous family groups. Socioecological arguments for the evolution of monogamy and paternal care, such as those proposed by Wilson (1975) or Clutton-Brock and Harvey (1978), are dismissed as inadequate.

Kleiman (1977b) has argued along similar lines, stating that ecological considerations give inadequate justification for the development of monogamous rather than polygynous groupings. Although such factors as the carrying capacity of the habitat are also regarded by Kleiman as relevant to a certain degree, the principal model of monogamy as a consequence of phyletic dwarfism is put forward and extended. Within the Callitrichidae, Kleiman showed that for those species with the highest litter-to-mother weight ratio (*Cebuella* and *Callithrix* spp., the smaller species), the father begins to carry the offspring from the very first day, whereas in the larger species (*Saguinus* spp. and *Leontopithecus*) the weight ratio is smaller and the male parent carries the infants only after they are a few days old. It should be pointed out, however, that this relationship within the Callitrichidae is based on little data. For several species, only a small number of adults and infants were measured, and a great deal of variability exists both in adult female and infant weights (see Napier and Napier, 1967; pers. obs.). In addition, there is a great deal of inter- and intra-individual variation in the timing and extent of parental involvement for many species (pers. obs.; Epple, 1975; chapters 1 and 14; H. Box, pers. comm.). Therefore, conclusions about relatively minor differences between species await more detailed confirmation.

But what of the principal hypothesis, that phyletic dwarfism in the Callitrichidae necessarily led to twinning, paternal involvement, and monogamy? Superficially, it appears to apply very well in the case of the Callitrichidae, but what happens when a broader view of other New World primates is taken?

Table 13-1 summarizes the data for several species of New World primates in which the adult female weighs, on the average, less than one kilogram. The species are arranged in descending order by body weight. Generally speaking, it can be seen that the weight ratio decreases as the mother's weight increases. There is, however, one important exception. *Callimico goeldii,* a primate considerably smaller than *Saimiri,* gives birth to singleton infants and also has a relatively low weight ratio. *Callimico* occupies a peculiar taxonomic position: it is intermediate between the Callitrichidae and the Cebidae and has at different times been described as a subfamily of each of these two families (Napier and Napier, 1967). More recently, Hershkovitz (1977) has ascribed to it the status of a distinct family, Callimiconidae. Whatever its taxonomic status, records from captive breeding

Table 13-1. Infant and Adult Female Weights in Some Smaller Species of New World Primate.

Species	Usual No. Infants	Mean Adult Female Weight (g.)	Mean Infant Weight (g.)	Litter: Mother Weight Ratio	Source
Cebuella pygmaea	2	122	14.6	24%	Hampton et al., 1972
	2	145	16	22%	Christen, 1974
Callithrix jacchus	2	345	32.5	19%	Personal observation, based on breeding records Jersey Zoological Park, 1974–77
		262	28	21%	Leutenegger, 1973
Saguinus fuscicollis	2	350	34	19%	Personal observation, same as above
Callithrix argentata	2	384	32	17%	Personal observation, same as above
Saguinus oedipus	2	476	41	17%	Personal observation, same as above
		510	40	16%	Christen, 1974
Saguinus midas	2	483	39	16%	Christen, 1974
Callimico goeldii	1	582	53	9%	Personal observation, same as above
		472	40	8%	Kleiman, 1977b
Callicebus moloch	1	680	?	?	Leutenegger, 1973
Saimiri sciureus	1	600–700 (5)	?	15%	Leutenegger, 1973
Leontopithecus rosalia	2	745	61	16%	Kleiman, 1977b
Aotus trivirgatus	1	950	135	14%	Kleiman, 1977b

colonies (Beck, in litt.; Hampton et al., 1972; pers. obs.) and from field studies (Pook and Pook, in press a) suggest that anything other than single offspring is rare. Only one case of twins has been recorded to the author's knowledge, by Hill (1966), and then only one twin was viable. *Callimico,* approximately the same size as some *Saguinus* species and smaller than *Leontopithecus,* appears to be under no "obstetrical obligation" to produce twins. If this is the case, it may be that other factors have brought about the preponderance of twinning in the Callitrichidae. Also, it should be noted that twinning is only the general rule in Callitrichidae and is by no means fixed. Hampton et al. (1972) found that twins accounted for just over half the births in *Cebuella pygmaea,* most of the rest being singletons. Singletons are also fairly common in *Saguinus* spp. (20% of 319 births, Gengozian et al., 1977; pers. obs.) and can occur in *Leontopithecus* (Kleiman, 1977b) and other species. On the other hand, breeding by well-established animals or by successive generations in captive colonies under optimal conditions can lead to an increasingly high proportion of triplet and even quadruplet births in *Callithrix jacchus* (Rothe, 1977; H. Box, pers. comm.). The number of young in a callitrichid litter appears to be fairly flexible, certainly when compared to the rarity with which twins are born to *Callimico* and other large primates, and there appears to be no physiological obligation on callitrichids to produce multiple births because of their size.

To summarize, as a general rule the litter-to-mother ratio does increase as the size of the mother decreases; but phyletic dwarfism rather than putting any obligation on the female callitrichids to produce multiple births, perhaps provided an opportunity for the smaller species to produce viable multiple offspring when conditions were suitable. This point will be discussed further later.

Let us now turn our attention to the second stage of the hypothesis, that multiple births have necessitated the development of parental care and monogamous family groups. As with the previous discussion, the argument seems sound when considering just the Callitrichidae; but if the viewpoint is broadened to include the Cebidae, what then are the reasons for the development of monogamy and paternal care in such genera as *Aotus, Callicebus, Pithecia,* and *Callimico?* All four genera give birth to only a single offspring, although in the first two the infant-to-mother weight ratio is still fairly high (Kleiman, 1977b; Eisenberg, 1977). None of the adults are large animals—the heaviest is *Pithecia,* weighing approximately 1400 g. (Napier and Napier, 1967). It follows that for all these genera, an infant still represents a considerable burden, especially as it grows; and so if the male parent is present, as is the case in a monogamous family group, it is obviously advantageous to the survival prospects of the offspring for the father to help considerably in caring for the infant. In *Aotus,* the male parent starts to carry the infant in the second week (Moynihan, 1964). *Callicebus* and *Pithecia* also show paternal care within a short time after birth (Eisenberg, 1977). The *Callimico* male does not usually start to carry the infant until it is three weeks old, but may then do most of the carrying (Pook, 1975; Heltne et al., 1973; Beck, in litt.). It is interesting that in this latter species the infant-to-mother weight ratio is unusually low (8-9%) and so the burden on the female is not so great. Larger primates, such as *Cebus* and *Ateles* have similar weight ratios (8.5% and 7%, respectively; Napier and Napier, 1967), and these species show no distinct pattern of paternal care.

The conclusion that can be drawn from this is that whereas all New World species that have multiple offspring exhibit monogamy and paternal care, the reverse it not always the case. This suggests that monogamy may have evolved independently from, and perhaps prior to, the production of multiple offspring. The problem of the evolutionary sequence in the development of these reproductive patterns is one that has been raised before briefly, by Clutton-Brock and Harvey (1978). They suggested that there was no obvious solution to this problem. However, it seems more logical that a monogamous social system should *precede* the production of multiple offspring, since it would confer little evolutionary advantage to a female if she were to produce twin infants that she could not rear herself, if there was not already in existence a monogamous relationship with a male that was predisposed to assist in the care of the young. However, there must obviously be some benefit in a monogamous relationship with paternal care, even when single infants are born, as shown by *Aotus, Callicebus, Pithecia,* and *Callimico.* Contrary to the propositions of Kleiman and Leutenegger, I would therefore argue that monogamy was a necessary prerequisite for, rather than a necessary consequence of, the production of multiple births.

In what circumstances then did monogamy evolve in the New World primates? The conditions favoring monogamy in mammals have previously been viewed as largely socioecological, often in terms of population distribution (e.g., Clutton-Brock and Harvey, 1978; Wilson, 1975). First, there are advantages in group living (better awareness of predators, more efficient food gathering, and utilization of resources such as sleeping sites, for example) which have ensured that nearly all primates live in social groups. Further than this, permanent group living in a fixed home range is also advantageous in many species, since this allows for animals to benefit from prolonged experience of the particular resources in a relatively small area.

However, there are limits imposed on the size of a group and its home range. One of the principal factors limiting group size is the amount of travel needed to gather enough food for the whole group. Obviously, the more animals in a group, the more food they require. If the food resources utilized by a particular species are distributed fairly evenly and not very densely, a larger group will have to travel farther to collect sufficient food. However, if the food items are distributed in uneven clumps or are more plentiful (if, for example, the species is folivorous), group and home range size may be determined more by other factors, such as the absolute amount of food available rather than its distribution.

Many of the smaller species of New World primate, such as most of the Callitrichidae, which are carnivorous and insectivorous as well as frugivorous, belong to the former category, in which an important part of their diet consists of food distributed evenly through the forest. Since they are small animals, it is energetically very expensive to travel any farther than is necessary. This has led to an optimal group size for callitrichids of approximately six animals (e.g., Kleiman, 1977a; Heltne et al., 1975). *Cebuella,* on the other hand, is a callitrichid that may be restricted in its distribution more by the distribution of a clumped resource, since it feeds largely on the exudate from a small number of tree species, which may occur infrequently (Ramirez et al., 1977). However, typical group sizes in favorable locations again appear to be restricted to about six because of the limited nature of the resource.

For similar ecological reasons, *Callicebus, Aotus,* and *Pithecia* have formed groups of two to four, and occasionally five, animals, inhabiting, at least in the case of *Callicebus,* very small home ranges (Mason, 1966; Pook and Pook, in press b). If it is optimal for these genera to live in stable groups of two to four animals, the only reproductive system available to them is monogamy; and in such circumstances, when the father is sure which are his offspring, it is obviously in his own genetic interests to ensure their survival. In the case of fairly small animals, this may be done by helping to alleviate the burden on the mother by carrying, cleaning, and helping to wean the young. In these genera, the option of producing twins is not advantageous if group size were to be restricted to four or five animals at the most. It is of interest to note that in other larger primates who live in small monogamous groups *(Hylobates* and *Symphalangus),* the mother needs little direct assistance in rearing the young, and that such obvious paternal care is absent in the case of *Hylobates* or delayed until the infant is 12 months old in the case of *Symphalangus* (Kleiman, 1977b; chapter 15).

Other New World genera form larger polygynous troops of 15 to 50 or more animals, although *Ateles* and *Cebus* may sometimes split into smaller temporary subgroups (Izawa, 1976). In such cases, the male parent is probably unaware which are his offspring and, when necessary, juveniles and other adult females may help out a little in caring for young infants; this is the case for *Saimiri* (Leutenegger, 1980), in which other females care for infants aged five weeks or more for short periods. The lack of paternal care at an early stage of the infant's development may have prevented *Saimiri* from twinning.

In the Callitrichidae, in which an optimal average group size is about six animals living in stable groups in fixed home ranges, there could perhaps have evolved a monogamous system with one or two breeding pairs of adults in the group, or a very restricted polygynous system of one male and two females. In any of these cases, a large investment in paternal care of the young would be profitable in helping to perpetuate the father's own genes.

This level of paternal investment made it possible for the Callitrichidae to take up the option of multiple offspring provided by phyletic dwarfism, thus greatly increasing the reproductive potential of the family. The production of more than one offspring at a time would have forced the callitrichids to adopt the more strictly monogamous family structure that has been found in field studies carried out to date (Kleiman, 1977a).

The only apparently anomolous case is that of *Callimico goeldii.* According to Leutenegger's (1973) weight relationship, it is probably small enough to have multiple births. A considerable amount of paternal care is present, even if not until the third week of age. Why, therefore, does this species, which has an optimal group size approximately the same as the callitrichids (Pook and Pook, in pr. a; Izawa, in litt.), not produce twins? The answer perhaps lies in the peculiar distribution of this species in the wild (Pook and Pook, in press a). Unlike the Callitrichidae, which have a continuous microdistribution, living in overlapping home ranges, *Callimico* groups usually live in isolation from conspecifics, sometimes two kilometers or more from the nearest group. This is probably because of their highly specialized habitat requirement related to their particular style of foraging (Pook and Pook, in pr. a).

Groups of *Callimico* may therefore live in genetic isolation for considerable periods of time, since opportunities for transfer of individual animals are severely limited compared to the Callitrichidae. Therefore, strictly monogamous family grouping may be disadvantageous, since it could lead to prolonged periods of inbreeding, or a lack of breeding should one of the adult pair die. From the limited evidence available so far—for example the presence of more than one infant in some wild groups (Pook and Pook, in press a; Izawa, in litt.) or more than one breeding female in a captive group (Carroll, in litt.)—it appears that there may frequently be more than one breeding female in wild groups. Whether there is more than one breeding male has not yet been determined. However, the genetic advantages of such a system for groups living in isolation must outweigh the advantages of producing twins, which would necessitate either a larger group size or strict monogamy.

In conclusion, there is a marked relationship between monogamy, paternal care, and multiple births in the family Callitrichidae. A consideration of this family alone may lead to the conclusion that phyletic dwarfism has led to multiple births, which have required close paternal care and therefore a monogamous family structure. A broader view of all the New World primates suggests, on the other hand, that monogamy may have evolved in response to ecological needs for restricted group sizes in particular species. Paternal care would then be advantageous, especially for smaller animals, and this would have favored the production of multiple offspring, made possible by phyletic dwarfism.

REFERENCES

Bridgwater, D. D. 1972. *Saving the Lion Marmoset.* Wheeling: The Wild Animal Propagation Trust.

Christen, A. 1974. Fortpflan zungsbiologie und Verhalten bei *Cebuella pygmaea* and *Tamarin tamarin. Z. fur Tierpsychol. Beiheft* 14: 1-78.

Clutton-Brock, T. H.; and Harvey, P. H. 1978. Mammals, resources and reproductive strategies. *Nature* 273: 191-95.

Eisenberg, J. F. 1977. Comparative ecology and reproduction of New world monkeys. In D. G. Kleiman (ed.), *The Biology and Conservation of the Callitrichidae,* pp. 13-22. Washington, D. C.: Smithsonian Institution Press.

Epple, G. 1975. Parental behaviour in *Saguinus fuscicollis* spp. (Callitrichidae). *Folia Primatologica* 24: 221-38.

Gengozian, N.; Batson, J. S.; and Smith, T. A. 1977. Breeding of tamarins (*Saguinus* spp.) in the laboratory. In D. G. Kleiman (ed.), *The Biology and Conservation of the Callitrichidae,* pp. 207-13. Washington, D. C.: Smithsonian Institution Press.

Hampton, S. H. 1975. Placental development in the marmoset. In S. Kondo; M. Kawai; and A. Ehara (eds.), *Proceedings of Vth International Congress of Primatology, vol.1, Contemporary Primatology,* pp.106-14. Basel: Karger.

Hampton, S. H.; Hampton, J. K.; and Levy, B. M. 1977. Husbandry of rare marmoset species. In D. D. Bridgwater (ed.), *Saving the Lion Marmoset,* pp. 70-85. Wheeling: The Wild Animal Propagation Trust.

Heltne, P.; Freese, C.; and Whitesides, G. 1975. A field survey of the nonhuman primates in Bolivia. Unpublished report to the Pan American Health Organization.

Heltne, P.; Turner, D.; and Wolfhandler, J. 1973. Maternal and paternal periods in the development of infant *Callimico goeldii. Am. J. Phys. Anthrop.* 38: 555-60.

Hershkovitz, P. 1977. *Living New World Monkeys (Platyrrhini) with an Introduction to the Primates,* vol.1. Chicago: University of Chicago Press.

Hill, W. C. O. 1966. On the neonatus of *Callimico goeldii* (Thomas). *Roy. Soc. (Edinburgh), Proc. B.,* 69: 321-33.

Izawa, K. 1976. Group sizes and compositions of monkeys in the Upper Amazon basin. *Primates* 17(3): 367-99.

Kleiman, D. G. 1977a. *The Biology and Conservation of the Callitrichidae.* Washington, D. C.: Smithsonian Institution Press.

———. 1977b. Monogamy in mammals. *Quart. Rev. Biol.* 52: 39-69.

Leutenegger, W. 1973. Maternal-fetal weight relationships in primates. *Folia Primatologica* 20: 280-93.

———.1980. Monogamy in callitrichids: a consequence of phyletic dwarfism? *Internat. J. Primatology* 1: 95-98.

Mason, W. 1966. Social organization of the South American monkey *Callicebus moloch:* A preliminary report. *Tulane Stud. Zool.* 13: 23-28.
Moynihan, M. 1964. Some behaviour patterns of platyrrhine monkeys. I. The night monkey *(Aotus trivirgatus). Smithsonian Misc. Coll.* 146(5): 1-84.
Napier, J. R. and Napier, P. H. 1967. *A Handbook of Living Primates.* London: Academic Press.
Pook, A. G. 1975. Breeding Goeldi's monkey *(Callimico goeldi)* at the Jersey Zoological Park. *Jersey Wildlife Pres. Trust Ann. Rep.* 12: 17-20.
Pook A. G., and Pook, G. A. In press, a. A field study of the socioecology of the Goeldi's monkey *(Callimico goeldii)* in northern Bolivia. *Folia Primatologica.*
———. In press, b. Polyspecific assocation between *Saguinus fuscicollis, Saguinus labiatus, Callimico goeldii* and other primates in north-western Bolivia. *Folia Primatologica.*
Ramirez, M. F.; Freese, C. H.; and Revilla, J. C. 1977. Feeding ecology of the pygmy marmoset, *Cebuella pygmaea,* in northeastern Peru. In D. G. Kleiman (ed.), *The Biology and Conservation of the Callitrichidae,* pp. 91-104. Washington, D. C.: Smithsonian Institution Press.
Rothe, H. 1977. Parturition and related behaviour in *Callithrix jacchus* (Ceboidea, Callitrichidae). In D. G. Kleiman (ed.), *The Biology and Conservation of the Callitrichidae,* pp. 193-206. Washington, D. C.: Smithsonian Institution Press.
Wilson, E. O. 1975. *Sociobiology: The New Synthesis.* Cambridge, MA: Harvard University Press.

14

Interactions Between Adult Males and Infants in Prosimians and New World Monkeys

Jerry L. Vogt
Department of Psychology
St. John's University
Collegeville, MN

INTRODUCTION

Discussions of the roles that adult male mammals play in the social structure and organization of their group usually revolve around such topics as dominance, territory defense, or sexual behavior. Typically the adult male is *not* thought of as an infant caretaker, as a parent playing an important or necessary part in the physical and/or social development of the offspring. Yet, adult male mammals do frequently show extensive male care. A recent review by Kleiman and Malcolm (1981) on the evolution of parental care in mammals estimates that about 10% of all mammalian genera show some form of direct male care of infants. In the Primate order, they calculated the percent of genera in which males showed some form of direct, positive interaction with infants to be even greater—nearly 40%. This figure was the highest of any individual mammalian order (Kleiman and Malcolm, 1981, p. 357). Thus, with the primates it might be supposed that interactions between adult males and infants are not only rather prevalent but also important for the proper development of the infant.

The purpose of this paper is to review the interactions between adult males and infant across two major groups of primates: the prosimians and the New World monkeys. Adult male-infant interactions of all types will be examined in terms of taxonomic distribution and will be related to the various ecological and social variables. Within each genus information on social interactions between adult males and infants will be noted from both field and laboratory sources, and the variability among and flexibility within species will be examined in order to reveal those environmental conditions supporting the appearance of active male care of infants.

ADULT MALES AND INFANTS

Infant Socialization in Broader Context

The primary agent of infant care in mammals is usually seen as the mother. And indeed she provides the developing offspring crucial and indispensable assistance, including physical protection and nutritional requirements. However, other social figures besides the mother are also important and essential to the infant's proper social development and growth, and this is particularly true in primates. For example, research by Harry Harlow and his colleagues (Harlow and Harlow, 1962; Harlow, 1969) has documented the role of peer relations in rhesus monkey social development, and studies by Rosenblum and Kaufman (1968) have indicated the importance of familiar adult females to bonnet macaque infants that have been separated from the mother. The evidence for important relationships between infants and conspecifics besides the mother has been reviewed by Spencer-Booth (1970) for all mammals and more recently by Hrdy (1976) for nonhuman primates. Other recent papers by McKenna (1979), Quiatt (1979), and Vogt and Hennessy (1982), have discussed in various ways the significance of the nonmaternal social environment in the infant development of different primate species.

One part of the primate infant's social milieu that has received some special attention is the adult male. Over the last 10 to 15 years, there has appeared an increasing emphasis on documenting and explaining the interactions between primate adult males and infants. In humans this research has explored the role of the father in various aspects of infant and child development (Lynn, 1974; Lamb, 1981 and chapter 16; Pedersen, 1980). In nonhuman primates, the literature on adult male-infant interactions has been reviewed by Mitchell (1969) and Mitchell and Brandt (1972), who graded the social interactions on various levels ranging from toleration to protection and care of the infant. Hrdy (1976) classified adult male interactions with infants as falling somewhere between exploitation and true care (i.e., between detracting from and contributing to the infant's development). And Kleiman and Malcolm (1981) have very recently examined the patterns of paternal care in all mammals by taking an evolutionary perspective. The present paper reviews adult male-infant interactions in each genus in two taxa of the primate order, the prosimians and the New World monkeys, in an attempt to delineate those social and environmental factors that appear to encourage male care of infants.

As mentioned in the previous section, a large proportion (nearly 40%) of the genera in the primate order, compared to other mammalian orders, exhibit some form of direct positive social interaction with infants. However, within the primate order, the prevalence of male care in different taxa appears to vary considerably. For some groups, notably all species in the callitrichid family, true parental care by the males is a salient and recognized feature. In the prosimians, on the other hand, although patterns of maternal care have become well documented (Klopfer and Boskoff, 1979), male care of infants appears to be rather infrequent. Among the species in the New World monkeys and prosimians, then, interactions between adult males and infants seem to be quite variable.

Definitions of Male Care

The interactions between adult male monkeys and infants encompass a wide variety of social behaviors. Mitchell and Brandt (1972) graded these interactions from the (low level) touching to the (high level) adopting of the infant by the male. This set of levels might be expanded and revised to become a range of behaviors extending from "less intense" interactions, such as avoidance, to "more intense" interactions such as protection. As can be seen in Figure 14-1, this proposed continuum starts with "no reported interactions," which might be seen as the most effective form of "avoidance," the next point on the scale. It then goes through "tolerate," "approach," and "touch" before getting to those behaviors involving active and/or extensive physical interaction by the male with the infant. These active behaviors are "groom/play," "sleep with," "retrieve," "feed," "baby sit," "protect," "carry," and "care for." The more intense end of the range is anchored by those interactions that might also be labeled exploitation (cf. Hrdy, 1976): "agonistic buffering," "aggression," and "infanticide."

In attempting to assess the ultimate effects of a male's interactions on the infant, Kleiman and Malcolm (1981) slightly revised Trivers's (1972) concept of parental investment for the following definition of male parental investment: "Any increase in a prereproductive mammal's fitness attributable to the presence or action of a male" (Kleiman and Malcolm, 1981, p. 348). They distinguish between direct and indirect investment; the former has an immediate physical influence on the infant's survivorship, while the latter has its effect in the infant's absence. Direct investment was also referred to as "male parental care " by Kleiman and Malcolm and appears to include a set of interactions from Figure 14-1. Since this paper is concerned with direct behavioral interactions between adult males and infants, Kleiman and

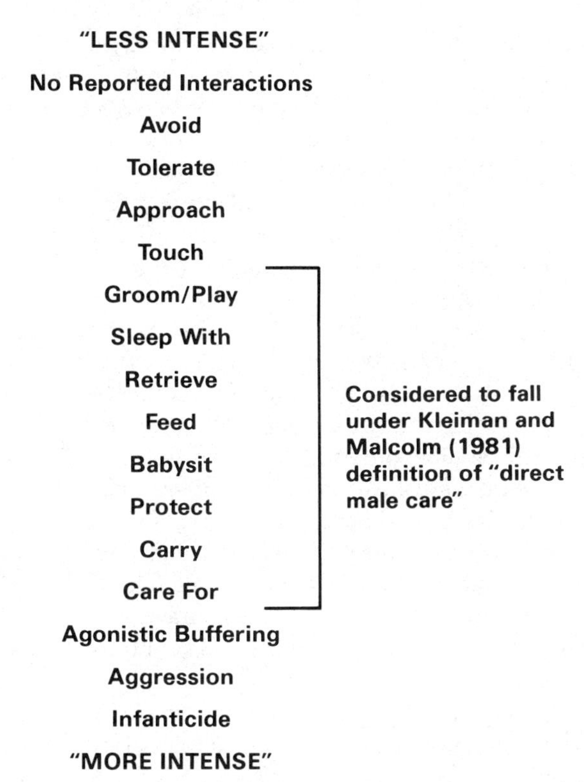

Fig. 14-1. Continuum of interactions, from less to more intense, between adult males and infants.

Malcolm's definition of direct parental care will be used to indicate which behaviors will have direct benefit for the infant. These are, as indicated in Figure 14-1, the following: groom/play, sleep with, retrieve, feed, babysit, protect, carry, care for. In this paper the term "male care" will thus refer to any of these.

Available Data

Before reviewing the reported interactions between adult males and infants, a few remarks should be made about the problems of and constraints on the available data on prosimians and New World monkeys. First, although the qualitative and quantitative aspects of the field data for prosimians (e.g., Charles-Dominique and Bearder, 1979) and New World monkeys (e.g., Defler, 1979) are improving and are now rather good, overall the field data for these two groups are not as good as the data for the Old World primates. Second, most field workers do not go into the field with the express purpose of examining adult male-infant interactions. Third, there are the problems common to all field work done on arboreal or nocturnal animals, and most of the prosimians and New World monkeys fall into either or both of these classifications. These types of animals are typically elusive, hard to see, distant, and/or fast moving. And fourth, a few species in both the prosimians and New World monkeys are highly endangered and there continues to be little information available on their behavior.

When animals are captive within the confines of a laboratory or zoo setting, the data on social behavior become far more accessible but there are other constraints. We must not presume that the captive environment does not distort the behavior of the animals and that the animals are arranged into proper social groups. Especially with regard to possibly infrequent interactions between adult males and infants, we must assure ourselves that any observed male care is *not* an artifact of the laboratory. It should be clear that the male was not forced to interact with the infant.

There are also the problems of negative reports and the lack of details. If authors indicate that they saw no male-infant interactions, male care of some form could still be occasionally occurring. And even when these interactions are reported, often there are no qualitative details given of the male's interactions with the infant.

Within these constraints on the available data, this paper reviews the interactions of adult males and infants in prosimians and in New World monkeys. There are separate sections for each of these two groups, and within each, the subsections are arranged by family. At the risk of losing some of the details of species differences, the data are grouped according to genus within each family. In almost all cases (except the Callitrichidae) there seemed not enough information to justify further splitting. The taxonomic classification used in this paper follows that of Bramblett (1976).

The available information is also summarized in tables. Table 14-1 lists the following for each genus in the prosimians and the New World primates: the social structure (solitary, monogamous, one-male harem, or multi-male); the reported interactions between adult males and infants; and the reference. Table 14-2 summarizes the information for each species of callitrichids in some greater detail. Since all species in this family are monogamous, it is hoped that this table will help clarify the differences among species, and between studies on the same species, with regard to when the adult male starts carrying the infants and how the male compares to the mother in amount of time carrying the infants. Also listed is the neonate-adult female ratio: the ratio of the weight of the infant(s) at birth to the weight of an adult female. Throughout this paper I will attempt to emphasize those environmental variables that correlate with different patterns of male care so as to allow us to better understanding of parental behavior and infant socialization in primates.

Table 14-1. Summary Table of Adult Male-Infant Interactions in Prosimians and New World Monkeys.

	Social Structure[1]	Source[2]	Reported Interactions of Adult Males with Infants	Reference
PROSIMIANS				
1. Tarsiidae				
Tarsius	M	F, C	Ignore young	Niemitz, 1974, 1979
2. Lorisidae				
Galago	M,S	F	Live and sleep with	Charles-Dominique and Bearder, 1979
		C	Groom if mother not present	Doyle et al., 1969
		C	Frequently groom, sometimes protect	Marney-Petix, pers. comm.
		C	Extreme interest	Sauer, 1967
		C	Play	Sauer and Sauer, 1962
		C	Play	Ehrlich, 1977
		C	Groom and hold, when presented with separated infants	Rosenson, 1972
Loris				
Arctocebus	S, M	F	No reported interactions	Charles-Dominique and Bearder, 1979 (all genera)
Perodicticus				
Nycticebus				
3. Daubentoniidae				
Daubentonia	S	F	No interactions	Petter, 1962, 1965
4. Indriidae				
Avahi	M	F	Attracted to infants, but no reported interactions; sleep?	Petter, 1965
Indri	M	F		
Propithecus	M	F	Approach, groom, occasionally hold	Jolly, 1966, 1972

[1]S = solitary; M = monogamous; H = polygynous (one-male, multi-female groups); P= polygamous (multi male, multi female groups).
NIA = No information available
[2]F = field, C = captive.

(continued)

Table 14-1. *(Continued)*

	Social Structure[1]	Source[2]	Reported Interactions of Adult Males with Infants	Reference
5. Cheirogaleidae				
Cheirogaleus	S	F	Females live apart from males except during brief mating interactions	Petter, 1962, 1965; Martin, 1973
Microcebus				Petter et al., 1975
Phaner				
6. Lemuridae				
Hapelemur	M	F	Little interaction; sleep?	Petter, 1965
				Petter and Peyrieras, 1974
				Pollok, 1979
		C?		Boskoff, cited in Kloper and Boskoff, 1979
Lepilemur	S	F	Solitary	Petter, 1962, 1965
			Infant parked in branch	Pollock, 1979
Lemur	P	C	Occasional groom or sniff, usually in 1st week	Klopfer and Klopfer, 1970
Varecia				Klopfer, 1972, 1974
				Klopfer and Dugard, 1976
				Harrington, 1978
Lemur		C	When mother removed for week, contact and groom	Klopfer, 1972
Varecia				Klopfer and Dugard, 1976
NEW WORLD MONKEYS				
1. Cebidae				
1.1 Aotinae		C	Takes active role and shares carrying about equally with female	Leibrecht and Kelley, 1977
Aotus	M	C?	Carries only after first week	Renquist, cited in Kleiman, 1977
		C	Carries in first few hours after birth	Meritt, 1980
		F	Primary carrier after first 10 days	Moynihan, 1964
1.2 Alouattinae				
Alouatta seniculus	H, P	F	Tolerant but no interest	Neville, 1972

A. palliata	H, P	F	Frequently investigated and several times "baby sat" for up to 30 minutes	Glander, 1975
		F	Play	Bernstein, 1964
		F	Adult males intruding into play terminated bouts	Baldwin and Baldwin, 1978
		F	Rescue and protect	Collias and Southwick, 1952
A. seniculus		F	Infanticide	Rudran, 1979
1.3 Cebinae				
Cebus	P	C	Play, sometimes carry	Bernstein, 1965
C. albifrons		F	Tolerant, sometimes carry	Defler, 1979
Saimiri	P	C	Frequently play-wrestle during and after week 5	Vandenbergh, 1966
		C	Approach but not retrieve separated infants	Rosenblum, 1972
		C	Respond to threat to infant with behavioral and hormonal changes	Vogt et al., 1982.
		C	Avoid females during parturition	Bowden et al., 1967
		C	In zoo subadult, but never adult, males observed carrying	Hunt et al., 1978
		C	Greatly increased contact when group females removed	Vaitl, 1977
		C	When infant trapped and giving distress calls, tried to retrieve and then threatened human	Baldwin, 1968
		C	Avoid infants for first 2 months, then allow developing infants to establish and maintain contact	Rosenblum and Coe, 1977
1.4 Atelinae				
Ateles geoffroyi	P	L	In zoo, no reports	Klein and Klein, 1971
Lagothrix	P		In zoo very interested in birth and showed protective behavior to human visitors and to other group members	Williams, 1967

[1]S = solitary; M = monogamous; H = polygynous (one-male, multi-female groups); P= polygamous (multi male, multi female groups).
NIA = No information available
[2]F = field, C = captive.

(continued)

Table 14-1. *(Continued)*

	Social Structure[1]	Source[2]	Reported Interactions of Adult Males with Infants	Reference
Brachyteles	NIA			
1.5 Pitheciinae				
Pithecia	P	C	In zoo, tried to steal newborn	Hanif, 1967
Chiropotes	NIA			
Cacajao	NIA			
2. Callitrichidae				
2.1 Callimiconinae				
Callicebus torquatus	M	F	Usually accompany and protect older infants	Kinzey et al., 1977
C. torquatus		F	Often carry	Sleeper, 1978
C. moloch		F	Frequently carry	Mason, 1966; Robinson, 1977
C. moloch		C	Male carried more than mother or siblings	Fragaszy et al., 1982
Callimico	M	C	Mother carries exclusively first few weeks; next few weeks male and mother share equally	Heltne et al., 1973; Pook, 1978
2.2 Callithrichinae				
Callithrix	M	C	Males share carrying equally with females	Ingram, 1977, 1978a
		C	Males carry frequently, but less than females	Box, 1975, 1977
		C	Investigate infant from birth	Rothe, 1978; Stevenson, 1976
		C	More likely to carry male infants	Ingram, 1978a
		C	More accepting of infants than are females	Ingram, 1977, 1978a
		C	Experience has little effect, but lab-born males carry more than wild-born	Ingram, 1978b
Saguinus oedipus	M	C	Carry as much as females	Moynihan, 1970
		F	Both males and females actively care for young	Dawson, 1978
S. nigricollis		F	Occasionally touch infant and show protective behavior but normally only mother seen carrying infant	Izawa, 1978
S. mystax		F	Mother primary carrier	Castro & Soini, 1978

S. oedipus	M	C	Frequently carry	Epple, 1967; Hampton et al., 1966
S. geoffroyi	M	C	Approached and retrieved still-wet neonate	Epple, 1970
S. oedipus	M	C	Almost exclusive care during first 5 weeks	Wendt, 1964
		C	Early infant development (5-6 weeks) more the responsibility of male; carries equally with female in weeks 7-14	Snowden, pers. comm.
S. fuscicollis	M	C	In each of 7 lab groups, infant carrying shared; in 5 of 7, dominant male carried most frequently	Epple, 1975; chapter 1
		C	For 3 sets of infants, males carried more than females at almost every week	Vogt et al., 1978
Cebuella	M	C	Carry most frequently, mothers carry only when nursing	Christen, 1974
		C	Primarily responsible after day 2	Pola and Snowden, 1975
Leontopithecus	M	C	More interested in infants than were females; mother carried exclusively for first few days (~9), then males exclusively except for nursing	Snyder, 1974
		C	In zoo, mothers primary carrier through week 3, then males in weeks 4-12; fathers carried sons more than daughters; experienced parents more successful (i.e., more infants survived) than inexperienced parents	Hoage, 1978

[1]S = solitary; M = monogamous; H = polygynous (one-male, multi-female groups); P= polygamous (multi male, multi female groups).
NIA = No information available
[2]F = field, C = captive.

Table 14-2 Selected Characteristics of Paternal Care in Callitrichids.

Species	Neonate-Adult[1] Female Ratio	Source[2]	Infant Care by Adult Male		Reference
			Starts at[3]	Vs. Mother[4]	
Callicebus	0.10				
C.?		C	4 hr	More than	Meritt, 1980
C. torquatus		F	n/a	More than	Kinzey et al., 1977
		F	n/a		Sleeper, 1978
C. moloch		F	day 1	Much more than	Mason, 1966
		F	n/a	Equal?	Robinson, 1977
		C	day 1	Much more than	Fragaszy et al., 1982
Callimico	0.07-0.10				
C. goeldi		C	wk 3	First 3 weeks much less than; next few weeks, equal	Heltne et al., 1973 Pook, 1978
Callithrix	0.23-0.27				
C. jacchus		C	day 1	Less than	Box, 1975, 1977
		C	day 1	Equal	Ingram, 1977, 1978
Saguinus	0.14-0.18				
S. oedipus		F	n/a	Equal	Dawson, 1978
		C	n/a	Equal	Moynihan, 1970

		C	n/a	Equal?	Epple, 1967; Hampton et al., 1966
		C	day 1	Much more during first 5 weeks	Wendt, 1964
		C	day 1?	More during first 5-6 weeks; equal in weeks 7-14	Snowden (pers. comm.)
S. nigricollis		F	n/a	Much less than	Izawa, 1978
S mystax		F	n/a	Less than	Castro & Soini, 1978
S. geoffroyi		C	day 1	Equal?	Epple, 1970
S. fuscicollis		C	day 1	Much more than	Vogt et al., 1978
			day 1	More than	Epple, 1975
Cebuella	0.22				
C. pygmaea		C	day 1	Much more than	Christen, 1974
		C	day 3?	After day 2, much more than	Pola and Snowden, 1975
Leontopithecus	0.19				
L. rosalia		C	day 10	None for first 9 days; after which much more than	Snyder, 1974
		C	wk 2	Much less than, weeks 1-3; much more than, weeks 4-12	Hoage, 1978

[1]Ratio of the weight of the infant(s) at birth to the weight of a (nonpregnant) adult female. All data from Kleiman (1977), except for *Callicebus*, which is from Fragaszy et al. (1982).
[2]C = captive; F = field.
[3]Time after parturition at which adult initiates infant caretaking. n/a = no information available.
[4]During the infant development, how caretaking of adult male compares to that of mother.

PROSIMIANS

Introduction

Behavioral studies of the prosimians have increased dramatically in the last few years. Several books (Martin et al., 1974; Tattersall and Sussman, 1975; Charles-Dominique, 1977; Doyle and Martin, 1979) as well as numerous journal articles indicate the quantity of new data on prosimian behavior from both field and laboratory sources. Enough information on maternal behavior in prosimians has accumulated to allow a recent review of this area by Klopfer and Boskoff (1979). They were able to relate the patterns of prosimian maternal care to social structure variables, and their review greatly facilitated my efforts in this chapter to examine and organize the reports on adult male-infant interactions in prosimians. The information in this paper is detailed in the section below ("Review of Each Family") and is summarized in the first part of Table 14-1. As can be seen in this table, it appears the role of prosimian males in infant development is rather limited. There is a wide range of social structures represented across the six families of the prosimians. However, even in those genera that are monogamous, there is only infrequent direct caretaking by the adult male.

Review of Each Family

Tarsiidae. The members of the Tarsiid family, the diminuitive tarsiers (*Tarsius* sp.), appear to live as male-female adult pairs (Charles-Dominique, 1977). However, there is some species variability since in *T. syrichta,* adults have the ability to form more complex social groupings, while in *T. bancanus,* an adult male and female have been observed to inhabit and defend a territory and to show a strong pair-bond (Niemitz, 1979). For this latter species there also are earlier reports of their asocial nature (Fogden, 1974). The limited observations on the social behavior of the tarsiers indicate, at least for *T. bancanus,* minimal interactions between adult males and infants. Niemitz (1974) introduced a hand-reared, wild-born infant *T. bancanus* to an adult male who showed no response whatsoever, even when the infant grabbed onto the male's tail. Niemitz (1974) concluded that the adult male "seemed to avoid the young tarsier's approaches" (p. 261).

Lorisidae. There are two subfamilies of lorisids, the lorisines and the galagines. The lorisines have the widest distribution of all prosimian groups and occur throughout the tropical rain forests of Africa and Asia (Petter and Petter-Rousseaux, 1979). Their social structure is considered to be unclear and they probably live singly or in pairs (Charles-Dominique and Bearder,

1979; Napier and Napier, 1967) with no reported interactions between adult males and infants. Much more is known of the galagines, which, like the lorisines, are nocturnal. The social structure of *Galago senegalensis* and *G. crassicaudatus* in the wild has been observed to be stable, sleeping groups of two to six, possibly consisting of family groups (Bearder and Doyle, 1974; Charles-Dominique and Bearder, 1979). The social structure of *G. demidovi* appears to be more solitary (Charles-Dominique, 1972). These limited reports on social behavior indicate that adult males play no role in the rearing of infants but do appear to live quietly with juveniles and to sleep occasionally with a maternal group (Charles-Dominique and Bearder, 1979).

Observations of the galagines under seminaturalistic or laboratory conditions have presented more information with regard to the interactions of males with infants. For *G. senegalensis* adult males are apparently tolerant but generally not attracted to infants. Doyle and colleagues (1969) reported that very occasionally the adult male of the group groomed an infant but only if the mother was not present. In the same species Marney-Petix (unpub. obs.) observed that of two males in a lab group, one individual frequently interacted with an older infant and the other male was only rarely seen with infants. And Sauer and Sauer (1962) observed an adult male *G. senegalensis* occasionally playing with infants. For *G. crassicaudatus* Ehrlich (1977) reported that an adult male played extensively with the infant, although in this group the infant spent as much time with the other animals, especially an adult female, as with the mother.

Rosenson (1972) performed a series of experiments to study the issue of the responsiveness of adult *Galago* to separated infants. In this research he presented a young (7 to 10-day-old) maternally separated infant to adult *G. crassicaudatus* of various backgrounds. The four adult males (all wild-born) in the test all groomed the infant and one attempted to hold the infant. The males, who were as responsive in the tests as were the adult females, were apparently quite persistent in their interest in the younger animals, since the infant's most frequent response was to reject the social advances of any of the adults.

Daubentoniidae. The one species of this family, *Daubentonia madagascariensis,* is considered to be solitary and highly endangered. There is little information on this elusive, slow-moving nocturnal prosimian, and the only observed groups have been females with an offspring (Petter, 1962, 1965; Petter and Petter, 1967; Petter and Peyrieras, 1974).

Indriidae. According to Klopfer and Boskoff (1979), the indriids are very difficult to observe in the wild and to maintain in captivity. Members of this family are arboreal folivores and live in small family groups (Petter, 1965), although larger groups may exist (Pollock, 1979). The wooly lemur *(Avahi laniger)* is a smaller nocturnal species of this family, while the indri *(I. indri)* is diurnal and the largest of the indriids. Groups of two to four individuals have been observed for both these species (Petter, 1965; Petter and Charles-Dominique, 1979; Pollock, 1979) and these groups are generally considered to be family groups (but see Pollock, 1979). Little information on social behavior is available for these species, but Petter (1965) reported that adult males were attracted to newborn infants.

For *Propithecus verreauxi,* another species in this family, there is much more information available (Petter, 1965; Klopfer and Klopfer, 1970; Jolly, 1966, 1972; Richard, 1974a, 1974b). Indeed, this diural folivore is considered by some (Pollock, 1979) to be the most intensively studied lemur. Members of this species exist in groups of three to eight individuals, which are presumed to be extended family groups (Jolly, 1966; Richard, 1974a, 1974b). One of the most salient features of the social structure of this species is the multiple parenting of the infants (Jolly, 1966). This appears to be the only prosimian species in which the adult males, albeit with other group members, have more than casual interactions with infants. The *P. verreauxi* infant grows up with extensive interactions with all other members of its group: the father, older siblings, etc. Males approach and groom the infant as frequently as the mother does and at times are even allowed to hold the infant (Jolly, 1966).

Cheirogaleidae. Members of this family include the smaller lemurs—the dwarf *(Cheirogaleus* sp.), mouse *(Microcebus murinus),* and fork-marked *(Phaner furcifer)* lemurs—all of which are nocturnal (Petter, 1962, 1965; Martin, 1973; Petter et al., 1975). Adult males show virtually no interaction with infants; females, either alone or in small groups, live apart from adult males except for brief interactions during the mating season (Pollock, 1979).

Lemuridae. Of the four genera within this family, two *(Lemur* and *Varecia)* are rather widely studied while the other two *(Hapelemur* and *Lepilemur)* are relatively unknown. For *Hapelemur* the social organization is a family group of three to six individuals (Petter, 1965; Petter and Peyrieras, 1974; Pollock, 1979). The limited information available on this genus implies that little interaction occurs between the adult male and the infant (Petter and Peyrieras, 1974; Boskoff, cited in Klopfer & Boskoff, 1979). The nocturnal *Lepilemur,* which is sometimes considered to be a separate taxonomic family (Petter and Petter-Rousseaux, 1979), exhibits a solitary social structure; individuals are observed alone except for mother-infant pairs (Petter, 1962, 1965; Pollock, 1979). The *Lepilemur* infant's contact, with even its mother is not as extensive as the mother-infant interactions in other prosimian species, since the infant is often left clinging to a branch while the mother feeds (Petter, 1965). Interactions with males are unreported.

Much more detailed research has been done on members of *Lemur* and *Varecia.* Extensive field studies have been reported on *L. catta* (Jolly, 1966; Jolly, 1972; Klopfer, 1972; Sussman, 1974, 1977), *L. fulvus* (Harrington, 1975), *L. macaco* (Jolly, 1966; Sussman, 1977), and *V. variegata* (Klopfer and Klopfer, 1970). In all of the species of *Lemur,* the social structure consists of groupings of 10 to 20 animals, which are larger than troops reported for other prosimians. *V. variegata* apparently lives in small groups of two to five individuals. The detailed laboratory studies on these species have provided the most complete information on maternal care in prosimians (Klopfer and Boskoff, 1979). This research indicates that during undisturbed conditions, the adult male's involvement with infants is limited to occasional grooming or sniffing of the infant, usually after the first week of life (Klopfer and Klopfer, 1970; Klopfer, 1972, 1974; Klopfer and Dugard, 1976; Harrington, 1978). However, in *L. fulvus* the presence of the adult male does serve to accelerate the infant's spatial independence from the mother; when compared with infants raised without the presence of adult males, infants raised with adult males were spatially further from the mother during development (Klopfer, 1972). *L. macaco* males under captive conditions attempted to approach infants from the first week but were not allowed contact until the infant was five weeks old (Harrington, 1978).

Despite the infrequent interaction with the infant, the adult male lemur appears capable of moderating the behavioral distress of a maternally separated infant. Studies with *L. macaco, L. fulvus,* and *V. variegata* have indicated that infants (five to seven months of age) separated for seven days from the mother showed minimal behavioral agitation if housed with another animal—even if that animal was an adult male. During the mother's absence, the infant's contact and grooming activity with the male was greatly increased, but when the mother returned, the infant's contact and proximity with the male declined to the very low preseparation values (Klopfer, 1972; Klopfer and Dugard, 1976).

Summary

Among the prosimians, there appear to be several genera in which the adult males show some form of direct care of infants. Those seven genera are: *Galago, Avahi, Indri, Propithecus, Hapelemur, Lemur,* and *Varecia.* In only one of these, *Propithecus,* is there any actual infant carrying by the male, although his participation in infant caretaking is still less than that of the mother (Jolly, 1966).

The information on the remaining 10 genera of prosimians *(Tarsius, Loris, Arctocebus, Perodicticus, Nycticebus, Daubentonia, Cheirogaleus, Microcebus, Phaner,* and *Lepilemur)* indicate that the adult male-infant interactions in these taxa are extremely limited or have not been observed. For some of these genera, especially those that are nocturnal (e.g., *Daubentonia*), there is little information available on social behavior and thus subsequent research may somewhat modify these groupings of those prosimians in which males do and do not display male care of infants. However, it does seem safe to conclude that no species within the prosimians exhibits direct male care involving protection or carrying of the infant, at least equivalent to that of the mother.

NEW WORLD MONKEYS

Introduction

The New World primates are divided into two families: the Cebidae and the Callitrichidae. As can be seen in the second part of Table 14-1, for the cebids there is great variability in the range of social structures and the types of adult male-infant interactions; in contrast, for *all* the callitrichids, the social organization is monogomous and the adult males share infant caretaking with the mother. The callitrichids also exhibit another notable trait; each species in this family, with the exception of *Callimico* and *Callicebus,* characteristically produces twin offspring, a feature not otherwise seen in any other primates (Eisenberg, 1978). Cebid monkeys, in contrast, give birth to singleton infants, and the social organization varies from monogamous family groups *(Aotus)* to one-male groups *(Alouatta)* to large multi-male, multi-female groups *(Saimiri)* (Eisenberg et al, 1972; Eisenberg, 1978). The only monogamous cebid species, the owl monkey *(Aotus trivirgatus),* is the only species in which the adult male reveals active care of the infant. For the remainder of the Cebidae, the role of the adult male in infant development is, to varying degrees, less pronounced. There are, however, recent reports to indicate that adult males in some cebid species will respond vigorously to infants.

Review of Each Family

Cebidae. Of the five subfamilies in the Cebidae, only the Aotinae show a monogamous social structure (Moynihan, 1964, 1976). The owl monkey is the only species in this subfamily and the only nocturnal primate in all of the New World. The adult male owl monkey takes a very active role in infant care (Leibrecht and Kelley, 1977). Some observers have reported that during the first few days after parturition, the infant is carried exclusively by the mother, but later in development the adult male is the only animal to transport the infant except during its nursing on the mother (Moynihan, 1964; Renquist cited by Kleiman, 1977). However, Meritt (1980) has observed that in zoo groups *Aotus* males will carry the infant a few hours after its birth.

The Alouattinae subfamily includes the howler monkeys (*Alouatta* sp.), and as their common name implies, these animals are characterized by their dramatic vocalizations. The social structure of howlers has been regarded as either one-male or age-graded male troops (Eisenberg et al., 1972; Bramblett, 1976). The field reports indicate that for *A. seniculus,* adult males are generally tolerant of infants but show little interest or curiosity (Neville, 1972). For *A. palliata* there appears to be somewhat more interaction between infants and nonmaternal group members. A very brief report of a 14-month field study by Glander (1975) noted that adult males "frequently investigated the infants and often the infants transferred to the males at this time" (p. 492). Adult males were also occasionally observed "baby sitting" an infant for as long as 30 minutes. There are also reported instances of *A. palliata* adult males playing with infants (Bernstein, 1964), although Baldwin and Baldwin (1978) reported for the same species that the very infrequent intrusions of adult males into infant-juvenile play groups terminated those play bouts. Adult male howler monkeys will also apparently show protective behavior. Collias and Southwick (1952) reported that an adult male *A. palliata* made a bridge with his body so that an infant could make a difficult crossing and that adult males were seen rescuing infants whose mothers had been shot.

Recent evidence has revealed even more dramatic interactions between adult male howlers and infants. Infanticide has been observed and documented by Rudran (1979) in *A. seniculus.* Apparently the male(s) taking over a harem of females will attempt to kill or severely injure young infants; the result is that the females will then cease lactation and become sexually receptive to the new males. Thus, this social process in howlers appears to be similar to the infanticide reported in langurs (Hrdy, 1977) and lions (Bertram, 1976). The observations by Rudran (1979) are the only reports of infanticide in a New World monkey species (Hrdy, 1979).

The Cebinae subfamily consists of the *Saimiri* and *Cebus* genera, and species in both of these are organized as multi-male, multi-female groups with polygamous mating systems. Bernstein (1965) reported that captive *Cebus* males were generally tolerant of infants and sometimes played with and carried them. In a field study on *C. albifrons,* Defler (1979) observed adult males to be highly tolerant of infants and sometimes to carry them for very brief distances. Squirrel monkey *(Saimiri)* adult males are not quite as tolerant of infants, although after the first few months of development the adult males will interact some with infants. In captivity, adult males avoid the mother during parturition (Bowden et al., 1967) and move or stay away from younger infants (Rosenblum and Coe, 1977). When these infants reach two months of age, however, the adult males begin to allow the young animal to establish contact rather than immediately fleeing from the approaching infant (Rosenblum and Coe, 1977). Although there are reports of adult males vigorously play-wrestling with infants older than five weeks (Vandenbergh, 1966), these interactions apparently do not include infant carrying. Observations of "aunting" in a zoo group of *Saimiri* by Hunt et al. (1978) noted that subadult males sometimes carried infants, but adult males were never observed infant carrying.

Under conditions of removal of the mother, adult male squirrel monkeys will exhibit some interactions with infants, especially if the infants are older. Vaitl (1977) removed the females from laboratory social groups and found that the infants (4 to 14 months of age) and adult males frequently contacted and played with each other. They even slept together, although all these interactions ceased as soon as the females were returned to the group. Using much younger infants, Rosenblum (1972) performed a laboratory study on adult responsiveness to separated infants by exposing various age-sex groups of *Saimiri sciureus* to a young (17 to 24 days old) maternally separated infant. He observed that, of all groups of animals tested, adult males were least likely to contact or retrieve the infant. Although the adult males approached the introduced infant as much as did any other age-sex class, the adult males would not fully retrieve (i.e., attempt to pick up) the infant, even if no other animals were present.

Adult male squirrel monkeys are not, however, totally insensitive to infants, as they appear to respond in a protective manner to threats toward the infant. Baldwin (1968) reported that under seminaturalistic conditions, subadult males protected, and even attempted to retrieve, an infant when it was threatened by a human stranger. Recent experimental evidence appear to support these observations on adult male protection. Vogt and colleagues (1980; 1982) have recently assessed the responsiveness of adult squirrel monkeys to mother-infant separation. In the first study (Vogt et al., 1980) a group of adult male and female *Saimiri,* living next to a group of mother-infant pairs, was exposed to several brief (30-minute) experimental conditions involving three types of social disruption. In two of these types, an infant was removed from its mother and its home cage and then either placed next to the adult group or placed in a separate room. The third social disruption condition involved placing an unfamiliar juvenile conspecific next to the adult group. Behavioral measures and adrenocorticoid activity were assessed for both adult males and females for each test period. The results indicated that only the two infant separation conditions produced significant behavioral and hormonal differences and that males differed from females during these. The females approached, while the males avoided, the separated infant. And compared with undisturbed baseline conditions, only the infant separation conditions resulted in significant elevations in plasma cortisol and then only for the females. The adult males failed to exhibit any adrenocorticoid responses to any of the manipulations.

These findings of Vogt et al. (1980) were pursued in a subsequent study that tested other factors in *Saimiri* adult male responsiveness to infants (Vogt et al., 1982). In this experiment, the adult male squirrel monkeys actually lived with the mother-infant dyads from the birth of the infants. Thus for several months before testing, the adult males were continually with the infants. Additionally, the test conditions included several different kinds of mother-infant separation, one of which involved one of the human experimenter's holding the separated infant in front of the group cage throughout the 30-minute test period. It was only this condition of holding the separated infant, and not any of the other types of mother-infant separation, that produced in the adult males elevated plasma cortisol values and behavioral changes indicative of arousal (Vogt et al., 1982). Thus, in a situation threatening to infant squirrel monkeys, adult male *S. sciureus* respond with behavioral and adrenocorticoid changes, while in a situation of mother-infant separation, adult males show avoidance and no cortisol elevations.

For the two remaining subfamilies in the Cebidae—the Atelinae and the Pitheciinae—there is rather less information available. Species in both subfamilies show the multi-male, multi-female social structure. Within the Atelinae, the spider monkey *(Ateles)* has been the subject of several major field studies (e.g., Richard, 1970; Klein and Klein, 1973; Eisenberg, 1976); yet typical group size and the details of social organization are unclear, and the reports on social behavior are very limited. Observations on a zoo colony of *A. geoffroyi* by Klein and Klein (1971) indicated no interactions between infants and adult males. A zoo report for *Lagothrix lagothricha,* another species in this subfamily, did note the interest of the adult male in a newborn (Williams, 1967). During parturition the only adult male in the group evidenced great curiosity and after the birth overt protective behavior when human visitors or other group members approached the young infant.

The Pitheciinae subfamily include three genera (*Chiropotes, Cacajao, Pithecia)* for which there is very little information available. Pithecia is presumed to be monogamous, with the adult male sharing infant carrying with the mother (Eisenberg, 1978). However, the only report of male interest in infants is by Hanif (1967). He noted that in a zoo group of saki monkeys *(Pithecia pithecia),* the adult attempted to steal a newborn infant.

Callitrichidae. Throughout the callitrichids, active male care of infants is an essential and salient feature of their social behavior and is observed in every species of this group. Thus, among the New World primates—and indeed among all primates—the Callitricihidae is the family in which adult male-infant interactions are most frequent and direct. There are other unique features of these monkeys: the monogamous social structure, the small size, and the production of twin offspring. It is not clear whether these features are phylogenetically recent developments or were also characteristic of the ancestral groups, although in a recent paper Leutenegger (1980) argues that monogamy, multiple births, and male care are the result of evolved small size (phyletic dwarfism) (see too chapter 13).

Although there are some species differences, the adult males of each species in this family exhibit a strong interest in the infants soon after birth and show active infant caretaking that, except for nursing, is comparable to that of the mother. Table 14-2 summarizes for each callitrichid species the available information on the male's initiation of carrying and the male's infant carrying compared with that of the mother. Also given is the neonate to adult female ratio: the ratio of the weight of the newborn infant(s) to the weight of an adult female.

The callitrichids are divided into two subfamilies, the Callimiconinae and the Callithrichinae (Bramblett, 1976). Within the former, there are two genera, *Callicebus* and *Callimico,* and the males in both of these exhibit extensive infant care. Field studies by Kinzey et al. (1977) on *Callicebus torquatus* and by Mason (1966, 1968), Robinson (1977), and Sleeper (1978) on *C. moloch* indicated that infants were frequently observed to be carried by the adult male. A recent report on zoo *Callicebus* noted that the adult male carried the infant over 60% of the time during the first 70 days (Meritt, 1980). A more detailed study on captive *C. moloch* has provided further quantitative data on parental care. Fragaszy et al. (1982) observed three infants over the first six months of life and found that each infant spent at least twice as much time with the adult males as with the mother during the first five months. Various interactions indicated that although the infants showed strong affiliative relationships with the mother and the sibling, the father appeared to be the primary focus of attention. If frightened or startled, the infants were observed to retreat to, or to be retrieved by, the adult male. Fragaszy et al. also noted that for the first four months the infants were primarily dependent upon an adult carrier. This period of the infants' dependence upon the adults appears to be longer than that observed in other New World monkeys characterized by male involvement in infant development, and its significance is not yet clear.

The other member of the Callimiconinae subfamily is Goeldi's monkey *(Callimico goeldii),* for which there are two studies on captive groups of animals (Heltne et al., 1973; Pook, 1978 and chapter 13). Each of these reports presents a picture of active male participation in infant development, but a slightly different pattern than for *Callicebus.* Both Heltne et al. and Pook reported that the mother was exclusively involved in infant care for the first few weeks, and for the subsequent few weeks the mother and father shared the carrying of the infant about equally. Thereafter each adult slowly started to reject the infant, as the older siblings started to show carrying of the infant for the first time. Thus, *Callimico* males appear to play an especially active parental role during the second month of the infant's life.

Besides the Callimiconinae subfamily, the other subfamily in the callitrichids is the Callitrichinae, which includes the four genera of marmosets and tamarins; *Callithrix, Saguinus, Cebuella,* and *Leontopithecus.* Although there are some species differences, the marmoset and tamarin males generally exhibit care of infants equivalent to adult females (see also chapter 1).

For the common marmosets *(Callithrix jacchus),* virtually all observers have reported that adult males are actively involved with infants (Fitzgerald, 1935; Epple, 1967). An early study indicated that except for nursing, only the adult male carried the infants (Stellar, 1960); more recent reports have indicated that adult male marmosets share infant carrying about equally with females (Ingram, 1977, 1978a) or that females carry more than males (Box, 1975, 1977). The interest of adult male *Callithrix* in infants is apparently present from at least parturition onward, as adult males, along with the rest of the group, show a strong interest in the birth process. This trend to investigate the newborn infant is stronger in the adult male than in the other animals (Rothe, 1978) and the adult male is the only animal able to contact the infant immediately after birth (Stevenson, 1976).

Ingram (1977, 1978a, 1978b) has examined more closely the interactions of *Callithrix* adult males and infants in order to reveal the effects of variables such as the infant's sex and the male's previous experience. She found that adult males were more likely to carry male infants than female infants (Ingram, 1978a). The adult males were more accepting of the infants than were adult females, but when the adult males did reject an infant, it was likely to be a male infant (Ingram, 1977). Also, adult males more that adult females were willing to allow infants to initiate contact and were responsible for proximity maintenance (Ingram, 1978a). And, finally, a male's caretaking experience with one set of infants had little effect upon his interest in subsequent offspring, although laboratory-born adult males did carry infants more frequently than did wild-born adult males (Ingram, 1978b).

The data on adult male-infant interactions for the various species of tamarins *(Saguinus)* reflect rather large intergroup and/or interspecific variability; there does, however, appear to be a trend for *Saguinus* adult males to care for infants at least as much as do the adult females. This tendency does contrast with the data for *Callithrix.* The field observations indicate that for *S. oedipus* (Moynihan, 1970; Dawson, 1978) both males and females actively participate in infant care, while for *S. mystax* (Castro and Soini, 1978) and *S. nigricollis* (Izawa, 1978) the mother appeared to be the primary caretaker. Izawa (1978) noted that the adult male occasionally touched the infants and showed protective behavior, but normally only the mother was seen carrying the infant.

Laboratory observations on *S. oedipus* by Hampton et al. (1966) noted that the adult males showed extraordinary interest in the infants and frequently solicited transfer of the young. Other observations on captive groups of the same species indicated that the adult male carried the infant about as frequently as the mother did (Epple, 1967) or, in one report (Wendt, 1964), that the adult male took almost exclusive care of the infant during the first five weeks of its life. There may be, however, further support for the hypothesis that early infant development—at least in terms of infant carrying—is more the responsibility of the (adult?) males. Preliminary results ($n = 8$ infants) of a long-term and careful study of captive *S. oedipus* by Snowden (pers. comm.) indicate that in the first 5 to 6 weeks of the infant's development, the males (father and subadult siblings) carried significantly more than the females, and that in weeks 7 to 14, the males and females shared infant carrying about equally. Thus, in the first month of the life of infant *S. oedipus,* males more than females may engage in caretaking.

There is one other tamarin species , *S. fuscicollis,* for which parental care patterns are well documented in the laboratory. It appears for these animals that the adult male carries the infants more frequently than does the female (but see chapter 1 for a different view). In a longitudinal study, Vogt et al. (1978) obtained data on three sets of infants from one family group of *S. fuscicollis* maintained under seminaturalistic conditions. They found that at almost every week through 12 weeks, when the infants are essentially no longer carried, the father carried more than the mother and the males more than the females. Overall, the males accounted for over 70% of the observed infant carrying (Vogt et al., 1978). In a cross-sectional study, Epple (1975) observed seven lab groups, for a total of eight infants, and found that in five of those groups, the adult male was the most frequent carrier of the infants. However, in all seven groups, infant carrying was shared by all members of a group (chapter 1). From these observations, Epple (1978) hypothesized that experience is rather important in adequate parental care but more so for mothers than for fathers, and that the care of siblings may be the crucial factor in the development of adequate parental care.

The remaining two genera of this subfamily are *Cebuella* and *Leontopithecus,* and for these less detailed information is available than for *Saguinus or Callithrix.* The pygmy marmoset *(Cebuella pygmaea)* is slightly smaller than the other marmosets and has only recently been the subject of field study. For example, Ramirez et al. (1978) studied the feeding ecology of *C.*

pygmaea in Peru and noted that the adult male did carry the twin infants. However, little detailed information on infant carrying or other social behavior in the field is currently available. It has been reported that, under captive conditions, the male of the adult pair performs the great majority of the infant carrying, and the mother tolerates the infants usually only during nursing (Christen, 1968, 1974; Pola and Snowden, 1975).

The golden lion marmoset *(Leontopithecus rosalia)* is a highly endangered species, and much of our information on the social behavior of this animal is the result of the breeding programs at zoos (e.g., Kleiman, 1978). Brief observations by Snyder (1972) and by Coimbra-Filho and Magnanini (1972) noted that the adult males were more interested in the infants than were the mothers, although there was an initial period of several days when the mother was almost exclusively in charge of the infants. A more recent study with animals at the National Zoo has presented some of the parameters of parental care in *Leontopithecus.* Hoage (1978) observed seven sets of *L. rosalia* infants ($n = 11$) in three groups. He found that, although the mothers were the principal infant carriers through week 3, the adult males took over this role in weeks 4 to 12. He also observed that in cases where heterosexual twins were born, the parents preferred carrying infants of their own sex. And with regard to previous parental experience, Hoage noted for *L. rosalia,* as did Epple (1978) for *S. fuscicollis,* that juveniles allowed exposure to infants were more effective parents when they became adults.

Summary

In the New World primates, the clear majority of genera (12 out of a total of 16) involve some sort of direct male care. These genera include all those in the Callitrichidae *(Callimico, Callicebus, Callithrix, Saguinus, Cebuella,* and *Leontopithecus)* plus six in the Cebidae *(Aotus, Alouatta, Cebus, Saimiri, Lagothrix,* and *Pithecia).* Of these 12 genera, at least seven (all the callitrichids plus *Aotus),* and possibly *Pithecia,* are monogamous, and the males care for the infants by carrying them during a substantial portion of their development. The males in the remaining four genera in the Cebidae exhibit less intensive forms of interaction with infants (e.g., play).

The New World genera in which the adult males seem not to show direct care of infants are *Ateles, Brachyteles, Chiropotes, and Cacajo.* Since the information on the social behavior in these four latter genera is very limited, future research may allow us to reassign some of these to the group in which direct male care is shown. Regardless, even if the present tentative division proves to be an accurate one, it is clearly the case that a majority of the genera in the New World primates can be characterized by direct male care of infants.

CONCLUSIONS

Across the prosimians and the New World primates, adult male care of infants is somewhat variable from genus to genus but is nevertheless present in a rather large proportion. Direct male care has been observed in 7 of 17 genera in the prosimians and in 12 of 16 genera in the New World monkeys. From this high incidence of male care, one might ask what social or ecological variables correlate with the adult males' interactions with infants. Of the social factors, the mating system—specifically, monogamy—seems to be that factor most directly related to the prevalence of direct interactions between adult males and infants. Indeed, of any variable, monogamy appears to be that factor most correlated with male care. For the New World monkeys, each species with a monogamous social organization (all of the callitrichids, which are listed in Table 14-2, plus *Aotus trivirgatus)* show intense infant care by the adult males. For the prosimians, the relationship between monogamy and male care of infants is not as strong. Some genera that are ostensibly labeled as monogamous, such as *Tarsius* and *Hapelemur,* reveal little male care of infants. But all those genera in which some adult male-infant interactions are observed are also classified as monogamous: *Indri* and *Propithecus.* The reasons male care is not as prevalent in the monogamous prosimians as it is in the monogamous New World primates are not clear at this time.

Within the New World monogamous species, there is some variability in the details of male care. As can be seen from Table 14-2, there are differences in the time at which the males start to carry the infants *(Callicebus* versus *Callimico)* and in the amount of infant carrying the males perform *(S. oedipus* versus *L. rosalia).* Moreover, these differences are not related to the neonate to adult female weight ratio. Possible other social or ecological factors are important in accounting for this variability in male care of infants (see, for example, chapter 13).

How important are ecological variables in relation to male care in prosimians and New World primates? A general qualitative assessment would indicate that these variables do not play a strong role. In her examination of monogamy in mammals, Kleiman (1977) found that none of several variables—daily activity patterns, feeding habits, breeding patterns, or neonatal maturation—was correlated with the presence of monogamy in mammals. Rather, three reproductive factors were related to monogamy: low reproductive potential, long maturation, and delayed sexual maturation. Several species in the New World monkeys, both those that are monogamous (e.g., *Saguinus fuscicollis* versus *Leontopithecus rosalia)* and nonmonogamous (e.g., *Cebus albifrons* versus *Saimiri sciureus*) differ in the details of adult male care of infants, and yet the species live under similar ecological conditions.

Perhaps it is best to conclude by indicating that there are probably several factors acting simultaneously that are important in predisposing males to care for infants. In a recent paper on the evolution of mammalian paternal care, Kleiman and Malcolm (1981) conclude that "it is probably not useful to look for global explanations in terms such as richness or harshness of the habitat. In many cases there are probably two or more predisposing factors that act in concert" (p. 377).

ACKNOWLEDGMENTS

During the writing of this chapter, I was supported by grant MH35827 from the National Institute of Mental Health. I would like to express my thanks to Mary Olk, Patty Wiczek, and Ann Arndt for reference and secretarial assistance.

REFERENCES

Baldwin, J. D. 1968. The social behavior of adult male squirrel monkeys *(Saimiri sciureus)* in a seminatural environment. *Folia Primatologica* 9: 281-314.

Baldwin, J. D., and Baldwin, J. I. 1978. Exploration and play in howler monkeys *(Alouatta palliata). Primates* 19(3): 411-22.

Bearder, S. K., and Doyle, G. A. 1974. Ecology of bushbabies, *Galago senegalensis* and *Galago crassicaudatus,* with some notes on their behavior in the field. In R. D. Martin; G. A. Doyle; and A. C. Walker (eds.), *Prosimian Biology,* pp. 109-30. Pittsburgh: University of Pittsburgh Press.

Bernstein, I. S. 1964. Role of the dominant male rhesus monkey in response to external challenges to the group. *Journal of Comparative and Physiological Psychology* 57: 404-06.

———. 1965. Activity patterns in a cebus monkey group. *Folia Primatologica* 3: 211-14.

Bertram, B. C. R. 1976. Kin selection in lions and in evolution. In P. P. G. Bateson, and R. A. Hinde (eds.), *Growing Points in Ethology,* pp. 281-302. London: Cambridge University Press.

Bowden, D.; Winter, P.; and Ploog, D. 1967. Pregnancy and delivery behavior in the squirrel monkey *(Saimiri sciureus)* and other primates. *Folia Primatologica* 5: 1-42.

Box, H. O. 1975. A social developmental study of young monkeys *(Callithrix jacchus)* within a captive family group. *Primates* 16: 419-35.

———. 1977. Quantitative data on the carrying of young captive monkeys *(Callithrix jacchus)* by other members of their family groups. *Primates* 18: 475-84.

Bramblett, C. A. 1976. *Patterns of Primate Behavior.* Palo Alto, CA: Mayfield Publishing Co.

Castro, R. and Soini P. 1978. Field studies on *Saguinas mystax* and other callitrichids in Amazon, Peru. In D. G. Kleiman (ed.), *The Biology and Conservation of the Callitrichidae,* pp. 73-78. Washington, D. C.: Smithsonian Institution Press.

Charles-Dominique, P. 1972. Ecologie et vie sociale de *Galago demidovi* (Fischer 1808, Prosimii). *Zeitschrift für Tierpsychologie* 9: 7-41.

———. 1977. *Ecology and Behavior of Nocturnal Primates.* New York: Columbia University Press.

Charles-Dominique, P., and Bearder, S. K. 1979. Field studies of lorisid behavior: Methodological aspects. In G. A. Doyle, and R. D. Martin (eds.), *The Study of Prosimian Behavior,* pp. 567-630. New York: Academic Press.

Christen, A. 1968. Haltung und brutbiologie von *Cebuella. Folia Primatologica* 8: 41-49.

———. 1974. Fortpflanzimgsbiologie und verhalten bei *Cebulla pygmaea* und *Tamarin tamarin. Zeitschrift fur Tierspsychologie* 14: 1-78.

Coimbra-Filho, A. F., and Magnanini, A. 1972. On the present status of *Leontopithecus* and some data about new behavioral aspects and management of *L. rosalia rosalia.* In D. Bridgewater (ed.), *Saving the Lion Marmoset,* pp. 59-69. Wheeling, WV: WAPT.

Collias, N., and Southwick, C. H. 1952. A field study of population density and social organization in howling monkeys. *Proceedings of the American Philosophical Society* 96: 143-56.

Dawson, G. A. 1978. Composition and stability of social groups of the tamarin, *Saguinas oedipus geoffroyi,* in Panama: Ecological and behavioral implications. In D. G. Kleiman (ed.), *The Biology and Conservation of the Callitrichidae,* pp. 23-38. Washington, D. C.: Smithsonian Institution Press.

Defler, T. R. 1979. On the ecology and behavior of *Cebus albifrons* in Eastern Columbia: II. Behavior. *Primates* 20: 491-502.

Doyle, G. A.; Anderson, A.; and Bearder, S. K. 1969. Maternal behavior in the lesser bushbaby *(Galago senegalensis moholi)* under semi-natural conditions. *Folia Primatologica* 11: 215-38.

Doyle, G. A., and Martin, R. D. 1979. *The Study of Prosimian Behavior.* New York: Academic Press.

Ehrlich, A. 1977. Social and individual behavior in captive greater Galagos. *Behaviour* 63: 192-214.

Eisenberg, J. F. 1976. Communication mechanisms and social integration in the black spider monkey, *Ateles fuciceps robustus,* and related species. *Smithsonian Contributions to Zoology* 213: 1-108.

Eisenberg, J. F. 1978.Comparative ecology and reproduction of New World monkeys. In D. G. Kleiman (ed.), *The Biology and Conservation of the Callitrichidae,* pp. 13-22. Washington, D. C.: Smithsonian Institution Press.

Eisenberg, J.F.; Muckenhirn, N.A.; and Rudran, R. 1972. The relation between ecology and social structure in primates. *Science* 176: 863-74.

Epple, G. 1967. Soziale Kommunikation bei *Callithrix jacchus* Erxleben, 1777. In D. Stark; R. Schneider; and J. J. Kuhn (eds.), *Neue Ergebnisse der Primatologie.* Stuttgart: Fischer Verlag.

———. 1970. Maintenance, breeding, and development of marmoset monkeys (Callithricidae) in captivity . *Folia Primatologica* 12: 56-76.

———. 1975. Parental behavior in *Saguinus fuscicollis* ssp. (Callithricidae). *Folia Primatologica* 24: 221-38.

———. 1978. Reproductive and social behavior of marmosets with special reference to captive breeding. *Primates in Medicine* 10: 50-62.

Fitzgerald, A. 1935. Rearing marmosets in captivity. *Journal of Mammology* 16: 181-88.

Fogden, M. 1974. A preliminary field-study of the western tarsier, *Tarsius bancanus* horsefield. In R. D. Martin; G. A. Doyle; and A. C. Walker (eds.), *Prosimian Biology,* pp. 151-65. Pittsburgh: University of Pittsburgh Press.

Fragaszy, D.M.; Schwarz, S.; and Shimosaka, D. 1982. Longitudinal observations of care and development of infant titis *(Callicebus moloch). American Journal of Primatology* 2:191-200.

Glander, K. E. 1975. Babysitting, infant sharing and adoptive behavior in mantled howling monkeys. *American Journal of Physical Anthropology* 41: 482.

Hampton, J. K.; Hampton, S.H.; and Landwehr, B. T. 1966. Observations on a successful breeding colony of the marmoset, *Oedipomidas oedipus. Folia Primatologica* 4: 265-87.

Hanif, M. 1967. Notes on breeding the white-headed saki monkey *(Pithecia pithecia)* at Georgetown Zoo. *International Zoo Yearbook* 7: 81-82.

Harlow, H. F. 1969. Age-mate or peer affectional system. In D.S. Lehrman; R.A. Hinde; and E. Shaw (eds.), *Advances in the Study of Behavior,* vol. 2, pp. 197-239. New York: Academic Press.

Harlow, H. F. and Harlow, M. K. 1962. Social deprivation in monkeys. *Scientific American* November.

Harrington, J. E. 1975. Field observations of social behavior of *Lemur fulvus fulvus.* In I. Tattersall and R. W. Sussman (eds.), *Lemur Biology,* pp. 259-80. New York: Plenum Press.

———. 1978. Development of behavior in *Lemur macaco* in the first nineteen weeks. *Folia Primatologica* 29: 107-28.

Heltne, P.G.; Turner, D.L.; and Wolhandler, J. 1973. Maternal and paternal periods in the development of infant *Callimico goeldii. American Journal of Physical Anthropology* 38: 555-60.

Hoage, R. J. 1978. Parental care in *Leontopithecus rosalia rosalia:* Sex and age differences in carrying behavior and the role of prior experience. In D. G. Kleiman (ed.), *The Biology and Conservation of Callitrichidae,* pp. 293-306. Washington, D. C.: Smithsonian Institution Press.

Hrdy, S.B. 1976. Care and exploitation of nonhuman primate infants by conspecifics other than the mother. In J.S. Rosenblatt; R.A. Hinde; E. Shaw; and C. Beer (eds.), *Advances in the Study of Behavior,* pp. 101-58. New York: Academic Press.

———. 1977. *The Langurs of Abu: Female and Male Strategies of Reproduction.* Cambridge, MA: Harvard University Press.

———. 1979. Infanticide among animals: A review, classification, and examination of the implications for the reproductive strategies of females. *Ethology and Sociobiology* 1: 13-40.

Hunt, S. M.; Gamache, K. M.; and Lockard, J.S. 1978. Babysitting behavior by age/sex classification in squirrel monkeys *(Saimiri sciureus). Primates* 19: 179-86.

Ingram, J. C. 1977. Interactions between parents and infants, and the development of independence in the common marmoset. *Animal Behavior* 25: 811-27.

———. 1978. Parent-infant interactions in the common marmoset *(Callithrix jacchus).* In D. G. Kleiman (ed.), *The Biology and Conservation of the Callitrichidae,* pp. 281-92. Washington, D.C.: Smithsonian Institution Press.

———. 1978. Preliminary comparisons of parental care of wild-caught and captive-born common marmosets. In H. Rothe; H.J. Wolters; and J.P. Hearn (eds.), *Biology and Behavior of Marmosets,* pp. 225-31. Gottingen, W. Germany: Eigenverlag Hartmut Rothe.

Izawa, K. 1978. A field study of the ecology and behavior of the black-mantle tamarin *(Saguinus nigricollis). Primates* 19:241-74.

Jolly, A. 1966. *Lemur Behavior.* Chicago: University of Chicago Press.

———. 1972. Troop continuity and troop spacing in *Propithecus verreauxi* and *Lemur catta* at Berenty (Madagascar). *Folia Primatologica* 17: 335-62.

Kinzey, W. G.; Rosenberger, A. L.: Heisler, P. S.; Prowse, D. L.; and Trilling, J. S. 1977. A preliminary field investigation of the yellow handed titi monkey, *Callicebus torquatus torquatus,* in Northern Peru. *Primates* 18: 159-81.

Kleiman, D. G. 1977. Monogamy in mammals. *The Quarterly Review of Biology* 52: 39-69.

———. 1978. Characteristics of reproduction and sociosexual interactions in pairs of lion tamarins *(Leontopithecus rosalia)* during the reproductive cycle. In D. G. Kleiman (ed.), *The Biology and Conservation of the Callitrichidae.* Washington, D. C.: Smithsonian Institution Press.

Kleiman, D. G. and Malcolm, J. R. 1981. The evolution of male parental investment in mammals. In D. J. Gubernick, and P. H. Klopfer (eds.), *Parental Care in Mammals,* pp. 347-88. New York: Plenum Press.

Klein, L., and Klein, D. 1971. Aspects of social behavior in a colony of spider monkeys *(Ateles geoffroyi)* at San Fransisco Zoo. *International Zoo Yearbook* 11: 175-81.

Klein, L. L., and Klein, D. J. 1973. Observations of two types of neotropical primate intertaxa associations. *American Journal of Physical Anthropology* 38: 649-53.

Klopfer, D. H. 1972. Patterns of maternal care in lemurs: II. Effects of group size and early separation. *Zeitschrift für Tierpsychologie* 30: 277-96.

———. 1974. Mother-young relations in lemurs. In R. D. Martin; G. A. Doyle; and A. C. Walker (eds.), *Prosimian Biology*, pp. 273-292. Pittsburgh: University of Pittsburgh Press.

Klopfer, D. H., and Boskoff, K. J. 1979. Maternal behavior in prosimians. In G. A. Doyle and R. D. Martin (eds.), *The Study of Prosimian Behavior*, pp. 123-57. New York: Academic Press.

Klopfer, D. H., and Dugard, J. 1976. Patterns of maternal care in lemurs: III. *Lemur variegatus. Zietschrift für Tierpsychologie* 40: 210-20.

Klopfer, D. H., and Klopfer, M. S. 1970. Patterns of maternal care in lemurs: I. Normative description. *Zietschrift für Tierpsychologie* 27: 984-96.

Lamb, M. E., (ed.). 1981. *The Role of the Father in Child Development.* 2nd ed. New York: Wiley.

Leibrecht, B. C., and Kelley, S. T. 1977. Some observations of behavior in breeding pairs of owl monkeys. Paper presented at the *American Society of Primatologists meeting*, Seattle, Washington.

Leutenegger, W. 1980. Monagamy in callitrichids: A consequence of phyletic dwarfism? *International Journal of Primatology* 1: 95-98.

Lynn, D. B. 1974. *The Father: His Role in Child Development.* Monterey, CA: Brooks/Cole.

Marney-Petix, V. C. Unpublished manuscript. Adult male-infant interactions in a captive Galago colony.

Martin, R. D. 1973. A review of the behavior and ecology of the lesser mouse lemur. In R. P. Michael, and J. H. Crook, (ed.), *Comparative Ecology and Behavior of Primates*, pp. 1-68. New York: Academic Press.

Martin, R. D.; Doyle, G. A.; and Walker, A. L. (eds.). 1974. *Prosimian Biology*. Pittsburgh: University of Pittsburgh.

Mason, W. A. 1966. Social organization of the South American monkey, *Callicebus moloch:* A preliminary report. *Tulane Studies in Zoology* 13: 23-28.

———. 1968. Use of space by callicebus groups. In P. Jay (ed.), *Primates: Studies in Adaptation and Variability*, pp. 200-16. New York: Holt, Rinehart and Winston.

McKenna, J. J. 1979. Aspects of infant socialization, attachment, and maternal caretaking patterns among primates: A cross-disciplinary review. *Yearbook of Physical Anthropology* 20: 250-86.

Meritt, D. A. 1980. Captive reproduction and husbandry of the Douroucouli *(Aotus trivirgatus)* and the titi monkey *(Callicebus spp). International Zoo Yearbook* 20: 52-59.

Mitchell, G. 1969. Paternalistic behavior in primates. *Psychological Bulletin* 71: 399-417.

Mitchell, G. D., and Brandt, E. M. 1972. Paternal behavior in primates. In F. E. Poirier (ed.), *Primate Socialization*, pp. 173-206. New York: Random House.

Moynihan, M. 1964. Some behavior patterns of platyrrhine monkeys: I. The night monkey *(Aotus trivirgatus). Smithsonian Miscellaneous Collections* 146 (5): 1-84.

———. 1970. Some behavior patterns of platyrrhine monkeys. II. *Saguinus geoffroyi* and some other tamarins. *Smithsonian Contributions to Zoology* 28: 1-77.

———. 1976. *The New World Primates.* Princeton: Princeton University Press.

Napier, J. R., and Napier, P. H. 1967. *A Handbook of Living Primates.* New York: Academic Press.

Neville, M. K. 1972. Social relations within troops of red howler monkeys *(Alouatta seniculus). Folia Primatologica* 18: 47-77.

Niemitz, C. 1974. A contribution to the postnatal behavioral development of *Tarsius bancanus*, Horsefield 1821, studied in two cases. *Folia Primatologica*, 21: 250-76.

———. 1979. Outline of the behavior of *Tarsius bancanus.* In G. A. Doyle and R. P. Martin (eds.), *The Study of Promisian Behavior*, pp. 631-60. New York: Academic Press.

Pedersen, F.A. 1980. *The Father-Infant Relationship.* New York: Praeger Publishers.

Petter, J. J. 1962. Ecological and behavioral studies of Madagascar lemurs in the field. *Annals of the New York Academy of Science* 102: 267-81.

———. 1965. The lemurs of Madagascar. In I. DeVore (ed.), *Primate Behavior: Field Studies of Monkeys and Apes,* pp. 292-321. New York: Holt, Rinehart and Winston.

Petter, J. J., and Charles-Dominique, P. 1979. Vocal communication in prosimians. In G. A. Doyle and R. D. Martin (eds.), *The Study of Prosimian Behavior,* pp. 247-306. New York: Academic Press.

Petter, J. J., and Petter, A. 1967. The aye-aye of Madagascar. In S. A. Altmann (ed.), *Social Communication Among Primates,* pp. 195-205. Chicago: University of Chicago Press.

Petter, J. J., and Petter-Rousseaux, A. 1979. Classification of the prosimians. In G. A. Doyle and R. D. Martin (eds.), *The Study of Prosimian Behavior,* pp. 1-42. New York: Academic Press.

Petter, J. J., and Peyrieras, A. 1974. A study of the population density and home range of *Indri indri* in Madagascar. In R. D. Martin, G. A. Doyle and A. C. Walker (eds.), *Prosimian Biology,* pp. 39-48. Pittsburgh: University of Pittsburgh Press.

Petter, J.; Shilling A.; and Pariente, G. 1975. Observations on the behavior and ecology of *Phaner fucifer.* In I. Tattersall and R. W. Sussman (eds.), *Lemur Biology* pp. 209-18. New York: Plenum.

Pola, Y. V., and Snowdon, C. T. 1975. The vocalizations of pygmy marmosets *(Cebuella pygmaea). Animal Behavior,* 23: 826-42.

Pollock, J. I. 1979. Spatial distribution and ranging behavior in lemurs. In G. A. Doyle, and R. P. Martin, (eds.), *The Study of Prosimian Behavior.* pp. 247-306. New York: Academic Press.

Pook, A. G. 1978. A comparison between the reproduction and parental behavior of the Goeldi's monkey *(Callimico)* and of the true marmosets *(Callitrichidae).* In H. Rothe, H. J. Wolters and J. P. Hearn (eds.), *Biology and Behavior of Marmosets* pp. 1-14. Gottingen, W. Germany: Eignenverlag Hartmut Rothe.

Quiatt, D. 1979. Aunts and mothers: Adaptive implications of allomaternal behavior of nonhuman primates. *American Anthropologist* 18: 310-19.

Ramirez, M. F.; Freese, C. H.; and Revilla, J. C. 1978. Feeding ecology of the pygmy marmoset, *Cebuella pygmaea,* in northeastern Peru. In D. G. Kleiman (ed.), *The Biology and Conservation of the Callitrichidae.* pp. 91-104. Washington, D. C.: Smithsonian Institution Press.

Richard, A. 1970. A comparative study of the activity patterns and behavior of *Alouatta villosa* and *Ateles geoffroyi. Folia Primatologica* 12: 241-63.

———. 1974a. Patterns of mating in *Propithecus verreauxi verreauxi.* In R. D. Martin, G. A. Doyle, and A. C. Walker (eds.), *Prosimian Biology,* pp. 49-74. Pittsburgh: University of Pittsburgh.

———. 1974b. Intra-specific variation in the social organization and ecology of *Propithecus verreaui. Folia Primatologica* 22: 178-207.

Robinson, J. 1977. Vocal regulation of spacing in the titi monkey *(Callicebus moloch).* PhD. dissertation, University of North Carolina, Chapel Hill.

Rosenblum, L. A. 1972. Sex and age differences in response to infant squirrel monkeys. *Brain, Behavior and Evolution* 5: 30-40.

Rosenblum, L. A., and Coe, C. L. 1977. The influence of social structure on squirrel monkey socialization. In S. Chevalier-Skolnikoff, and F. E. Poirier (eds.), *Primate Bio-Social Development,* pp. 479-500. New York: Garland.

Rosenblum, L. A., and Kaufman, I. C. 1968. Variations in infant development and response to maternal loss in monkeys. *American Journal of Orthopsychiatry* 38: 418-26.

Rosenson, L. M. 1972. Interactions between infant greater bushbabies *(Galago crassicaudatus)* and adults other than their mothers under experimental conditions. *Zeitshrift für Tierpsychologie* 31: 240-69.

Rothe, H. 1978. Parturiton and related behavior in *Callithrix jacchus* (Ceboidea, Callitrichidae). In D. G. Kleiman (ed.), *The Biology and Conservation of the Callitrichidae* pp. 193-206. Washington, D. C.: Smithsonian Institution Press.

Rudran, R. 1979. Infanticide in red howlers *(Alouatta seniculus)* of Northern Venezuela. Paper presented at the VIIth International Congress of Primatology, Bangalore, India. January 8-12.

Sauer, E. G., Mother-infant relationship in Galagos and the oral-child transport among primates. *Folia Primatologica* 7: 127-49.

Sauer, E. G., and Sauer, E. M. 1962. The southwest African bushbaby of the *Galago senegalensis* group. *Journal of the South West Africa Scientific Society* 16: 5-36.

Sleeper, P. 1978. The implications of monogamy for parental roles and paternal care in the yellow-handed titi monkey *(Callicerpus torquatus).* Paper presented at *Animal Behavior Society,* Seattle, Washington, June 6-10.

Snyder, P. A. 1974. Behavior of *Leontopithecus rosalia* (Golden lion marmoset) and related species: A review. *Journal of Human Evolution* 3: 109-22.

Spencer-Booth, Y. 1970. The relationship between mammalian young and conspecifics other than mothers and peers: A review. In D. S. Lehrman, R. A. Hinde and E. Shaw (eds.), *Advances in the Study of Behavior* pp. 120-194. New York: Academic Press.

Stellar, E., 1960. The marmoset as a laboratory animal: Maintenance, general observations of behavior, and simple learning. *Journal of Comparative and Physiological Psychology* 53: 1-10.

Stevenson, M. F. 1970. Birth and perinatal behavior in family groups of the common marmoset *(Callithrix jacchus jacchus)* compared to other primates. *Journal of Human Evolution* 5: 265-81.

Sussman, R. W. 1974. Ecological distinctions in sympatric species of *Lemur.* In R. D. Martin; G. A. Doyle; and A. C. Walker (eds.), *Prosimian Biology* pp. 75-108. Pittsburgh: University of Pittsburgh Press.

Sussman, R. W. 1977. Feeding behavior of *Lemur catta and Lemur fulvus.* In T. H. Clutton-Brock (ed.), *Primate Ecology* pp. 193-230. New York: Academic Press.

Tattersall, I., and Sussman, R. W. 1975. *Lemur Biology.* New York: Plenum Press.

Trivers, R. L. 1972. Parental investment and sexual selection. In B. Campbell (ed.), *Sexual Selection and the Descent of Man,* pp. 136-79. Chicago: Aldine Publishing Co.

Vaitl, E. A. 1977. Social context as a structuring mechanism in captive groups of squirrel monkeys *(Saimiri sciureus). Primates* 18: 861-74.

Vandenbergh, J. G. 1966. Behavioral observations of an infant squirrel monkey. *Psychological Reports* 18: 683-88.

Vogt, J. L.: Carlson, H.: and Menzel, E. 1978. Social behavior of a marmoset *(Saguinus fusciocollis)* group I: Parental care and infant development. *Primates* 19: 715-26.

Vogt, J. L.; Coe, C. L.; Lowe, E.; and Levine, S. 1980. Behavioral and pituitary-adrenal response of adult squirrel monkeys to mother-infant separation. *Psychoneuroendocrinology* 5: 181-90.

Vogt, J. L., and Hennessy, M. B., 1982. Infant separation in monkeys: Studies on social figures other than the mother. In H. E. Fitzgerald; J. Mulins; and P. Gage (eds.), *Studies of Development in Nonhuman Primates* vol. 3 of the Child Nurturance Series. New York; Plenum Press.

Vogt, J. L.; Siperstein, L; and Levine, S. 1982. Responsiveness of male adult squirrel monkeys to mother-infant separation. *American Journal of Primatoloty* 3:161-66

Wendt, H. 1964. Erfolgreiche Zucht des Baumwoltkipfchens oder Pincheaffchens *Leontocebus (Oedipomidas) eodipus* (Linne, 1758), in Gefangenschaft. Saugetierkdl. Mitt. 12: 49-52. Cited in Epple (1975).

Williams, L. 1967. Breeding Humbolt's woolly monkey *(Lagothrix lagotricha)* at Murrayton woolly monkey sanctuary. *International Zoo Yearbook* 17: 86-89.

15

Adult Male-Infant Interactions in Old World Monkeys and Apes

David M. Taub
Yemassee Primate Center and
Medical University of South Carolina

William K. Redican
San Francisco, California

INTRODUCTION

As witnessed by this volume, which is devoted solely to the subject of male-infant interactions among primates, there has been in recent years an inexorable growth in interest of the male's role in parenting (Gubernick and Klopfer, 1981), especially as this phenomenon occurs among primates (Lamb, 1981; Lamb and Goldberg, 1981). For example, there have been several doctoral dissertation studies devoted almost exclusively to the investigation of certain aspects of male-infant relationships (e.g., Taub, 1978; Stein, 1981), and previous to the present volume, there have been numerous reviews summarizing the state of "primate paternalism," including those by Mitchell (1969), Mitchell and Brandt (1972), Blaffer-Hrdy (1976), Redican (1976) and Redican and Taub (1981).

Several contributing factors seem to have played a role in this burgeoning interest. First, there is now considerable interest in the association between caretaking and other cooperative or altruistic behavior and genetic relatedness and kinship relations. Thus, the male primate has come under closer scrutiny to test some currently vital theories of social behavior that have sprung from new approaches to evolutionary biology (Wilson, 1975). Second, it is probably correct to suggest that primatologists have followed the lead of their colleagues in the human social sciences, where there has been a shift away from viewing the female as the exclusive caretaker, as indeed this present volume demonstrates (see e.g., Lamb chapter 16). Third, there has been a growing accumulation of data from field and laboratory investigations on this behavior and it is too easy to forget that primatology is a relatively new discipline, having coalesced only in the past few decades. So in recent years there has been an opportunity to examine a broader range of questions than had been possible with earlier methods and conceptual orientations, and it is to be hoped that an understanding of the male's role as a caretaker has benefited greatly from that development.

The objectives of this chapter are to present a broad overview of patterns of male caretaking among Old World monkeys and apes. It is beyond the scope of this chapter to attempt any evaluation, resolution, and reconciliation (if such is even now possible) of all the conflicting interpretations of the diverse male-infant behavioral systems (although such an effort is surely needed; Taub, in prep. 1983). Neither is it necessarily appropriate that we should attempt to interpret the occurrence of a structurally (and probably functionally) diverse set of behavioral phenomena within the context of current sociobiological theories (see Kurland and Gaulin, chapter 11). This chapter is designed to be summarizing and descriptive, a complimentary companion to the previous and following "summary" chapters, and little repetition will be given to new materials presented in other chapters of this volume.

OLD WORLD MONKEYS

General

The majority of the patterned relationships among males and infants occur mainly among various species of macaques and baboons, and the bulk of this chapter will be concerned with those for whom sufficient studies provide relative detail. There are a number of minor reports of male-infant interactions among some cercopithecoid primate species that deserve passing mention: many male primates protect the group and episodes in which males may intervene on behalf of infants when they are threatened have been reported for patas monkeys *(Erythrocebus patas)* (Hall, 1965; Kummer, 1971); pigtail macaques *(Macaca nemestrina)* (Kaufmann and Rosenblum, 1969); black and white colous *(Colobus spp)* (Booth, 1962); and baboons (*Papio* spp) (Hall, 1963; Rowell, 1969; Saayman, 1971; Packer, 1980). Under captive and experimental conditions males of crab-eating macaques *(Macaca fascicularis)* (Auerbach and Taub, 1979; Gifford, 1967), rhesus macaques *(Macaca mulatta)* (Redican, 1975; Soumi, 1979); mangabeys (*Cercocebus* spp) (Chalmers, 1968; Bernstein, 1976); and hamadryas baboons *(Papio hamadryas)* (Kummer, 1967; Bernstein, 1975) will show varying degrees of interest in and care for immature conspecifics. Among the one-male "harem" groups of hamadryas baboons and gelada monkeys *(Theropithecus gelada)* "bachelor" or second males may carry and play with infants (Kummer, 1967, 1971; Dunbar and Dunbar, 1975; Mori, 1979a, b). Among the former it is suggested that this may be the incipient formation of a harem, and in the latter, where male rather than female infants appear to be the preferred companions, perhaps a form of "agonistic buffering." Kaufman and Rosenblum (1969) report that when mothers were removed from the group, adult male bonnet macaques *(Macaca radiata)* became quite solicitous to the "orphaned" infants, and carried and held them on occasion. This phenomenon of infant separation leading to increased male interest in the abandoned infant has also been reported for captive rhesus monkeys (Spencer-Booth and Hinde, 1967). Males in captive bonnets groups have been reported to hold and carry infants in the presence of higher-ranking males, suggesting perhaps a form of "agonistic buffering" (J. Silk, per. comm., 1981), while among wild bonnet macaques, adult males are described as sometimes carrying, playing with and protecting infants (Simonds, 1965, 1974; Sugiyama, 1971). Affiliative interactions were more common between the adult males and juveniles; in fact, adult males were reported to avoid infants with neonatal coats (Simonds, 1974), but as the coat color changed males did interact with the infants.

Macaques

Barbary Macaques. Among the polygynously mating Old World monkeys,the Barbary macaque *(Macaca sylvanus)* clearly shows the greatest degree of male caretaking activity (Taub, chapter 2). In recent times, Lahiri and Southwick (1966) were the first to draw attention to the special interest males of this species have in infants. Reporting on a 12-week study of a one-male zoo group, they reported that 7.5% to 8% of the infants' time was spent with the dominant male, who groomed or carried them an average of two to four times per hour. They further described four ways in which the male acquired the infant and/or contributed care to it. Besides the dominant male, all group members showed intense interest in infants, and no aggression was ever directed to them.

The first indication that adult males of this species are strongly attracted to and interact with infants apparently comes from the British naturalist Lydekker (1894) based on observations made on the Barbary macaques resident on Gibraltar. Since that time, many detailed studies have been done on the animals maintained there by the British Army. MacRoberts (1970) observed special relations between the two adult males (M1 and M2) and several infants of one group, and he provided a brief, descriptive sketch of a variety of male-infant interactions (especially of the male carrying the infant on his dorsum), which he called "adoption" behaviors, while he labeled these special male-infant associations as "protector-protege" relations. Later Burton (1972; Burton and Bick, 1972) chronicled extensive male-infant interactions among members of this colony, providing a lengthy, descriptive account of the male's dynamic role in the socializing process of infants. Adult and subadult males are described by Burton (1972) as carrying, grooming, playing with, holding (for "periods much exceeding 15 minutes," p. 37), approaching, retrieving, protecting, and assisting in navagating difficult terrain. Males are continually involved with infants in a wide variety of contexts through a rich diversity of behaviors from as early as the first day of life. Interestingly, Burton noted that subadult males became a predominant influence in the infant's early life, a focus observed also by Taub (chapter 2) for wild Moroccan groups. Burton (1972) summarized the adult male's role in infant socialization as follows (p. 55): "(1) to encourage the infant to develop motor abilities that permit social interactions; (2) to reorient the infant one as it matures, away from himself and toward other troop members; (3) to reinforce socially acceptable behaviors appropriate to the age group by not interfering, or by giving positive reward (chatter, embrace, and so on); (4) to extinguish or negate inappropriate behaviors by punishment (threat chase and so forth)." Two observations from this colony that appear to be at variance with observations of Moroccan population should be mentioned: Burton (1972) describes a case of male infanticide forcing the removal of this male from this colony—no such behavior has ever been observed among wild populations; she has also described aggressive behaviors exhibited by the male to the infants, behaviors not seen during Taub's study, but P. Melhman (pers. comm., 1982) has observed that on occasion males in the Rif region of Morocco may aggress infants.

When this species was first studied systematically in its natural habitat (central Morocco) a decade ago, Deag and Crook (1971; Deag, 1974, 1980) were immediately struck by the magnitude, intensity, and diversity of male-infant relationships. In a preliminary, descriptive report, extensive and unusual social interactions between males and unweaned monkeys were described (Deag and Crook, 1971), and they distinguished two types of male-infant interactions: male care and "agonistic buffering." The former, also referred to as "type a" behaviors, included all forms of dyadic interactions involving a single male and a single infant—thus male care subsumed such traditionally "maternal" behaviors as holding, carrying, retrieving, protecting, huddling, and grooming. A second category of male-infant interactions, one in which males used babies to regulate their social relations with other males was termed "agonitic buffering" (also referred to as "type b" behaviors) (Deag and Crook, 1971). Most of the work from this study has focused on the latter type of behavior, and its function has been said to be a means of enabling a subordinate male to approach and remain near a dominant male with a reduced likelihood of attack. Quantified data form this study (Deag, 1974, 1980) almost exclusively concern "type b" class of male-infant interactions, and are used to support the earlier "agonistic buffering" hypothesis advanced in the qualitative, descriptive report. For example, there were larger clusterings of animals in general, and of other males in particular, around a male that had a baby in his possession than when he did not; males made more friendly approaches to one another when in the presence of infants; and in general, subordinate males carried infants to males higher than themselves in social rank.

Two months of observation in late 1971 of isolated, wild groups in the Rif region of northern Morocco by Whiten and Rumsey (1972) showed males of this isolated region were also actively involved with infants, and they classified at least seven observations as falling into the category of "agonistic buffering" (1972, p. 423). Patrick Melhman is at the present time halfway through conducting a two-year study of the same Riffian populations, and he reports observing all the types of male-infant interaction exhibited by this species in the Middle Atlas region of Morocco (pers. comm., 1982; pers. obs. by DMT 1982).

Taub's subsequent study of wild, Moroccan groups in 1973-74 confirmed many of Deag and Crook's earlier observations, firmly establishing that intensive and elaborate interactions between males and infants are typical and characteristic of this African macaque, and that such extensive male care-taking was not an artifact of captivity or provisioning (Taub, 1978, 1980, chapter 2). Choosing to avoid such interpretive terminology as "agonistic buffering," Taub divided the caretaking interactions of males into two classes based on their structure: dyadic (equivalent to Deag and Crook's "male care") and triadic (equivalent to their "type b" class) male-infant interactions. Presented in chapter 2 of this volume is a detailed examination of the dyadic caretaking interactions from Taub's study, and these data will not be reviewed here. Relative to the triadic type of male-infant interaction, Taub has reexamined the "agonistic buffering" hypothesis in light of data from his field study (1980) and found cause both to reject the dichotomy between male care and "agonistic buffering" and to reevaluate the function of triadic interactions. Social rank did correlate in some general ways with the pattern of triadic interactions, but for the most part the data could not adequately accommodate the earlier hypothesis that "agonistic buffering" functions to regulate dominance/subordination relations among males. Males did not choose other males equally often to interact with (a pattern also found by Deag), but rather each male had a different set of 3 other males (out of 11 possible) that he preferred for a triadic interaction. Each male showed striking preferences for certain infants in triadic encounters, and these infants were the same ones preferred by that male in dyadic caretaking activities. Finally, males that preferred each other for a triadic interaction showed a mutual preference for the same infants. Taub (1980) concluded that "males choose to participate in triadic encounters by means of a shared, common and special caretaking relationship with the same infant" (p. 187). Thus Taub shifted his focus of interpretation of triadic relationships away from social status regulation to a shared caretaking network.

During a summer study, Kurland and Taub (in prep.), using a focal male protocol, focused attention on the aggressive behavior of males with and without infants in their possession. Preliminary analysis has shown some patterns that are at variance with Deag's observations, but a complete interpretation of these data await completion of data analysis. In this regard, however, it is interesting to note the results of a recent study by Smith and Pfeffer-Smith (1982). Studying a free-ranging, but captive island group, the Smiths found that lower-ranking animals approached higher-ranking animals for triadic interactions in 77% of the cases, but in 11%, the initiator was higher ranking. They found no difference in the frequency of nonagonistic interactions between juvenile, subadult, and adult males with and without infants present; interactions preceding the triad were characterized by no overt interaction between participants (in 3 of 18 episodes, or 17%, aggression actually followed the triadic interaction); affiliative male-male behavior after a triad occurred in 5 of 18 (28%) episodes.

Stumptail Macaques. Early studies of stumptail macaques *(M. acrtoides)* suggested that males were not particularly involved with or interested in infants, although Bertrand (1969) reported that they were tolerant and protective of infants. Other early reports (Jones and Trollope, 1968; Bernstein, 1970; Brandt et al., 1970) did describe some male-infant interactions that involved physical contact, in which infants actively sought proximity to males; they also described males as being "very aware" of infants.

In contrast to these earlier studies, recent investigations suggest that, at least among the multi-male/female groups of polygynously mating Cercopithecines, stumptail macaque males rank second to Barbary macaques in the extent of their interactions with infants. Gouzoules, reporting on a corral group at the Yerkes Primate Center, noted that adult males "showed interest in, and interacted with infants almost as much as females did" (1975, p. 413). During the first six months of the infant's life, adult and subadult males engaged in a variety of interactions, including holding, carrying, touching, genital manipulation, retrieval, and protection. During the first six months for the six study group infants, there were 836 and 268 episodes of "social interaction" between the two adult and two subadult males, respectively. Social rank of both the males and the infant's mother appeared to be important in influencing the patterns of male-infant interactions in this group. Thus Gouzoules found that adult males interacted most frequently with infants (contrasting with Brandt et al., where subadult males were more involved), but found that subadult males too showed similar, extensive interactions with infants, although they preferred to interact with older (four- to six-month-old) infants and infants of lower-ranking females.

In a study of early development of nine infants living in two captive social groups, Hendy-Neely and Rhine (1977) reported on 14 categories of contact between males and infants in the first two months after birth. Only one infant interacted with males before day 10, but thereafter momentary touching by males occurred more often than other behaviors, such as grooming and clinging, which were most frequently associated with maternal attachment. In fact, five categories of momentary touching accounted for a large majority of all interactions among males and infants. There were significant differences between the two adult males of each group in the distribution of their interactions with infants. These differences tended to correlate positively with both male age and social rank (similar to the results of Gouzoules). The dominant male of each group tended to be the most heavily involved with infants generally, but they also tended to focus the majority of their interest on a few specific infants. There was, consequently, a wide variation among the nine infants in the amount of male attention each received, and these differences were attributed to differences in the affiliative patterns of the adult group members. That is, the mother's dominance rank and whether she was "permissive" or "restrictive" in allowing males access to her infant appeared to influence strongly patterns of male care. In a follow-up paper (Rhine and Hendy-Neely, 1978), patterns of infant interaction with nonmaternal adult females and immatures during the first two months of life were compared with the patterns of interaction between these same infants and the adult males. With the exception of "play" behavior, the adult males interacted with infants more frequently than any other group members (except mothers), and did so most often through behaviors most commonly associated with the mother-infant interactional system (e.g., touching, carrying, retrieving).

In chapter 4 of this volume, Smith and Peffer-Smith report the results of a study of a captive stumptail group containing 6 adult males. They observed that 17% of all social interactions were between immatures (birth to two years) and adult males. Adult male social rank, but not age, had a significant effect on the rate of male-immature interaction, and interestingly, genealogically related males and immatures were found to interact at a higher rate than nonrelated individuals. The authors concluded that adult males served as important social foci for immature members of the social group.

A free-ranging, island colony of stumptail macaques has been studied extensively by Estrada and his colleagues (Estrada et al., 1977; Estrada and Sandoval, 1977, Estrada, chapter 3). Previously they reported on 17 different caretaking behaviors between males and infants (12 categories of contact and five categories of proximity and vocalizations), including carrying, huddling, protecting, grooming, and bridging. Infants less than six months of age received significantly more care from males than did older infants, and there were similar age-related differences in the type or quality of male care received: e.g., younger infants received more tactile stimulation from males than did older infants, with whom play interactions predominated. Several differences in the distribution of male care correlated with the sex of the infants: although there were two male and five female infants, male infants received 76% of all contact behaviors and 88% of the proximity and vocalization scores. Similarly, there were infant sex-related differences in the type of care infants received from males: "While males were touched, their genitals were manipulated and they were bridged, females interacted more in play, were touched and their genitals were manipulated in this order" (Estrada and Sandoval, 1977, p. 802-803). Infant age correlated weakest with the contact categories of male behavior, but again, there were clear-cut male preferences for interactions with male infants. Infant dominance rank did not correlate with the amount of male care received, but the dominance rank of the adult males was inversely correlated with the amount of male care exhibited. Proximity of the mother also correlated positively with the frequency of male care, suggesting that the mother's presence could have acted as an inhibiting influence in the male's ability to physically interact with infants.

Although males of all ages were involved with infants in caretaking interactions, juvenile males showed the greatest amount (in contrast to other studies): average per-male interactions with infants was 518 episoded per juvenile males (all infants pooled) versus 289 for adult males. However, the differences in the distribution by male age-category can be ascribed to the fact that there were striking individual male differences in interest in infants, with one juvenile male (Ch) accounting for fully 35% of all male-infant interactions recorded (out of nine males total). Not only did juvenile males show considerable interest in all infants, they also showed "substantial amounts of care behavior to their infant siblings" (Estrada and Sandoval, 1977, p. 803). Indeed, when the data are separated out on an individual male-to-sibling basis, the three juvenile males with sibling infants showed a significant preference for interactions with them over all other infants combined: 67% each for males Pa and Ch and 52% for Jo. Consequent to the significant differences among individual males in their interest in specific infants, there were also striking individual differences among the seven infants in the amount of care each received from males, so that male infants Ti and DJ received between them 84% of all male-infant interactions.

In chapter 3 of this volume, Estrada presents further data from this free-ranging colony, which expands and clarifies the distribution of male-infant care in this species. Summarizing the salient features of the male-infant care system, he notes that males are active and important participants in the early life of infants, with juveniles showing more intensive interest than other males; males display preferences for specific infants, although these patterns may be influenced by social rank and mother's proximity; and male-infant preferences seem to be strongly influenced by the degree of genetic relatedness between the participants.

Japanese macaques. It was among wild Japanese macaques *(M. fuscata)* that came the first systematic, albeit qualitative, account of social relations between males and infants among Cercopithecine species. From a study of the Takasakiyama populations, Itani (1959) described male care of infants as being extensive and similar to maternal care (except for suckling), including such behaviors as carrying, hugging, holding, grooming, and playing. He observed intensive male caretaking directed almost exclusively to one- and two-year-old individuals (yearlings accounted for 74% of such interactions), and only during the May to August birth season, when the yearlings were being supplanted from maternal care by the arrival of the newborns; and he suggested that the burden of caretaking was shifted to fully mature males at a time crucial for infant survival. Itani reported that only fully mature "leader" and "subleader" males greater than 10 years of age exhibited this seasonally intense interest in infants, and further that this male caretaking behavior was not common to all troops of Japanese macaques.

Subsequent studies of the Takasakiyama populations (Hasegawa and Hiraiwa, 1980; Hiraiwa, 1981) have substantiated these patterns of male care, although no systematic quantitative data have yet been provided on these populations that would allow a more definitive evaluation of the patterns of allopaternal behavior in this species. In regard to Itani's original speculation about the functional importance of the critical period in infant development in which this behavior appears and is confined, a report on "adoption" of orphaned infants among these troops is particularly noteworthy (Hasegawa and Hiraiwa, 1980). In this study, it was found that adult males showed intensive interest in and carried , groomed, and defended infants whose mothers had died. These adult males became the primary caretakers of these orphaned infants, surpassing siblings, other relatives, nonkin peers, and nonkin adult females in the extent of their interaction with infants. Moreover, the orphaned infants themselves preferred these adult males to their own immature kin.

Studying a confined, corralled troop at the Oregon Primate Center, Alexander (1970) has reported adult male behavior similar to that described for the Takasakiyama troops, but with some differences in their pattern of distribution. Seventy-five percent of all sexually mature males showed an increase of all types of affiliative male-immature (one to four years) interactions (especially male-initiated interactions) both prior to and during the April to June birth season. However, the younger (five to seven years) adult males interacted with juveniles more than the older males, and subordinate males were equally as likely to be involved with immatures as were the high-ranking males. Even though, as with the Japanese data, males-immature interactions tended to be seasonally dependent (a function of androgen withdrawal, suggested Alexander), at least one (third-ranking) adult male showed a year round involvement with infants, directing his behavior to several juveniles, especially one four-year-old female (although this male too showed a seasonal peak in his interactions with immatures). Although the amount of play and affiliative care immatures received from males was small in comparison with other group members and in spite of the seasonal fluctuations, Alexander postulated that these males "may play a very significant role in the socialization of the young" (Alexander, 1970, p. 282). It is interesting to note in this context that much of the immatures experiences with aggressive behavior also are provided by the adult males.

Patterns of male-infant interest among the transplanted, free-ranging Arashiyama troop in Texas differ in some fundamental ways from those reported for the Takasakiyama and Oregon troops. In chapter 6, Gouzoules presents a detailed picture of male-infant interactions observed in the Arashiyama West troop. Although these males did not exhibit the "babysitting" of juveniles during the birth season previously described, a third of the adult males (9 of 29) did develop long-term, persistent patterns of caretaking with individual infants. As with other reports, these males interacted with infants in a diversity of affiliative interactions, such as carrying, clutching, holding, grooming, and proximity. Dominance rank and maternal kinship relatedness (see also Kurland, 1977; Grewall, 1980) were among the most important variables accounting for the distribution of male-infant interactions.

Rhesus Macaques. Wild adult male rhesus macaques *(M. mulatta)* generally tend to be indifferent to infants, although on occasion they may be quite sensitive to their approaches and contact, and interact with them in positive and affiliative ways (Southwick, et al., 1965; Kaufman, 1968; Lindburg, 1971). In a study of the Cayo Santiago colony, Breuggeman (1973) found that the age of the male and the frequency of caretaking were positively correlated, but that three-year-old males tended to exhibit the highest relative levels of interest in infants. All males exhibited more frequent interest in infants during the mating season, and the caretaking males tended to prefer intereacting with male rather than female infants. Vessey and Meikle (chapter 5) and Meikle (1980) observed very few instances of adult males interacting with infants during a long-term study of the free-ranging colony at La Parguera, Puerto Rico. The primary type of association was close proximity, and direct interactions between adult males and infants accounted for less than 1% of all interactions. Nevertheless, they did observe a case of adoption, in which an adult male carried and cared for a female infant whose mother had died, and one case in which the alpha male carried the infant of a primiparous female who tended to abandon it. Berman (pers. comm., 1981) has observed a similar case of adoption by a four-year-old male of his orphaned male infant sibling in the Cayo Santiago colony, and Taylor et al. (1978) have also reported two cases of an alpha male who adopted a female infant after the infant's mother had died.

Captive, group-living rhesus males also tend to show relatively little interest in neonates (Rowell et al., 1964; Spencer-Booth, 1968), but other studies of captive rhesus males have indicated that under the proper conditions, males may show remarkably intense levels of interest and caretaking of infants. For example, in a series of papers, Redican and coworkers (Redican, 1975, 1978; Redican and Mitchell, 1974) present data from a longitudinal study investigating the behavior of adult male and infant rhesus living together in a laboratory environment, compared with control groups of mother-infant pairs reared under similar conditions. One of the fundamental differences between mother-infant and male-infant pairs was the pattern of physical contact. Mothers were consistently more contact-oriented toward infants than were adult males, being in more frequent and longer contact (especially ventral contact), particularly during early months, and both mothers and their infants made contact as often as they broke it. The generally lower level of contact in the male-infant pairs was reflected in the patterns of establishing and breaking contact, as males consistently broke contact more often than they established it, while the converse was true for their infants. Adult males restrained and retrieved infants very rarely, but actively protected them by directly attacking the source of danger and/or interposing themselves between it and the infant. However, distress vocalizations and facial expressions were far more frequent in male-reared infants. Play between adult males and infants was far more intense and reciprocal than that between mothers and infants, and adult males groomed infants to the same extent that the mother did. Thus some measures of proximity, such as ventral-ventral contact, were due primarily to the mother and not the infants; whereas in the male-infant dyads, the adult males were the principal agents for certain behavioral measures and the infants for others, and there was evidence that measures of attachment increased over time in the male-infant pairs.

The magnitude of sex differences in many attachment behaviors was greater in male-infant pairs, with contact more pronounced with male infants. Mothers tended to play with female infants, whereas adult males did so with male infants. In general, mothers interacted more positively with female infants, and adult males with male infants. No consistent sex or rearing differences were found in exploration and general activity measures for infants.

From the infant's viewpoint, perhaps the primary index of whether adult males are equally suitable objects of attachment is the infant's response to temporary separation. In terms of attachment measures such as vocalization, approach, locomotion, and self-directed behavior, responses to separation and reunion were similar in both mother-reared and male-reared infants. During separation, both groups of infants showed comparable frequencies of distress vocalization, decreases in solitary play and exploration, and increases in self-directed behaviors. It was inferred that both mothers and adult males were comparably potent objects of attachment for infants, and, in general, the results of these studies indicate a significant potential for adult males to form attachments with infants and for infants to form attachments with adult males. Clearly the dimensions of parental caretaking reflected the opportunities available to the individual, so that in the absence of mothers, adult males were seen to interact with infants in a highly affiliative manner rarely observed in the wild.

In a series of similar investigations at the Wisconsin Primate Center, triadic or "nuclear family" social groups (each composed of a father, mother, and infant) were studied (Harlow, 1971; Soumi, 1977, 1979; Soumi et al., 1973). Infants but not adults were allowed access to members of other triads. In social preference tests, infants raised in triads preferred their mothers to other adult females, their father to other adult males, and their mothers to their fathers. Fathers initiated few play sessions but responded to most of the infants' initiations, playing with male infants more frequently than with female infants. However, infants spent less than 5% of their time interacting with all adult males, including their fathers; this was at a rate only slightly higher than that reported for free-ranging animals.

Soumi (1977) reported that adult males showed quite stable behavioral profiles over the three-year period of the study, in strong contrast to those of the adult females and infants. During the relatively limited time males spent interacting with infants, the clear majority of interactions were play. Play and defense were the only categories in which adult males interacted more often with offspring than with other available animals. Patterns of sex differences were complex, so that adult males defended each sex equally, but they invited play and responded to invitations to play more often with male infants. Major differences in form and frequency of interactions with male versus female infants greater than four months were also evident. Data on mother-infant interactions, in contrast, suggested that son-daughter differences at the same age were present but not substantial. Thus, in keeping with several other studies, males differentially responded to infants on the basis of gender to a greater extent than did females, and furthermore, at least until adolescence, this gender-based disparity increased with adult males but decreased with adult females as the infants grew older.

Baboons

As with macaques, an expansion of field data in recent years has brought increasingly frequent reports of significant interactions between males and infants in baboon species possessing a multi-male/multi-female social organization. Interest in infants by baboon males was first documented in an early field study by DeVore (1963), who emphasized the importance to the infants' development of their relationships with adult males. DeVore described several types of care and protection afforded infants by males: all adult males responded to infant stress vocalizations, and also intervened on their behalf when aggressed by conspecifics. He noted a considerable variation among males in the interest they showed to infants; juvenile and young adult males showed little interest in infants, but fully mature males of the central dominance hierarchy frequently approached and manipulated infants.

In a field study of anubis baboons *(Papio anubis)* at the Gombe Stream Reserve, Tanzania, Ransom and Ransom (1971) qualitatively described four types of male-infant relations: (1) "special" relations derivative from a special bond or attachment between the cartaking male and the infant's mother; (2) an intensification of the male's generalized tendency to protect members of his major subgroup (independent of any male-infant bonding); (3) relations where young males took an interest in the infants of young, low-ranking females; and (4) dominance-mediated relationships that appeared to be based on the adult male's ability to increase his effectiveness in interactions with other males through close contact with an infant—the three highest-ranking adult males were the most frequently involved in this "agonistic buffering" type of male-infant behavior, especially when in proximity to new, immigrant adult males. Ransom and Ransom described several case histories, pointing out that most caretaking and "babysitting" occurred in the first or special maternal-mediated type of relationship.

In a more quantitative study of male-infant interactions among Gombe baboons, Packer (1980) distinguished three types of both care and "exploitation" of infants by adult males. Interactions Packer observed to be caretaking included defense against and rescue from predators (mainly chimpanzees), defense against aggression by conspecifics, and grooming. Analyzing several variables that might form the basis on which males chose particular infants to interact with, Packer found that infant age correlated strongest, with males preferring younger infants. Males resident in the troop at the time of possible conception (i.e., probable "fathers") were predominantly and almost exclusively involved with infants in all forms of male-infant interactions; moreover, infants spent significantly more time with males present in their troop at their birth. Neither infant sex nor maternal rank correlated with the patterns of male-infant interactions.

During Packer's two-year study, there were 26 male-male dyads (involving 17 different males) in which one male carried an infant during encounters with another male. Resident males (potential fathers) were especially active in this triadic type of behavior, and also in defending infants against males newly migrant to the troop, and infants frequently reacted very negatively toward these immigrant males. Males that most frequently "exploited" infants in this way were also those most frequently involved in caretaking interactions with infants. Packer interpreted the male system of infant caretaking and exploitation as a case of "mutualism" or "delayed return altruism," in which both the male giving care and the infant receiving care ultimately benefit more or less equally.

Thus, among the Gombe anubis baboons, all types of male-infant interactions were more common between infants and males residing in the troop at the time of their probable conception than with all other males. Busse and Hamilton (1981) and Busse (chapter 8) have reported similar patterns of male-infant interactions among chacma baboons *(P. ursinus)* in Botswana, where infants were carried by resident males against more dominant, recent immigrant males. Nine of 10 carrying males were resident in the troop when the carried infants were conceived, and these males carried the infants in 109 of 112 (97%) recorded triadic interactions. The recent immigrants could not have fathered the infants carried against them by the resident males, who in all likelihood were themselves the sires of the infants as they were among the higher-ranking resident males at the time of the conception. Busse and Hamilton thus argued that instead of using the infants to "buffer" their relations with other males, it was rather a mechanism whereby the probable fathers protected their probable offspring against injury or possible infanticide by unrelated male immigrants, and they classified this triadic form of infant carrying as "parental care." Busse provides additional data and further refines the functional interpretation of the male-infant interactions among chacma baboons in chapter 8 of this volume.

The patterns of male-infant interactions among anibus (Gombe) and chacma (Botswana) baboons, while generally concordant, are at variance with data from another anubis baboon study (Popp, 1978). Popp described triadic male-infant-male interactions as "kidnapping," where adult and subadult males took infants during agonistic encounters with other males. But Popp found that the kidnapper was less likely to be the father of the kidnapped infant than was the opponent, and in the 19 adult male dyads observed, there were no reversals in kidnapper versus opponent. Therefore, the "user" male appeared to gain advantage by placing an unrelated infant between himself and an opponent who was probably related to the infant. High-ranking males, who were mostly large, were kidnappers, but they only kidnapped in response to the few even larger, more dominant males in their troop. While not a frequent occurrence (39 cases with a mean duration of less than one minute each), this triadic male-infant interaction appears to be a form of "agonistic buffering" in the sense that infants appear to be objects that are exploited when needed by adult males. Since almost 40% of the kidnapping episodes occurred in a sexual consort context, Popp suggested it was a means whereby a competing male could harass the consorting pair with little risk of reprisal by the probable sire of the infant being used. However, in a reanalysis of Popp's data, Stein (1981) casts considerable doubt on Popp's assumptions relating to probable paternity of the kidnapper versus the opponent; i.e., according to Stein, the opponent had "higher mating success" (i.e., would be the probable father) in barely half the 39 male-infant carrying episodes. If true, then almost one-half the interactions would be analogous to the situation described by Packer (1980) and Busse and Hamilton (1981) where males are indeed carrying their probable offspring.

Stein's study of Amboseli baboons (1980, 1981; chapter 9; Stein and Stacey, 1981) is the most comprehensive yet of male-infant interactions among baboons, leading him to conclude that this system most often had an "agonistic buffering" effect. Stein describes several categories of male-infant interactions, including contract, proximity, connection (i.e. carrying, embracing, holding), transition, and grooming. Adult males almost never carried infants except during triadic (i.e., "agonistic buffering") episodes, and adult males spent a larger proportion of their time in proximity, in contact, and connected with infants if the male was involved in an agonistic encounter with another adult male than if not, which was ascribed to the higher frequency with which infants and males approached each other during inter-male conflicts (while a male was involved in an inter-male aggressive encounter, half the interactions with infants were initiated by the infant). Adult males aggressed at a greater rate when they were connected with infants than when they were not, but the amount of aggression received by males with and without infants did not differ. Although adult males may interact with several infants, they preferentially carried infants most likely to be related to them and the infants with whom they had special, affiliative relations: infants were carried by their potential father 34% of the time and by nonrelated males, 15%; whereas infants were carried against potential fathers 35% of the time and against nonfathers, 31%. Dyads in which these special male-infant relations were established were those where the infant and adult males were in the same subgroup, the male had copulated with the infant's mother at the time of its conception, and the male was high ranking. Maternal rank and infant gender were not significant factors, although younger infants were used more often.

In comparing the patterns of male-infant interaction between the multi-male Alto's group and the single-male Limp's group, Stein and Stacey found that the infant in the latter group did not have as much contact, proximity, connection and grooming with the adult male as did infants in the multi-male group, which would be consistent with Stein's emphasis on the "agonistic buffering" function of male-infant interactions among Amboseli baboons.

In a comprehensive study of mother-infant relations among Amboseli baboons, Altmann (1980) also observed several cases of males caretaking specific infants. Although individual males had wide variations in the degree of interest shown in infants, the most intense caretaking males tended to be those that had associated with or had special relationships with the infant's mother. Particular males had clear preferences for associating with certain females and, by extension, these males often associated with these females' infants, increasing the probability that these males may be caretaking their own progeny. These special male-female-infant relationships usually involved only fully mature, higher-ranking males.

Studying troops of anubis baboons at Gilgil, Kenya, Smuts (1981) found patterns of male-infant interactions similar to those in the Altmann, Ransom and Ransom, and Strum (chapter 7) studies; i.e., males who had "special" relationships with certain females also tended to have special relationships with the infants of these females. Smuts distinguished six types of affiliative male-infant interactions: class 1, or prolonged types, including holding, carrying, and grooming; and class 2, or brief types, including vocalizations to, approaches, and greetings. An average of 85% of class 1 and 97% of class 2 affiliative interactions (there were 109 episodes of class 2 and 38 of class 1 behaviors for a total of 147 male-infant interactions), and close proximity to adult males by infants when away from mother tended to be restricted to the males with special relations to an infant's mother. Smuts found that these "special" relationships with the mothers were the most important variables in explaining the distribution of male-infant interactions; kinship was also important, but affiliative interactions between infants and probable fathers were rare unless these males also had special relations with the infant's mother. As with all other baboons exhibiting male-infant interactions, the Gilgil baboons exhibited the triadic or "agonistic buffering" type: in 22 of 24 carrying episodes during a tense situation involving another male, the carrier was a probable or potential relative of the infant, and all opponents (except one) were unrelated; and in 41 of 43 such episodes, the carrier had a special relationship with the mother.

Strum's (chapter 7) detailed study of another Gilgil troop of baboons expands our understanding of the male-infant interactions among anubis baboons. In chapter 7, she presents detailed data on the distribution of male-infant interactions, and discusses and evaluates a number of variables (e.g., natal coat color, infant age, etc.) relating to the patterns of infant choice by males. As the title of her chapter suggests, she evaluates why male baboons use infants, and offers some interesting and unusual interpretations in an attempt to reconcile often conflicting data from several baboons studies.

Taking into account the diversity of reports on baboon male-infant associations, it is difficult to reconcile all points of conflict, but nevertheless it appears that real caretaking does occur [e.g., Packer (1980), Ransom and Ransom (1971), Busse and Hamilton (1981), Smuts (1982)] but it occurs rather infrequently and most often appears to be a consequence of a special relationship between the caretaking male and the infant's mother, the probability that the caretaking male is related to the infant, or both. The primary characteristic of most male-infant interactions appears to be a modulation of relations with the male troop members (Stein, 1981), so that the focus of attention is less the infant than it is another male with whom the caretaking male is interacting, although males tend to be related to infants they "use" against males who are less likely to be related to the infant.

APES

The apes encompass virtually every form of social organization common to mammals, including monogamy, promiscuity, and a form of one-male polygyny, although their cognitive and emotional traits allow for a lability and complexity of social organization that render categorization rather difficult. To the extent that coherent descriptions can be offered, at least the same sorts of questions and predictions raised for nonpongid primate groups can be examined in light of issues such as genetic relatedness and resource availability. The greatest ease in categorizing is afforded in the case of the smaller apes, gibbons *(Hylobates spp)* and siamangs *(Symphalangus)*. As is well known, they are classically monogamous. The gorilla presents itself largely as a one-male/multi-female polygynous system. Although the chimpanzee almost confounds classification, it is most comfortably assigned a polygamous (promiscuous) category. The orangutan is most often referred to as solitary.

Given the above admittedly not absolute range of social systems, certain predictions can be advanced regarding parental investment by males. Because of their monogamous mating system, male gibbons and siamangs would be expected to be preeminently involved in care of offspring. Gorillas, given the relative stability of their one-male polygynous system, should occupy an intermediate position, and the chimpanzee and orangutan should be expected to exhibit the least investment in male parental care among the pongids.

Gibbons and Siamangs

The smallest of the apes, the gibbon and siamang are arboreal, monogamous, and territorial (Carpenter, 1940; Chivers, 1971, 1972; Ellefson, 1968; McCann, 1933; McClure, 1964; Tenaza, 1975; Tenaza and Hamilton, 1971). As with monogamous New World monkeys, male investment in caretaking is extensive. But male callitrichids are clearly more active caretakers than male hylobatids (see chapters 1, 13, and 14). Several considerations may be instrumental in effecting this difference. First, primate mothers bearing relatively large offspring (i.e., whose weight exceeds 20 to 25% of their mothers') are more likely to share carrying responsibilities with the male or older offspring. The neonate-to-mother weight ratio of callitrichids is as great as 0.27, whereas that of hylobatids is as much as five times less (Kleiman, 1977). Parental duties are correspondingly more shared in callitrichids than in hylobatids. In addition, gibbons and siamangs bear singleton offspring, as opposed to twins or triplets for callitrichids. And the interbirth interval is considerably longer in the case of hylobatids. Both of these latter reproductive characteristics also promote a more shared parental investment in callitrichids. Although direct comparison would be difficult, it is possible also that different predation pressures could play a significant role—gibbon males, at least, devote considerable time to vigalence (Tenaza in Kleiman, 1977).

Male gibbons do not typically carry their offspring (Tenaza, 1975). But they have, in the wild, been observed to inspect and to groom neonates, and a male in a captive group was seen to carry a small juvenile for the greater part of a day (Carpenter, 1940). As infant gibbons mature, they become more independent of their mother and interact more frequently with their father (Berkson, 1966). Nearing sexual maturity, however, they are threatened and aggressed, principally by the father, until they eventually become peripheral to the natal group and thus ultimately establish independent monogamous units (Carpenter, 1940; Ellefson, 1968, 1974). Carpenter (1964, p. 224-25) proposed that this is an intrasexual antagonism—that fathers aggress sons and mothers aggress daughters. Further study is needed to substantiate the hypothesis, although preliminary studies suggest it holds for males (Tenaza, 1975).

A clearly more extensive degree of male caretaking has been documented for the siamang. Chivers, (1971, 1972) reported that infant siamangs are dependent on the mother for the first 12 to 16 months of life, but from that point onward, they are carried by the father until independence is attained during the third year. Siamang fathers groom as well as sleep with juveniles; mothers groom and sleep with infants. Maturing siamang offspring are also peripheralized from the natal group, but the process appears to be less severe than in the gibbon (Chivers, 1971, 1972; Fox, 1971, 1974). As in the other hylobatids, adult males take the more active role in peripheralizing offspring. One peripheralized young male was observed to establish a home range directly adjoining his parents' (Aldrich-Blake and Chivers, 1973)—a finding that may prove of interest in terms of altruistic or cooperative defense against predators by individuals sharing a high complement of genes, as documented for one or more rodent taxa. It may also pertain to another phenomenon—adoption—documented thus far, among the hylobatids, in the case of the gibbon. Brockelman, Ross, and Pantuwatana (1974) reported the transfer of two infant gibbons, in an island colony, from one family unit to another. In one case the mother had died; in the other, she had not. Clearly it would be helpful to know the degree of relatedness between individuals involved in the adoption.

As with the gibbon, there is a need for more data before it can be safely concluded that peripheralization takes place along intrasexual lines. There is also a need for data on the extent, if any, of "helpers-at-the-nest" activity by older offspring. Such data would certainly facilitate comparisons with callitrichid systems.

There are thus striking parallels in male caretaking between the groups of monogamous primates (see chapter 14): all are monogamous in their mating, all are classically territorial, all exhibit relatively frequent and intensive male parental care, and all engage in some degree of active peripheralization of offspring at sexual maturity. Degree of male involvement appears to vary according to the reproductive burden on the mother: male callitrichids may be involved from birth onward, assuming a major role in infant transport and engaging in intensive caretaking activities such as food sharing; in contrast, monogamous male apes become involved later in infancy and appear not to engage in specialized caretaking behaviors.

Chimpanzees

The intricacies of the chimpanzee (*Pan* spp) social system have become progressively apparent as more and more fieldwork has been carried out.

Clearly, the genus cannot be described in terms of a single category of reproductive systems. Tutin and McGinnis (1981), for example, distinguished three distinct mating patterns: (1) opportunistic, noncompetitive mating between a receptive female and 2 to 12 males; (2) possessive short-term relationships between a receptive female and a single male whereby lower-ranking males are excluded; and (3) consortive relationships between a male and a female (and her dependent offspring) in which they remain distinct from the remainder of the community (for a median duration of seven days). The majority (73%) of copulations observed by Tutin were opportunistic, but the existence of alternative consorting relationships is of considerable importance in terms of predicting patterns of paternal investment. That is, where confidence of paternity is enhanced as a result of prolonged mating associations, male parental investment should be correspondingly increased. At least for some, there is evidence that males may show special attention to the offspring of females with whom they had consorted.

The social structure of the chimpanzee is one in which flexible but independent multi-male/multi-female communities exist. Males traverse an area that includes the ranges of a number of females, but the several males of a given community collectively defend their territory against neighboring communities (Harcourt, 1981; Tutin and McGinnis, 1981) with intense agonism (Goodall et al., 1979). Males reside permanently within their natal community; all adolescent or subadult females migrate, although some return after becoming pregnant (Teleki et al., 1976; Pusey, 1979). Accordingly, males are more closely related within a community than are females— the opposite situation found in one-male/multi-female groups of other primates.

The implications of this social situation on male parental care are several. First, one would predict little or no parental investment by males in unrelated infants—those of other communities. Indeed, adult males have been observed even to attack, kill, and eat infants of neighboring communities (Suzuki, 1971; Bygott, 1972). More generally, one would predict that, within the community, male parental care would exist at relatively low levels but dispersed rather uniformly among the males, given the opportunistic mating and high inter-male relatedness. In view of the diffuseness of such a prediction, it is difficult to interpret available data. However, the extremes are readily discounted: chimpanzee males clearly do not transport and care for offspring in the sustained mode characteristic of monogamous taxa. But it is likewise too severe to suggest the other extreme, as did Tutin and McGinnis (1981, p. 262): that males "make no direct contribution to the rearing of their offspring." A range of caretaking behaviors is exhibited, but not at intensive levels.

The range of male-infant behavioral involvement includes both direct and indirect aspects. Infants remain in virtually unbroken contact with their mothers until 3½ to 5½ months of age (van Lawick-Goodall, 1967, 1968), and little direct interaction between infants and community males is seen. Males do show interest in young infants, as inferred by visual orientation, reaching toward and touching of infants (van Lawick-Goodall, 1968). As infants become mobile in the second half of year one, males may reach out to greet them and are tolerant of their interference with ongoing activities. For example, infants may jump onto the male's lap, climb about on his body, or grasp at his food—all of which is tolerated or responded to with care. Young infants are also comforted, protected, and assisted by adult males (Nissen, 1931; van Lawick-Goodall, 1968).

A set of behaviors is described by a consensus of observers as "interference" (van Lawick-Goodall, 1968; Tutin, 1979; Tutin and McGinnis, 1981). That is, immature males and females often make contact with a copulating pair (usually the male) during copulation with intromission—but neither before nor after—typically touching the adult's face. This is perhaps the most frequently cited form of male-infant contact in this genus. However, since the behaviors in question are typically fearful and submissive rather than aggressive, it may be unwise to describe the behavior as interference. Moreover, copulating males usually tolerate contact by infants and juveniles in such contexts and may even reach out to reassure them (Tutin and McGinnis, 1981, p. 252). Thus, the behavior may more closely resemble exploration or curiosity. It remains for field workers to more closely describe such interactions. Interference does suggest Oedipal dynamics, which, were they present, would be of considerable comparative interest to psychodynamically oriented psychologists. Such a dynamic would predict that a male offspring would attempt to interfere with copulations between his mother and an adult male or males. In this regard, data on the relative distribution of male versus female infants exhibiting interference, as well as the object of interference as related to sex of infant, would be useful. Any response of the adult male, particularly an aggressive one, could also be correlated with gender of the infant. Tutin and McGinnis (1981), for example, noted that copulating males became increasingly aggressive toward an interfering male after he reached puberty.

During the chimpanzee infant's second year, play increases and tolerance declines concurrently. Infants are rejected more aggressively and they may learn to avoid adult males. Adult males may thrust against infants during play and greeting. In the third or fourth year, infants begin to direct submissive behavior toward males, although they continue to be protected by them. And as weaning proceeds after the fourth year, they take their place more and more autonomously in the community (van Lawick-Goodall, 1968).

Indirect forms of male parental investment are also evident. Males are more mobile than females encumbered with dependent offspring (van Lawick-Goodall, 1968; Wrangham, 1979). Accordingly, they are more likely to discover sources of seasonally abundant but scattered food supplies. They signal by vocalizing and drumming, which attracts mothers and dependent offspring (Reynolds and Reynolds, 1965). In addition, males are well known to be more active in hunting, the spoils of which are shared with other community members.

As with some Cercopithecine species, the opportunity afforded males to interact affiliatively with infants in a laboratory housing context appears to elicit more parental interest than is observed among wild groups. In chapter 10, Davis describes male-infant interactions among one captive group of chimpanzees, which suggests a capability of greater male interest in infants than has been reported in the wild.

Gorillas

The mountain gorilla *(G. g. beringei),* of which most data are available of the three gorilla subspecies in the wild, is characterized by a social system that is functionally one-male/multi-female. That is, although more than one mature male may be present in a group, only one is actively reproductive with fertile females (Harcourt et al., 1981). The groups studies by Harcourt et al., for example, included one silverback male at least 18 years old, two to six adult females less than 8 years old, and four to eight immature individuals. Like chimpanzees, but unlike most other primates, adolescent females rather than males transfer from natal groups and join established groups or solitary siverback males. Males emigrate and remain solitary or join with females in forming new groups (Harcourt, 1978; Veit, 1982). Adult males within a group appear to be closely related and familiar (Harcourt, 1979), whereas males of different groups—given a lack of group transfer by males—are far less likely to be so characterized (Harcourt et al., 1981).

In accord with the expected pattern for one-male systems, male parental investment is relatively limited. Indeed, Veit (1982, p. 58) stated that "obvious male parental investment includes only the maintenance of a home range and the exclusion of foreign males"—both indirect benefits clearly. Foreign (i.e., unrelated) males can indeed be dangerous to immature individuals, as Fossey (cited in Redican, 1976, and Redican and Taub, 1981) demonstrated: one silverback male killed an infant of another group, yet left unaggressed a subsequent infant of the same mother when she joined his group.

Yet it would be less than perspicacious to limit male parental investment to the maintenance of the home range and the exclusion of unrelated males. According to Fossey (ibid.), silverbacks are characteristically tolerant of infants. And low-intensity play bouts have been documented between silverbacks and infants. Fossey also observed a prolonged nurturant relationship between a silverback male and female infant, in which the male groomed, slept with, protected, and remained in close proximity and contact for extended period of time with the infants. Very unlike chimpanzee infants, immature gorillas rarely "interfere" during the copulation of adults (Harcourt et al., 1981).

Two reports of parental behavior among captive lowland gorillas *(G. g. gorilla)* are worthy of note. Studying adult male-infant interactions from the third to tenth months of life in a captive group at the Yerkes Primate Center, Wilson et al. (1977) noted that the silverback male of the group would often take (and attempt to take without success) one of the male infants from its mother (who always retrieved it after a short period in the male's possession). This male also tolerated touches from the other, less preferred infant male, but he did not carry this infant. "Lest contact was observed between the infant female and the silverback male" (abstr). Tilford and Nadler (1978) studying what appears to be the same group of lowland gorillas, provide a somewhat more quantitative picture of male-infant interactions early in the infant gorilla's life. They define four categories of male-infant behavior: social approach, attempt to contact, touch, and hold/carry. During the six weeks of their study, the adult male "approached, attempted to contact and touched all the infants" (p. 221), but only carried and held one male and one female infant (of the one male and two female infants in the group). Interactions, while not frequent (n = 220 total for the six-week study period), were generally focused toward the infant male, Akbar, as he accounted for 86% (n = 190) of all the social interactions scored. The male's interactions with this preferred male infant were more or less constant from study weeks 1 to 5 (infant ages ca. 17 to 22 weeks) at an average of about 27 episodes per week, but rose rapidly on week 6 (the last week of the study) to 57 episodes. The authors suggest that infant age and sex, as well as the male's social relationship with the infant's mother, are important variables in determining this pattern of specificity with regard to the male's "parental" interactions with infants.

Orang Utans

Asserting that an animal is solitary is somewhat like attempting to prove a null hypothesis: that animals do not associate. Thus, the common description of the orang utan as a solitary taxon is likely to encounter difficulty as more data are collected. Orang utans do fulfill Eisenberg's (1966, p. 249) description of solitary patterning insofar as they generally do not seek to be with conspecifics and they remain out of contact or try to do so (Horr, 1971;

MacKinnon, 1971, 1974; Rodman, 1973; Rijksen, 1975). Solitary taxa are social only during mating and parental caretaking. However, orang utans have also been seen to cluster in temporary associations. Rijksen (1978), for example, documented 141 occasions in which associations of up to eight Sumatran orang utans formed. Such associations were observed more rarely among Bornean orang utans (Rodman, 1973). Indeed, there are suggestions of a full-fledged integrated social organization, at least among Sumatran orang utans, particularly in terms of adolescent social groupings (Rijksen, 1978).

In whatever classification the orang utan may ultimately fall, it remains the case that it is more solitary than most other primates. Adult Bornean males and females consort for several days only, during which time most conceptions occur, as in *Pan* (Tutin, 1975.) At other times, adults chase, avoid, or fight with members of the same sex; interactions between the sexes are strikingly nonaggressive (Galdikas, 1981). Females care for dependent young for an extraordinarily long period of five or more years (Galdikas, 1981; Rijksen, 1978).

Not surprisingly, given this type of social organization, contact between adult males and infants is virtually undocumented for this taxon in the wild, (although when the solitary nature of orang utans is circumscribed by the conditions of captivity, males do show tolerance for and interest in infants [Wilson et al., 1977]). Given this pattern, Ullrich's (1970) account (cited by Caine and Mitchell, 1979, p. 120) of a captive orang utan pair is all the more remarkable. The male was present as the female gave birth: "Assuming a calm and nonsexual demeanor, the animal positioned himself at the female's raised perineum. He then put his mouth over the emerging newborn's head and pulled gently. Once the head was fully freed, the male used his hands to deliver the rest of the neonate's body...he held the infant until the female had righted herself; at that time he relinquished the newborn to its mother." The animal behaviorist can either rejoice or despair over such unexpected deviations from what has been presumed to be the norm. At the least, they remind us that the unexpressed potential for behavioral plasticity is immense in such a primate.

REFERENCES

Aldrich-Blake, F. P. G., and Chivers, D. J. 1973. On the genesis of a group of siamang. *American Journal of Physical Anthropology* 38: 631-36.

Altmann, J. 1980. *Baboon Mothers and Infants.* Cambridge, MA: Harvard University Press.

Alexander, B. K. 1970. Parental behavior of adult male Japanese monkeys. *Behaviour* 36: 270-85.

Auerbach, K. G., and Taub, D. M. 1979. Paternal behavior in a captive "harem" group of cynomolgus macaques *(Macaca fascicularis). Laboratory Primate Newsletter* 18 (2): 7-11.

Berkson, G. 1966. Development of an infant in a captive gibbon group. *Journal of Genetic Psychology* 108: 311-25.

Bernstein, I. S. 1970. "Paternal" behavior in nonhuman primates. *American Zoologist* 10: 480.

Bernstein, I. S. 1975. Activity patterns in a gelada monkey group. *Folia Primatologica* 23: 50-71.

———. 1976. Dominance, aggression and reproduction in primate societies. *Journal of Theoretical Biology* 60: 459-72.

Bertrand, M. 1969. The behavioral repertoire of the stumptail macaque. *Bibliotheca Primatologica* 11: 1-273.

Blaffer-Hrdy, S. 1976. Care and exploitation of nonhuman primate infants by conspecifics other than the mother. In J. S. Rosenblatt; R. A. Hinde; E. Shaw; and C. Beer (eds.), *Advances in the Study of Behavior*, vol. 6, pp. 101-58. New York: Academic Press.

Booth, C. 1962. Some observations on behavior of cercopithecus monkeys. *New York Academy of Sciences* 102: 477-87.

Brandt, E.; Irons, R.; and Mitchell, G. 1970. Paternalistic behavior in four species of macaques. *Brain, Behavior and Evolution* 3: 415-20.

Breuggeman, J. A. 1973. Parental care in a group of free-ranging rhesus monkeys *(Macaca mulatta). Folia Primatolgica* 20: 178-210.

Brockelman, W. Y.; Ross, B. A.; and Pantuwatana, S. 1974. Social interactions of adult gibbons *(Hylobates lar)* in an experimental colony. *Gibbon and Siamang* 3: 137-56.

Burton, F. D. 1972. The integration of biology and behavior in the socialization on *Macaca sylvana* of Gibraltar. In F. E. Poirier (ed.). *Primate Socialization,* pp. 29-62. New York: Random House.

Burton, F. O., and Bick, M. J. A. 1972. A drift in time can define a deme: The implications of tradition drift in primate societies for hominid evolution. *Journal of Human Evolution* 1: 53-59.

Busse, C., and Hamilton, W. J. 1981. Infant carrying by male chacma baboons. *Science* 212: (4500): 1281-83.

Bygott, J. D. 1972. Cannibalism among wild chimpanzees. *Nature* 238: 410-11.

Caine, N., and Mitchell, G. 1979. Behavior of primates present during parturition. In J. Erwin et al., (eds.), *Captivity and Behavior*, pp. 112-24. New York: Van Nostrand Reinhold.

Carpenter, C. R. 1940. A field study in Siam of the behavior and social relations of the gibbon *(Hylobates lar). Comparative Psychology Monographs* 16: 1-212.

———. 1964. *Naturalistic Behavior of Nonhuman Primates.* University Park: Pennsylvania State University Press.

Chivers, D. 1971. Spatial relations within the siamang group. *Proceedings of the Third International Congress of Primatology, Zurich, 1970.* 3: 14-21. Basel: S. Karger.

———. 1972. The siamang and the gibbon in the Malya Peninsula. *Gibbon and Siamang* 1: 103-35.

Chalmers, N. 1968. The social behavior of free-living mangabeys in Uganda. *Folia Primatologica* 8: 263-81.

Deag, J. M. 1974. A study of the social behaviour and ecology of the wild Barbary macaque *Macaca sylvana* L. Ph.D. dissertation, University of Bristol, England.

———. 1980. Interactions between males and unweaned Barbary macaques: Testing the agonistic buffering hypothesis. *Behaviour* 75: 54-81.

Deag, J. M., and Crook, J. 1971. Social behavior and "agonistic buffering" in the wild barbary macaque *Macaca sylvana* L. *Folia Primatologica* 15: 183-200.

DeVore, I. 1963. Mother-infant relations in free-ranging baboons. In H. L. Reingold (ed.), *Maternal Behavior in Mammals,* pp. 305-35. New York: Wiley.

Dunbar, R. I. M., and Dunbar, E. P. 1975. Social Dynamics of Gelada Babbons, *Contributions to Primatology* vol. 6. Basel: Karger.

Eisenberg, J. F. 1966. The social organization of mammals. *Handbuch der Zoologie: Eine Naturgeschichte der Stamme des Tierreiches* 8(39): 1-92.

Ellefson, J. 1968. Territorial behavior in the common white-handed gibbon, *Hylobates lar.* In P. C. Jay (ed.), *Primates: Studies in Adaptation and Variability,* pp. 180-99. New York: Holt.

———. 1974. A natural history of white-handed gibbons in the Malay Peninsula. *Gibbon and Siamang* 3: 1-136.

Estrada, A., and Sandoval, J. M. 1977. Social relations in a free-ranging troop of stumptail macaques *(Macaca arctoides):* Male-care behavior I. *Primates* 18: 793-813.

Estrada, A.; Estrada, R.; and Ervin, R. 1977. Establishment of a free-ranging colony of stumptail macaques *(Macaca arctoides):* Social relations I. *Primates* 18: 647-76.

Fox, G. 1972. Some comparisons between siamang and gibbon behaviour. *Folia Primatologica* 18: 122-39.

———. 1974. Peripheralization behavior in a captive siamang family. *American Journal of Physical Anthropology* 41: 479.

Galdikas, B. M. F. 1981. Orangutan reproduction in the wild. In C. E. Graham (ed.) *Reproductive Biology of the Great Apes: Comparative and Biomedical Perspectives* pp. 281-300. New York: Academic Press.

Gifford, D. P. 1967. The expression of male interest in the infant in five species of macaque. *Kroeber Anthropological Society Papers* 36: 32-40.

Goodall, J. et. al. 1979. Perspectives on human evolution. In D. A. Hamburg and E. R. McCown (eds), *The Great Apes,* vol. 5. pp. 13-53. Menlo Park, CA: Benjamin Cummings.

Grewall, B. S. 1980. Social relationships between adult central males and kinship groups of Japanese monkeys at Arashiyama with some aspects of organization. *Primates* 21 (2): 161-80.

Gouzoules, H. 1975. Maternal rank and early social interactions of stumptail macaques, *Macaca arctoides. Primates* 16: 405-18.

Gubernick, D. J., and Klopfer, P. H. (eds). 1981. *Parental Care in Mammals.* New York: Plenum Press.

Hall, K. R. L. 1963. Some problems in the analysis and comparison of monkey and ape behavior. In S. L. Washburn (ed.), *Classification and Human Evolution,* pp. 263-300. New York: Viking Fund Publications, Wenner-Gren Foundation.

———. 1965. Behaviour and ecology of the wild patas monkey, *Erythrocebus patas,* in Uganda. *Journal of Zoology* 148: 15-87.

Harcourt, A. H. 1977. Social relationships of wild mountain gorilla. Ph.D. dissertation, Cambridge University, England.

———. 1979. Social relationships between adult male and female mountain gorillas in the wild. *Animal Behaviour* 27 (2): 325-42.

Harcourt, A. H.; Stewart, J. J.; and Fossey, D. 1981. Gorilla reproduction in the wild. In C. E. Graham (ed.), *Reproductive Biology of the Great Apes: Comparative and Biomedical Perspectives,* pp. 265-79. New York: Academic Press.

Harlow, M. K. 1971. Nuclear family apparatus. *Behavior Research Methods and Instrumentation* 3: 301-04.

Hasegawa, T., and Hiraiwa, M. 1980. Social interactions of orphans observed in a free-ranging troop of Japanese monkeys. *Folia Primatologica* 33: 129-58.

Hendy-Neely, H., and Rhine, R. R. 1977. Social development of stumptail macaques *(Macaca arctoides):* Momentary touching and other interactions with adult males during the infant's first 60 days of life. *Primates* 18: 589-600.

Hiraiwa, M. 1981. Maternal and alloparental care in a troop of free-ranging Japanese monkeys. *Primates* 22 (3): 309-29.

Horr, D. A. 1972. The Borneo orang utan. *Borneo Research Bulletin* 4: 46-50.

Itani, J. 1959. Paternal care in the wild Japanese monkey, *Macaca fuscata fuscata. Primates* 2: 61-93.

Jones, N. G. B., and Trollope, J. 1968. Social behaviour of stump-tailed macaques in captivity. *Primates* 9: 365-94.

Kaufman, I. C., and Rosenblum, L. A. 1969. The waning of the mother-infant bond in two species of macaque. In B. M. Foss (ed.), *Determinants of Infant Behaviour* vol. 4, pp. 41-59, London: Methuen.

Kaufmann, J. H. 1966. Behavior of infant rhesus monkeys and their mothers in a free-ranging band. *Zoologica* 51: 17-27.

Kleiman, D. G. 1977. Monogamy in mammals. *The Quarterly Review of Biology* 52: 39-69.

Kummer, H. 1967. Tripartite relations in Hamadryas baboons. In S. A. Altmann (ed.), *Social Communication among Primates,* pp. 63-71. Chicago: University of Chicago Press.

———. 1971. *Primate Societies.* Chicago: Aldine-Atherton.

Kurland, J. 1977. Kin selection in the Japanese monkey. *Contributions to Primatology* 12: 1-145.

Kurland, J., and Taub, D. M. In preparation. Does "agonistic buffering" buffer agonism. Testing the agonistic buffering hypothesis II.

Lahiri, R. K., and Southwick, C. H. 1966. Parental care in *Macaca sylvana. Folia Primatogica* 4: 257-64.

Lamb, M. E. (ed.), *The Role of the Father in Child Development,* 2nd ed. New York: Wiley.

Lamb, M. E., and Goldberg, W. A. 1981. The father-child relationship: A synthesis of biological, evolutionary and social perspectives. In R. Gandelman, and L. W. Hoffman (eds.), *Perspectives on Parental Behavior.* Hillsdale NJ: Lawrence Erlbaum Associates.

Lindburg, D. G. 1971. The rhesus monkey in North India: An ecological and behavioral study. In L. A. Rosenblum (ed.), *Primate Behavior: Developments in Field and Laboratory Research* vol. 2, pp. 1-106. New York: Academic Press.

Lydekker, R. 1894, *The Royal Natural History,* vol. 1. London: Frederick Warne.

McCann, C. 1933. Notes on the colouration and habits of the white-browed gibbon or hoolock *(Hylobates hoolock* Harl.). *Journal of the Bombay Natural History Society* 36: 395-405.

McClure; H. E. 1964. Some observations of primates in climax diptocarp forest near Kuala Lumpur, Malaya. *Primates* 3-4: 39-58.

MacKinnon, J. 1971. The orang utan in Sabah today. *Oryx* 11: 141-91.

———. The behaviour and ecology of wild Orang-utans *(Pongo pygmaeus). Animal Behaviour* 22: 3-74.

MacRoberts, M. H. 1970. The social organization of Barbary apes *(Macaca sylvana)* on Gibraltar. *American Journal of Physical Anthropology* 133: 83-99.

Meikle, D. B. 1980. Sex ratio and kin selection in rhesus monkeys. Ph.D. dissertation, Bowling Green State University, Ohio.

Mitchell, G. D. 1969. Paternalistic behavior in primates. *Psychological Bulletin* 71: 399-417.

Mitchell, G., and Brandt, E. 1972. Paternal Behavior in primates. In F. Poirier (ed.), *Primate Socialization* pp. 173-206. New York: Random House.

Mori, U. 1979a. Development of sociability and social status. In M. Kawai (ed.), *Ecological and Sociological Studies of Gelada Baboons, Contributions to Primatology* vol. 16, pp. 125-54. Basel: S. Karger.

———. 1979b. Individual relationships within a unit. In M. Kawai (ed.), *Ecological and Sociological Studies of Gelada Baboons, Contributions to Primatology* vol. 16: pp. 93-124. Basel: S. Karger.

Nissen, H. W. 1931. A field study of the chimpanzee. *Comparative Psychology Monographs* 8: 1-22.

Packer, C. 1980. Male care and exploitation of infants in *Papio anubis. Animal Behaviour* 28: 512-20.

Popp, J. L. 1978. Male baboons and evolutionary principles. Ph.D. dissertation, Harvard University.

Pusey, A. 1979. Intercommunity transfer of chimpanzees in Gombe National Park. In D. A. Hamburg and E. R. McCown (eds.), 1979. *Perspective on Human Evolution* vol. 5, *The Great Apes*, pp. 465-79. Menlo Park, CA: Benjamin/Cummings.

Ransom, T. W., and Ransom, B. S. 1971. Adult male-infant relations among baboons *(Papio anubis). Folia Primatologica* 16: 179-95.

Redican, W. K. 1975. A longitudinal study of behavioral interactions between male and infant rhesus monkeys *(Macaca mulatta).* Ph.D. dissertation, University of California, Davis.

———. 1976. Adult male-infant interactions in nonhuman primates. In M. E. Lamb (ed.), *The Role of the Father in Child Development*, pp. 345-85. New York: Wiley.

Redican, W. K., and Mitchell, G. 1974. Play between adult male and infant rhesus monkeys. *American Zoologist* 14: 295-302.

Redican, W.K., and Taub, D. M. 1981. Adult male-infant interactions in nonhuman primates. In M. E. Lamb (ed.), *The Role of the Father in Child Development*, 2nd ed., pp. 203-58. New York: Wiley.

Reynolds,V., and Reynolds, F. 1965. Chimpanzees of the Budongo Forest. In I. DeVore (ed.) *Primate Behavior: Field Studies of Monkeys and Apes*, p. 368-424. New York: Holt, Rinehart and Winston.

Rhine, R. J., and Hendy-Neely, H. 1978. Social development of stumptail macaques *(Macaca arctoides):* Momentary touching, play and other interactions with aunts and immatures during the infants' first 60 days of life. *Primates* 19: 115-23.

Rijsken, H. D. 1975. A field study of Sumatran orang utans *(Pongo pygmaeus abelii).* In S. Kondo; M. Kawai; and A. Ehara (eds.), *Contemporary Primatology*, pp. 373-79. Japan: Nagoya; Basel: S. Karger.

———. A field study on Sumatran orangutans: Ecology, behaviour and conservation. *Mededelingen Landbouwhogeschool Wageningen (Nederland)* 78-2, vi + 420 pp.

Rodman, P. S. 1973. Population composition and adaptive organization among orang-utans of the Kutai Reserve. In J. H. Crook, and R. P. Michael (eds.), *Comparative Ecology and Behaviour of Primates*, pp. 171-209. London: Academic Press.

Rowell, T. E. 1969. Long-term changes in a population of Ugandan baboons. *Folia Primatologica* 11: 241-54.

Rowell, T. E.; Hinde, R. A.; and Spencer-Booth, Y. 1964. "Aunt"-infant interactions in captive rhesus monkeys. *Animal Behaviour* 12: 219-26.

Saayman, G. S. 1971. Behaviour of the adult males in a troop of free-ranging chacma baboons *(Papio ursinus). Folia Primatologica* 15: 36-57.

Simonds, P. E. 1965. The bonnet macaque in South India. In I. DeVore (ed.), *Primate Behavior: Field Studies of Monkeys and Apes*, pp. 175-95. New York: Holt, Rinehart and Winston.

———. 1974. Sex differences in bonnet macaque networks and social structure. *Archives of Sexual Behavior* 3: 151-66.

Smith, E. O., and Peffer-Smith, P. G. 1982. Triadic interaction in captive Barbary macaques *(Macaca sylvanus* Linnaeus 1758): "Agonistic buffering"? *American Journal of Primatology* 2(1):99-108.

Smuts, B. 1982. Special relationships between adult male and female olive baboons *(Papio anubis).* Ph.D. dissertation, Stanford University.

Southwick, C. H.; Beg, M. A.; and Siddiqi, M. R. 1965. Rhesus monkeys in North India. In I. DeVore (ed.), *Primate Behavior: Field Studies of Monkeys and Apes*, pp. 111-59. New York: Holt, Rinehart and Winston.

Spencer-Booth, Y. 1968. The behaviour of group companions toward rhesus monkey infants. *Animal Behaviour*, 16: 541-77.

Spencer-Booth, Y., and Hinde, R. A. 1967. The effects of separating rhesus monkey infants from their mothers for six days. *Journal of Child Psychology and Psychiatry and Allied Disciplines* 7: 179-97.

Stein, D. M. 1980. Adult male baboons' affiliative relations with infants. Paper presented at 3rd meeting, *American Society of Primatologists,* Winston-Salem, N. C.

Stein, D. M. 1981. The nature and function of social interactions between infant and adult male yellow baboons *(Papio cynocephalus).* Ph.D. dissertation, University of Chicago.

Stein, D. M., and Stacey, P. B. 1981. A comparison of infant-adult male relations in a one-male group with those in a multimale group for yellow baboons *(Papio cynocephalus). Folia Primatologica* 36: 264-76.

Sugiyama, Y. 1971. Characteristics of the social life of bonnet macaques *(Macaca radiata). Primates,* 12: 247-66.

Suomi, S. J. 1977. Adult male-infant interactions among monkeys living in nuclear families. *Child Development* 48: 1255-70.

———. 1979. Differential development of various social relationships by rhesus monkey infants. In M. Lewis and L. A. Rosenblum (ed.), *The Social Network of the Child,* pp. 219-44. New York: Plenum Press.

Suomi, S. J.; Eisele, C. D.; Grady, S. A.; and Tripp, R. L. 1973. Social preferences of monkeys reared in an enriched laboratory environment. *Child Development.* 44: 451-60.

Suzuki, A. 1971. Carnivority and cannibalism observed among forest-living chimpanzees. *Journal of the Anthropological Society of Nippon* 79: 30-48.

Taub, D. M. 1975. "Paternalism" in free-ranging Barbary macaques, *Macaca sylvanus. American Journal of Physical Anthropology* 42: 333-34.

———. 1977. Geographic distribution and habitat diversity of the Barbary macaque, *Macaca sylvanus* L. *Folia Primatologica* 27: 108-33.

———. 1978. Aspects of the biology of the wild Barbary macaque (Primates, Cercopithecinae, *Macaca sylvanus* L. 1758): Biogeography, the mating system and male-infant associations. Ph.D. dissertation, University of California, Davis.

———. 1980. Testing the "agonistic buffering" hypothesis. I. The dynamics of participation in the triadic interactions. *Behavioral Ecology and Sociobiology* 6: 187-97.

———. In preparation. Male-infant interactions in Old World baboons and macaques: A summary and reevaluation. *American Society of Zoologists/Animal Behavior Society Symposium, Philadelphia, PA. 1983.*

Taylor, H.; Teas, J.; Richie, T.; Southwick, C.; and Shrestha, P. 1978. Social interactions between adult male and infant rhesus monkeys in Nepal. *Primates,* 19: 343-51.

Teleki, G.; Hunt, E.; and Pfifferling, J. H. 1976. Demographic observations (1963-1973) on the chimpanzees of the Gombe National Park, Tanzania, *Journal of Human Evolution* 5: 559-98.

Tenaza, R. 1975. Territory and monogamy among Kloss' gibbons *(Hylobates klossii)* in Siberut Island, Indonesia. *Folia Primatologica* 24: 60-80.

Tilford, B. A., and Nadler, R. D. 1978. Male parental behavior in a captive group of lowland gorillas *(Gorilla gorilla gorilla). Folia Primatologica* 29: 218-28.

Tutin, C. E. G. 1975. Sexual behaviour and mating patterns in a community of wild chimpanzees *(Pan troglodytes schweinfurthii).* Ph.D. dissertation, University of Edinburgh, Edinburgh, Scotland.

———. 1979. Responses of chimpanzees to copulation, with special reference to interference by immature individuals. *Animal Behaviour* 27: 845-54.

Tutin, C. E. G., and McGinnis, P. R. 1981. Chimpanzee reproduction in the wild. In C. E. Graham (ed.), *Reproductive Biology of the Great Apes: Comparative and Biomedical Perspectives,* p. 239-64. New York: Academic Press.

Ullrich, W. 1970. Geburt aund naturliche Geburtshilfe beim Orang Utan, *Der Zoologische Garten* 39: 284-89.

van Lawick-Goodall, J. 1967. Mother-offspring relationships in free-ranging chimpanzees. In D. Morris (ed.), *Primate Ethology* pp. 365-436. Chicago: Aldine.

———. 1968. The behaviour of free-living chimpanzees in the Gombe Stream Reserve. *Animal Behaviour Monographs* 1: 161-311.

Veit, P. G. 1982. Gorilla society. *Natural History* 91(3): 48-59.

Whiten, A., and Rumsey, T. J. 1973. "Agonistic buffering" in the wild Barbary macaque, *Macaca sylvana* L. *Primates* 14: 421-25.

Wilson, E. O. 1975. *Sociobiology: The New Synthesis.* Cambridge, MA: Belknap Press, Harvard University.

Wilson, Mark E. et al., 1977 (abstract only). Characteristics of paternal behavior in captive orang utans *(Pongo pygmaeus abelii)* and lowland gorillas *(Gorilla gorilla gorilla).* Paper presented at first annual meeting, *American Society of Primatologists,* Seattle, WA. April 16-19, 1977.

Wrangham, R. W. 1979. Sex differences in chimpanzee dispersion. In D. A. Hamburg, and E. R. McCown (eds.), *Perspectives on Human Evolution,* vol. 5: *The Great Apes.* Menlo Park, CA: Benjamin/Cummings.

16

Observational Studies of Father-Child Relationships in Humans[1]

Michael E. Lamb
Departments of Psychology, Pediatrics, and Psychiatry
University of Utah
Salt Lake City, Utah

INTRODUCTION

Researchers and theorists have expended much more effort in the investigation and elucidation of the father-child relationship among humans than among all other species put together. In fact, so much has been written about paternal influences and the father's role since Freud initiated theorizing about sociopersonality development at the turn of the century that it is necessary to impose some limitations on the scope of the chapter. First, I do not provide a review of theoretical perspectives; I simply introduce those propositions that help make sense of the data I present. Second, I deal solely with the results of studies in which fathers and children have actually been observed together, since these data are most comparable to those gathered by behavioral scientists concerned with male-infant relationships among nonhuman primates. Because most of the observational research has involved fathers and infants, relatively little can be said about the relationships between fathers and older children. However, since older children have been the focus of most investigations of paternal influences, I refer to well-established conclusions where they help to place in context or clarify the

---

[1]This chapter is based on and expanded from a chapter entitled "The development of father-infant relationships" published in Michael E. Lamb (ed.), *The role of the father in child development.* New York: Wiley, 1981.

findings of observational studies. This is especially pertinent where effects are concerned, because most observational research has been process rather than outcome oriented. Readers who want a more thorough review of the relevant theories and findings are referred to a recent volume containing chapters by the most prominent workers in the area today (Lamb, 1981).

As in the case of nonhuman primates (Redican, 1975; Redican and Mitchell, 1973, 1974), quasiexperimental strategies have proved useful for distinguishing between potential competence and average performance (e.g., Parke and Tinsley, 1981). Paternal caretaking competence is the focus of the next section of this chapter ("Paternal Caretaking Competence"); here I conclude that men are quite as capable of caretaking as women are, although they tend, almost universally, to yield to women and thus play a minor role in the physical care of their children. The relative lack of involvement with children is also evident in studies designed to determine how much time, on average, fathers spend with their children. These studies are reviewed in the next section, "Time-Use Studies of Fathers." Thereafter, I review studies concerned with the formation of attachments between infants and their fathers ("The Development of Attachments"). This topic has been a popular focus among researchers, largely because theorists have consistently argued that such relationships do not form in infancy. The differences between maternal and paternal styles of interaction are next described ("Characteristics of Father-Infant Interactions"), followed by the sex-differentiated treatment of sons and daughters ("Sex Differences in Paternal Behavior"). The chapter closes with a section ("Parental Influences on Development") in which indirect and direct paternal influences are described and efforts are made to illustrate important indirect effects—particularly those that may be analogous to processes of influence that occur in groups of nonhuman primates. The "discovery" of father-infant relationships has brought with it the realization that infants are raised within complex social systems in which each person affects every other person as well as the relationships among the others.

PATERNAL CARETAKING COMPETENCE

Most research designed to assess the parenting capacities of adult men and fathers has focused on sensitivity or responsiveness to infant signals and needs. The reason for this is straightforward: it is widely believed that parental sensitivity is crucial to understanding both how attachments are formed in infancy and how individual differences in infant-parent attachment are shaped. According to the ethological attachment theorists (Ainsworth, 1973; Bowlby, 1969; Lamb, 1978c, 1981d; Lamb and Easterbrooks, 1981), human infants are biologically predisposed to emit signals (e.g., cries, smiles) to which adults are biologically predisposed to respond. When adults consistently respond promptly and appropriately to infant signals, infants come to perceive the adults concerned as predictable and reliable. Such trust in the adults' reliability constitutes the basis of secure infant-parent attachment (Ainsworth, Bell, and Stayton, 1974; Lamb, 1981a, 1981b). By contrast, when adults do not respond sensitively (i.e., promptly and appropriately), insecure attachments result, and when they respond rarely, no attachment at all may develop. Consequently, it is very important to determine whether fathers are appropriately responsive to their infants; when they are not, the likelihood of infant-father attachments forming would be low. The issue is rendered especially important by claims (e.g., Klaus, Trause, and Kennell, 1975) that the biological predisposition to respond to infant signals is stronger in females than in males. This speculation implies that males are biologically prevented from exerting major direct influences on the socioemotional development of their infant offspring. These claims, however, are based largely on evidence concerning hormonal influences on parental behavior in rodents (Lamb, 1975b) and this animal model may be quite inappropriate for generalization to humans (Lamb and Goldberg, 1981).

In an early interview study, Greenberg and Morris (1974) reported that most fathers were elated by the birth of their infants and experienced extremely positive emotions, which Greenberg and Morris labeled "engrossment." In an observational study, meanwhile, Parke and O'Leary (1976) observed primiparous mothers and fathers interacting with their newborn infants in a hospital maternity ward. They found that the fathers were neither inept nor disinterested in interaction with the newborns. Indeed, all but a couple of the many measures showed the fathers and mothers to be equivalently involved in interaction.

In his subsequent research, Parke has considered not only the affectionate manner in which fathers behave, but also their behavioral responsiveness or sensitivity—that is, their propensity to attend to their infant's cues and emit appropriate responses. When observed feeding their infants, for example, both fathers and mothers responded to infant cues either with social bids or by adjusting the pace of the feeding (Parke and Sawin, 1980). Although the fathers were capable of behaving sensitively, however, they tended to yield responsibility for child-tending chores to their wives when not asked to demonstrate their competence for the observers.

Alternative ways of studying parental responsiveness to infant signals have been pursued by Feldman and Nash and by Frodi and Lamb. Their studies, however, have involved observations of parents with unfamiliar infants rather than with their own. Feldman and Nash (1977, 1978; Nash and Feldman, 1981) observed parents individually while they sat in a waiting room containing an infant and its mother. The subjects were observed interacting with the baby by concealed observers. These studies showed that sex differences in "baby responsiveness" waxed and waned depending upon the subject's age and social status. Thus females were more responsive than males in early adolescence and in early parenthood, whereas there were no sex differences among eight-year-olds, childless couples, and unmarried college students. Feldman and Nash concluded that sex differences in responsiveness to infants are experientially rather than biologically determined: they are evident when individuals are under increased social pressure to respond in a conventionally sex-typed fashion. In response to such social pressures, suggested these researchers, mothers become more responsive to infants than fathers are.

By contrast, the studies conducted by Frodi and Lamb revealed no sex differences in responsiveness to infants, but their measures were of psychophysiological rather than behavioral responsiveness. In the first two studies in this series, the heart rate, blood pressure, and skin conductance of mothers and fathers were monitored while the parents observed quiescent, smiling, or crying infants on a television monitor (Frodi, Lamb, Leavitt, and Donovan, 1978; Frodi, Lamb, Leavitt, Donovan, Neff, and Sherry, 1978). Crying and smiling infants elicited characteristic and distinct physiological response patterns in *both* mothers and fathers: crying elicited autonomic arousal and reports of irritation or aversion, whereas smiles elicited little change in the level of autonomic activation and pleasant emotion. In a later study, Frodi and Lamb (1978) found no sex differences in physiological responses among either 8- or 14-year-olds, whereas in a waiting room situation, 14-year-old females were more behaviorally responsive to infants than males were. Like Feldman and Nash, we concluded that sex differences in responsiveness to infants emerged in response to societal pressures and expectations rather than to biologically based predispositions.

This conclusion appears to be consistent with all the data currently available. Nevertheless, the fact that men *can* be as responsive as women does not mean that mothers and fathers typically are equivalently responsive. Fathers are not always highly responsive and all fathers are probably not equivalently responsive. Each father's responsiveness probably varies depending on the degree to which he assumes responsibility for infant care: caretaking experience appears to facilitate parental responsiveness (Zelazo et al., 1977). Furthermore, responsiveness is probably dependent on a variety of individual characteristics or traits, the attitudes and support of significant others, the family's circumstances, and the characteristic of the infant (Lamb and Easterbrooks, 1981).

TIME-USE STUDIES OF FATHERS

Formative significance depends on both the quality and quantity of adult-infant interaction. In the preceding section, I reviewed evidence concerning the quality of father-infant interaction, concluding that fathers were able to respond sensitively, although they did not always do so—

especially when caretaking might be involved. In this section, I review studies designed to determine how much time fathers tend to spend with their children.

Several researchers have attempted to determine how much time the average father spends with his infant. As one might expect, estimates vary widely, even when one considers only traditional families in which mothers assume primary and full-time responsibility for child care and fathers assume primary responsibility for economic provision and support. One early study found that according to maternal reports, fathers of 8- to 9½-month-old infants were home between 5 and 47 hours per week at times when the infants were awake (Pedersen and Robson, 1969). The average was 26 hours. The fathers reportedly spent between 45 minutes and 26 hours per week actually interacting with their infants. A few years later, Kotelchuck (1975) determined from interviews with the parents of 6- to 21-month-olds that mothers spent an average of 9 waking hours per day with their children, while fathers spent 3.2 hours. The parents interviewed by Golinkoff and Ames (1979) reported figures of 8.33 and 3.16 hours. By contrast, fathers interviewed by Lewis and Weinraub (1974) reported much less interaction than this: the average was 15 to 20 minutes per day.

Presumably the variability among these estimates is attributable to socioeconomic and subcultural differences in the populations studied. It may also differ depending upon the amount of encouragement and support fathers receive: Lind (1974), for example, found that Swedish fathers who were taught how to care for their newborns and were encouraged to do so were more involved with their infants three months later. Finally, it is important to distinguish between time spent in the same house and time spent in actual and intensive interaction: the mothers and children studied by Clarke-Stewart (1973) spent 90% of the observational sessions in the same room, but only 10 to 15% of the time in interaction with one another.

Most investigators would agree that traditional fathers spend relatively little time each day interacting with their infants, regardless of their class status. Unfortunately, although most theorists believe that at least a minimal amount of regular interaction is necessary if attachments are to form, no one has yet determined the minimum necessary amount. It is unlikely that such an estimate will ever be possible, given the fact that the quality of interaction appears to be more important than its quantity. Furthermore, Schaffer's (1963) research on hospitalized infants suggested that the amount and quality of social interaction in general may facilitate the later formation of bonds to people other than the sources of stimulation. Conceivably, therefore, the quality of mother-infant interaction may facilitate the formation of father-infant attachments and affect the amount of interaction necessary for attachments to form.

Two secular trends may render obsolete the statistics reviewed in this section. Best documented is the increasing tendency of mothers—including the mothers of infants and young children—to seek paid employment outside the home. Roughly 36% of the mothers of children under three are so employed, although many work part-time (Glick, 1979), and 52% of the married mothers of 6 to 17-year-olds have paying jobs. Employment rates are even higher among single mothers. Maternal employment means that mothers spend less time with their children (reducing the disparity between the levels of involvement of mothers and fathers) and is also associated with a surprisingly modest increase in the amount of paternal involvement (Hoffman, 1977; Pleck, 1981). A second trend, documented by Pleck (1981) and Sheehy (1979), suggests an increasing interest on the part of young men today in family relationships rather than occupational responsibilities. This increasing expressiveness on the part of young fathers will presumably result in closer and warmer father-child relationships, although there is little evidence that fathers have generally assumed increased responsibility for their children. Even when economic barriers are removed, as in Sweden's national parental leave policy, few fathers actually assume major responsibility for child care (Lamb, Frodi, Hwang, and Frodi, 1982). When they do so, finally, their venture into child care is often shortlived, as Russell (1978, 1982) found in his study of Australian fathers who had primary responsibility for child care.

THE DEVELOPMENT OF ATTACHMENTS

The first attempt to determine empirically whether and when infants formed attachments to their fathers was made by Schaffer and Emerson (1964). Mothers reported that their infants began to protest separations from them at seven to nine months of age; many also reported that their infants began to protest separations from their fathers at about the same time. By 18 months of age, 71% of the subjects appeared to be attached to (i.e., protested separation from) both parents. According to Schaffer and Emerson (1964) the babies form attachments to those with whom they interacted regularly; involvement in caretaking seemed unimportant.

As in Schaffer and Emerson, the next published study (Pedersen and Robson, 1969) also relied upon maternal reports, although here the focus was on responses to reunion following separation, rather than to protest concerning the onset of separation. The majority of the mothers (75%) reported that their infants responded positively and enthusiastically when their fathers' returned from work and this led Pedersen and Robson to conclude that these infants were attached to their fathers. Among the boys, intensity of greeting behavior was correlated with the frequency of paternal caretaking, paternal patience with infant fussing, and the intensity of father-infant play. Among daughters, however, intensity of greeting behavior was only correlated with reported paternal "apprehension over well-being."

Observational studies of father-infant attachment began in the 1970s—a decade marked by a sharp shift from interviews to observations as the most popular sources of data. For his doctoral dissertation, Kotelchuck (1972) developed an experimental procedure that permitted him to observe the reactions of 6-, 9-, 12-, 15-, 18-, and 21-month-old infants to brief separations from mothers, fathers, and strangers. Although no usable data were obtained from the 6- and 9-month-olds (perhaps because the sequence of 13 three-minute episodes was too confusing or exhausting for them), it seemed that the older infants were attached to both of their parents. They predictably protested when left alone by either parent, explored little while the parents were absent, and greeted them positively when they returned. Few infants protested separation from either parent when the other parent remained with them. A majority (55%) of the infants were more concerned about separation from, and thus seemed to prefer, their mothers, but 25% preferred their fathers and 20% showed no preference for either parent. Kotelchuck and his colleagues proceeded to conduct a series of investigations using the same experimental procedure (Spelke et al., 1973; Ross et al., 1975; Lester et al., 1974, Zelazo et al., 1977). The results of these studies confirmed the conclusions of the initial study.

Cohen and Campos (1974) observed, as Kotelchuck and his colleagues did, infants' responses to brief separations from each of their parents, but they also recorded the infants' propensity to seek comfort from the people (mother, father, or stranger) remaining with them. Distress did not discriminate between mothers and fathers, but on measures such as frequency of approach, speed of approach, time in proximity, and use of the parent as a "secure base" from which to interact with a stranger, 10-, 13-, and 16-month-old infants showed preferences for their mothers over their fathers, as well as clear preferences for father over strangers. This finding contrasts with that of Feldman and Ingham (1975) and Willemsen et al. (1974). Both of these latter studies revealed no preferences for either parent over the other. In a study of two-year-olds, Lamb (1976c) found no preferences for either parent on measures of separation protest or greeting behavior.

Instead of a separation-reunion paradigm, Lewis and his colleagues observed one- and two-year-old infants in 15-minute free play sessions—once with each parent. One-year-olds touched, stayed near, and vocalized to their mothers more than their fathers, whereas no comparable preferences were evident among two-year-olds (Ban and Lewis, 1974; Lewis and Ban, 1971; Lewis and Weinraub, 1974; Lewis, Weinraub, and Ban, 1972).

By the mid-1970s, therefore, there were numerous indications that children indeed developed attachments to their fathers in infancy, but there were no data available concerning the period between six and nine months of age, during which infants form attachments to their mothers (Bowlby, 1969). Consequently, it was not known whether mother-infant and father-infant relationships developed contemporaneously or serially. There was also controversy concerning the existence of preferences for mothers over fathers—some studies reported such preferences while others did not. Finally, no researchers had examined father-infant interaction in the unstructured home environment rather than in the laboratory.

It was in this context that I initiated a naturalistic longitudinal study of mother- and father-infant attachment in 1974. As in all other studies reviewed, the participant families were all traditional in role structure, with mothers as full-time caretakers and fathers as full-time breadwinners. Two-hour-long observations revealed that 7-, 8-, 12-, and 13-month-old infants showed no preference for either parent over the other on attachment behavior measures (measures of their propensity to stay near, approach, touch, cry to, and ask to be held by specific adults), although these measures all showed preferences for parents over a relatively unfamiliar adult (Lamb, 1976b, 1977c). Measures of separation protest and greeting behavior likewise showed no preferences for either parent over the other (Lamb, 1979). During the second year of life, however, the situation changed. Boys began to show significant preferences for their fathers on the attachment behavior measures, whereas girls as a group showed no significant and consistent preference for either parent (Lamb, 1977a). By the end of the second year, all but one of the nine boys showed consistent preferences for their fathers on at least four of the five attachment behavior measures (Lamb, 1977b).

The results of this longitudinal study indicated that most infants formed attachments to both their parents at about the same time, and that by the second year of life boys preferred to interact with their fathers. On superficial examination, these findings appear to contradict Bowlby's (1969) claim that there is a hierarchy among attachment figures, with the primary caretaker becoming the preferred attachment figure. My data, however, were not really appropriate for testing this hypothesis. According to attachment theory, preferences among attachment figures may not be evident when infants do not need comfort from or protection by attachment figures. Under stress, however, infants should inhibit affiliation with nonattachment figures and focus their attachment behavior more narrowly on primary attachment figures.

The participants in my longitudinal study were also observed in more stressful laboratory contexts at 8, 12, and 18 months (Lamb, 1976a, 1976d, 1976f). At eight months, the infants did not show consistent preferences for either parent, even when they were distressed (Lamb, 1976d). The results of the 12- and 18-month observations were clearcut and consistent with

Bowlby's hypothesis, however. In stress-free episodes, infants at these ages behaved much as they did at home; the attachment behavior measures showed no significant preferences for either parent. When the infants were distressed, the display of attachment behaviors increased and the infants organized their behavior similarly around whichever parent was present. When both parents were present, however, distressed infants turned to their mothers preferentially (Lamb, 1976a, 1976f). A study involving another sample of two-year-olds found that infants of this age did not manifest preferences, even when distressed (Lamb, 1976c). Evidently, the predicted hierarchy among attachment figures is marked only during a relatively brief period. Notice too that the boys showed strong preferences for their fathers in free play at the very ages that they turned to their mothers preferentially when distressed. This indicates that mothers were still deemed more reliable sources of comfort and security, even though fathers had become more desirable partners for playful interaction.

The fact that stress affects the display of preferences may help explain some of the inconsistency evident in the results of the studies reviewed earlier. Even when primary caretaking mothers become primary attachment figures, preferences for them may not be apparent unless the infants are distressed. Studies in which infants either are not distressed or cannot choose between the parents when distressed are less likely to uncover reliable preferences than studies in which distressed infants can choose between two potential sources of comfort and security.

CHARACTERISTICS OF FATHER-INFANT INTERACTION

Even in the first trimester, fathers and mothers appear to engage in different types of interactions with their infants. When videotaped in face-to-face interaction with their 2- to 25-week-old infants, for example, fathers tended to provide staccato bursts of both physical and social stimulation, whereas mothers' actions tended to be more rhythmic and containing (Yogman et al., 1977). Mothers addressed their babies with soft, repetitive, imitative sounds, whereas fathers touched their infants with rhythmic pats (Yogman et al., 1977). During visits to hospitalized premature infants, mothers were responsive to social cues, fathers to gross motor cues (Marton and Minde, 1980).

Most of the data concerning the characteristics of interaction between parents and infants older than four months of age have been gathered in the course of naturalistic home observations. In my longitudinal study, I found that fathers tended to engage in more physically stimulating and unpredictable or "idiosyncratic" play than mothers did (Lamb, 1976b, 1977c). Since these types of play elicited more positive responses from the infants, the average response to play bids by fathers was more positive than the average response to maternal bids. Power and Parke (1979) and Clarke-Stewart (1978) later confirmed that mothers and fathers engaged in different types of play. On the other hand, Belsky (1979) did not find any differences of this kind. Since he used different observational codes and studied older infants, however, direct comparison of the findings is not possible.

When I examined the reasons mothers and fathers picked up and held their 7- to 13-month-old infants, I found that mothers were more likely to hold infants in the course of caretaking, whereas fathers were more likely to do so in order to play with the babies or in response to the infants' requests to be held (Lamb, 1976c, 1977c). These findings were replicated by Belsky (1979). Not surprisingly, infants responded more positively to being held by their fathers than by their mothers (Lamb, 1976a, 1977c).

Data gathered by interview confirm that fathers are identified with playful interactions while mothers are associated with caretaking. According to Kotelchuck's (1975) informants, mothers spent an average of 1.45 hours per day feeding their 6- to 21-month-olds, whereas fathers averaged 15 minutes. Mothers spent 55 minutes per day cleaning their infants and 2.3 hours playing with them, while the comparable figures for fathers were 9 minutes and 1.2 hours. Mothers thus spent a greater proportion of their total interaction time caretaking, while fathers spent a greater proportion of their interaction time in playful social interaction. Relative to the total amount of interaction, Clarke-Stewart's (1978) data likewise suggested that fathers were consistently notable for their involvement in play, and that their relative involvement in caretaking increased over time. Although Rendina and Dickerscheid (1976) did not record maternal behavior (making a comparison of maternal and paternal behavior impossible), it is clear that fathers spent most of their time in playful interaction; on average, only 3.8% of the time was spent caretaking. A study of English families reported similar findings: from maternal interviews, Richards et al. (1975) found that at both 30 and 60 weeks, the most common father-infant activity in 90% of the families was play. Routine involvement in caretaking was rare: only 35% regularly fed their infants at 30 weeks and 46% at 60 weeks. Diaper changing and bathing were the least common paternal activities.

Further evidence that fathers are especially notable for their involvement in play comes from a small ($N = 14$) but intensive longitudinal study of 15- to 30-month-olds undertaken by Clarke-Stewart (1978, 1980). In this study, it was found that fathers gave more verbal directions and positive reinforcement than mothers did. Fathers were rated higher than mothers on ability to engage children in play. For their part, babies showed more enjoyment and involvement when playing with fathers than with mothers. No differences were found in parental responsiveness. When the parents were asked (as part of a laboratory task) to choose an activity in which to engage their infants, mothers tended to choose intellectual activities, whereas fathers selected playful "social-physical activities." Given the playfulness of much father-infant interaction, it is little wonder that young children come to prefer play with their fathers when they have the choice (Clarke-Stewart, 1978; Lamb, 1976c, 1977c; Lynn and Cross, 1974).

Diaries kept by the mothers of Clarke-Stewart's (1978) subjects also suggested important age changes not explored in the other studies. At 15 months, mothers and infants engaged in more play than did fathers and infants; at 20 months, mothers and fathers spent equivalent amounts of time in play; and by 30 months, fathers were spending more time than mothers in play. There was also a change with age in relative responsibility for caretaking. At 15 months, mothers spent far more time than fathers in caretaking, but by 30 months, both parents devoted equivalent amounts of time to this, in part because the total amount of caretaking was much less.

The results of a recent study by Pedersen, Cain, Zaslow, and Anderson (1982) suggested that that the patterns of involvement may differ when both parents work full-time outside the home. When observed with their infants, employed mothers stimulated their infants more than nonemployed mothers did, and they were far more active than their husbands were. In accordance with the findings just reviewed, fathers with nonemployed wives played with their infants more than the mothers did, but this pattern was reversed in the families with employed mothers. Maternal responsibility for caretaking did not differ depending on the mother's working status.

Although observational studies have not been conducted, questionnaire studies confirm the everyday observation that fathers retain their association with play and adventure and mothers their association with caretaking and nurturance even after their infants become young children (Bronfenbrenner, 1961; Devereux, Bronfenbrenner, and Rodgers, 1969; Devereux, Bronfenbrenner, and Suci, 1962; Devereux, Shouval, Bronfenbrenner, Rodgers, Kav-Venaki, and Kiely, 1974; Kohn, 1959; Radke, 1946; Kagan, Hosken, and Watson, 1961; Nadelman, 1976). These studies also indicate that fathers are perceived as more threatening, rigid, and demanding than mothers. Interestingly, children's conceptions about "average" families are usually more stereotyped than attitudes about their own parents (Nadelman, 1976).

Despite the pervasiveness and ubiquity of the distinctive paternal and maternal interactional styles, their origins remain uncertain. Brazelton and his colleagues (1979) have argued that they reflect biologically determined differences, and that because even very young infants innately "expect" fathers and mothers to behave differently, they behave in ways that elicit "appropriate" paternal behavior, even from fathers who assume a primary caretaking role rather than the more typical playmate role. This claim is implausible, since it attributes to young infants greater cognitive and perceptual skills than they are known to possess (see Lamb and Sherrod, 1981, for reviews of the relevant literature). It seems more likely that maternal and paternal styles are consequences of the socially prescribed roles assumed by mothers and fathers in traditional families. Unfortunately, there have been few opportunities to compare the behavior of primary and secondary caretaking fathers. According to Field (1978), primary caretaking fathers behaved more like mothers than did secondary caretaking fathers, although there were still differences between primary caretaking mothers and fathers. Particularly noteworthy was the fact that playful and noncontaining interactions were more common among fathers regardless of their caretaking responsibilities. In an attempt to address similar issues, I am currently involved in a longitudinal study of parent-infant interaction in families in which fathers are minimally, moderately, or centrally involved in child care during the first year of life. By studying couples who share caretaking responsibilities in a variety of ways, we should be able to determine whether biological gender, social role, or both determine the distinctive parental styles. Naturalistic observations of the interactions between three-month-olds and their parents showed that sex of parent was a more powerful determinant of parental behavior than was family type (Lamb, Frodi, Hwang, Frodi, and Steinberg, 1982). Similar conclusions were suggested by observations conducted when the infants were 8 and 16 months old (Lamb, Frodi, Hwang, and Frodi, 1982; Lamb, Frodi, Hwang, Frodi, and Steinberg, 1982; Lamb, Frodi, Frodi, and Hwang, 1982).

Within traditional families, in any event, it is clear that mothers and fathers generally do represent different types of experiences for their children. This increases the likelihood that both parents have independent and significant influences on infant development. What aspects of development might this affect? There has been a great deal of speculation that fathers influence the development of sex role and gender identity, especially in boys, and this evidence is reviewed in the next section.

SEX OF CHILD DIFFERENCES IN PATERNAL BEHAVIOR

Although both men and women have a preference for sons, this preference is especially marked among men (e.g., Arnold et al., 1975; Hoffman, 1977). It is perhaps not surprising, therefore, that fathers tend to interact preferentially with sons from shortly after delivery. Parke and O'Leary (1976) found that fathers vocalized to, touched, and responded to their first-born sons more frequently than their first-born daughters. In a later study of three-week-olds and three-month-olds, Parke and Sawin (1975) found that fathers looked at sons more than at daughters, and provided them with more visual and tactile stimulation than they provided daughters. Mothers, by contrast, stimulated girls more than boys. Parke and Sawin (1980) found that fathers were more likely to diaper and feed three-month-old sons than three-month-old daughters. Rendina and Dickerscheid (1976) confirmed that the fathers of 6- and 13-month-olds watched sons more than daughters. In Israeli kibbutzim, fathers spent more time visiting their four-month-old sons than their four-month-old daughters (Gewirtz and Gewirtz, 1968).

During the second year of life, sex-differentiated treatment appears to intensify (Lamb, 1977b). Fathers verbalize to their sons more than to their daughters during the second year (Lamb, 1977a, 1977b) and they report spending more time (about 30 minutes more per day) playing with first-born sons than with first-born daughters (Kotelchuck, 1976). Even among the nomadic !Kung Bushmen, fathers spend more time with sons than with daughters (West and Konner, 1976).

From early in their sons' lives, therefore, fathers make themselves especially salient to male offspring. Although I found that mothers and fathers treated boys and girls similarly in the first phase of my longitudinal study (Lamb, 1977c), there was a dramatic change early in the second year (Lamb, 1977a, 1977b). Fathers began to direct more social behavior to sons than to daughters, and this appeared to encourage sex differences in the infants behavior. As noted earlier, boys focused their attachment behaviors on fathers during the second year, whereas girls continued to show no consistent preference for either parent. Thus the fathers behaved in a manner that encouraged preferential relationships with sons.

As noted earlier, furthermore, fathers and mothers tend to behave in a sex-differentiated fashion, which should facilitate the fathers' influence on the development of gender identity. Unfortunately, there is no substantive support for this speculation, but circumstantial support comes from two sources. First, there is Money's claim that gender identity must be established in the first two to three years of life if it is to be established securely (Money and Ehrhardt, 1972). Second, there is the evidence that father absence is most likely to affect the development of masculinity in boys when the absence occurs early in the children's lives (Lamb, 1976e, 1981d; Biller, 1971, 1974; Hetherington and Deur, 1971). Nevertheless, the tenuousness of these data underscores the need for further research in this area. In addition, we do not know whether and how fathers affect the development of infant daughters (Lamb, Owen, and Chase-Lansdale, 1979).

Observational data are again lacking concerning fathers' relationships with older children. Correlational studies indicate that warm, masculine fathers have more masculine sons and more feminine daughters (Biller and Borstelmann, 1967; Hetherington, 1967; Reuter and Biller, 1973; Heilbrun,

1965; Sears, Rau, and Alpert, 1965; Johnson, 1963), but the warmth of the relationship appears critical since, on the whole, sons do not resemble their fathers (Hetherington, 1965; Hetherington and Brackbill, 1963; Lynn and Maaske, 1970; Sears, Rau, and Alpert, 1965; cf. Heilbrun, 1965). Several studies of parental attitudes reveal that parents expect fathers to adopt more sex-differentiated roles in relation to sons and daughters than mothers do. For example, Fagot (1974) found that the parents of boys believed there was a special paternal role (as play partner and role model), whereas the parents of girls did not think that mothers and fathers had different roles. Fathers remain more interested in and involved with sons than with daughters through adolescence (Kemper and Reichler, 1976; Bronfenbrenner, 1961; Kohn and Carroll, 1960). They are also much more concerned than mothers are about adherence to conventional sex-stereotyped norms by both sons and daughters (Kohn, 1969, 1979; Bronfenbrenner, 1961; Fagot, 1978; Sears, Maccoby, and Levin, 1957; Goodenough, 1957; Heilbrun, 1965). Other studies—both observational and questionnaire-based—reveal interactions between sex of parent and sex of child when disciplinary strategies are compared. Bearison (1979) reported that parents were more person- (rather than position-) oriented when attempting to control the behavior of same-sex rather than opposite-sex children. Cross-sex leniency has also been reported (Langlois and Downs, 1980; Tasch, 1952; Lansky, 1967; Noller, 1980; Atkinson and Endsley, 1976; Aberle and Naegele, 1952; Rothbart and Maccoby, 1966).

PATERNAL INFLUENCES ON DEVELOPMENT

Instead of sex role development, several recent studies have explored paternal influences on cognitive and motivational development. In part, this focus is attributable to evidence concerning the impact of mother-infant interaction of cognitive development (see Stevenson and Lamb, 1981, for a review) and in part it is attributable to evidence that male infants raised without fathers are less cognitively competent than infants raised in two-parent families (Pedersen, Rubenstein, and Yarrow, 1979). This conclusion appears consistent with Wachs, Uzgiris, and Hunt's (1971) finding that increased paternal involvement is associated with better performance on the Uzgiris-Hunt scales.

In her observational study of 15- to 30-month-olds, Clarke-Stewart (1978) found that intellectual competence was correlated with measures of maternal stimulation (both material and verbal), intellectual acceleration, and expressiveness, as well as with measures of the fathers' engagement in play, their positive ratings of the children, the amount they interacted, and the fathers' aspirations for the infants' independence.

Some speculations concerning paternal influences focus less on the specific differences between maternal and paternal styles than on the fact that they differ in many ways. In the early months, for example, it may be easier for infants to learn to recognize the distinctive features of one parent when they are exposed relatively frequently to a distinctly different person. Because mothers and fathers represent different types of interaction, furthermore, infants are likely to develop different expectations of them (Lamb, 1981d), which should, in turn, increase their awareness of different social styles and perhaps facilitate a perceptive sensitivity to such subtle differences. This would contribute to the development of social competence. Pedersen et al. (1979) found that the degree of paternal involvement (as reported by mothers) was positively correlated with the social responsiveness of five-month-olds.

Finally, we should consider the formative significance of parental sensitivity once again. There is much speculation that maternal sensitivity determines the security of the mother-infant attachment (Ainsworth, Blehar, Waters, and Wall, 1978) and that the security of this relationship determines how the child will later behave in a variety of contexts. As yet, no published studies have examined the effects of paternal sensitivity on the security of father-infant attachments. However, three studies have shown that many infants have secure relationships with one parent and insecure relationships with the other, suggesting that the security of each attachment is independently determined—presumably by the sensitivity of the parent concerned (Lamb, 1978b; Grossmann et al., 1980; Main and Weston, 1981; Owen, Chase-Lansdale, and Lamb, in prep.). Main and Weston (1981) found that the security of both mother-infant and father-infant attachments affected infants' responses to an unfamiliar person (dressed as a clown). It seems likely that the relationship with the primary caretaker is more influential than that with the secondary caretaker and that the child who has at least one secure attachment relationship is better off than a child who has no secure attachment.

Direct and Indirect Effects

As students of infancy have come to recognize that there are multiple influences (e.g., maternal, paternal, biological) on infant development, they have come to appreciate that many influences are indirectly mediated (Belsky, 1981; Lewis, Feiring, and Weinraub, 1981; Lewis and Weinraub, 1976). According to Lewis and Weinraub (1976), *most* paternal influences on infant development are indirectly mediated via the father's impact on their wives. Although this suggestion remains unsubstantiated, several studies confirm that indirect influences are very important. Pedersen, Anderson, and Cain (1977), for example, showed that the affective quality of parent-infant interaction could be predicted from observational measures of mutual criticism in the spousal relationship. Price (1977) reported that the ability of mothers to enjoy and be affectionate with their infants was asssociated with the quality of the marital relationship and Minde, Marton, Manning, and Hines (1979) reported that the quality of the marital relationship predicted the frequency of maternal visits to hospitalized premature infants. The suggestion that marital conflict produces unsatisfactory parent-child relations (Pedersen et al., 1977) is consistent with evidence (reviewed by Lamb, 1981d, and Rutter, 1979) indicating that marital conflict has a more harmful impact on socioemotional development than does parent-child separation or father absence.

As Parke, Power, and Gottman (1979) point out, there are many ways in which indirect effects may be mediated, and this makes it difficult to explore the patterns of influence within the family. It is not sufficient to compute correlations between characteristics of the individual family members and of the relationships among them. Instead, researchers need to develop clear theoretical frameworks from which to derive precise hypotheses that are then subjected to empirical scrutiny. Belsky (1981) and Parke et al. (1979) have begun to develop such frameworks and hypotheses.

One way in which fathers may indirectly affect their children's development, despite limited amounts of direct interaction, was described by Bowlby (1951) in an early discussion of the significance of mother-infant attachment. Bowlby wrote in his introduction:

> Fathers . . . provide for their wives to enable them to devote themselves unrestrictedly to the care of the infant and toddler and by providing love and companionship, they support her emotionally and help her maintain that harmonious contented mood in the aura of which the infant thrives. In what follows, therefore, while continual reference will be made to the mother-child relation, little will be said of the father-child relation; his value as the economic and emotional support of the mother will be assumed. (Bowlby, 1951, p. 13)

Empirical studies of indirect paternal effects are rare. Hennenborg and Logan (1975) and Anderson and Standley (1976) reported that women whose husbands were supportive during labor were themselves less distressed. According to Pedersen (1975), mothers whose husbands evaluated their maternal skill positively were more effective in feeding four-month-old infants, although the direction of effects is especially difficult to determine in this case. In a recent study, Belsky (1980) found significant correlations between the frequency of the parents' comments about the baby and the amount of father-infant interaction, as well as between the frequency of comments *not* about the baby and the frequency of ignoring. Finally, Lytton (1979) reported that the father's presence increased the likelihood that mother would respond positively to her toddler's compliance, reduced the number of maternal efforts to exert control, and increased the mother's effectiveness as a disciplinarian.

An increasing awareness of the interface between spousal and parent-child relationships has accompanied recognition of the fact that both parents, their children, and the marriage all develop and change with time. The changes within any individual or relationship may affect all other persons and relationships for there exists a complex network of influences and interrelations that we have yet to explore in any depth.

CONCLUSION

Although many important questions remain unanswered, at least some issues have been resolved. First, there is substantial evidence that infants form attachments to both mothers and fathers at about the same point during the first year of life. Second, there appears to exist a hierarchy among attachment figures, such that most infants prefer their mothers over their fathers. These preferences probably developed because the mothers were primary caretakers; the preference patterns may well disappear or be reversed when fathers share caretaking responsibilities or become primary caretakers, but the necessary research has not yet been done. No research has been done on changes in parental preferences as children grow older.

Third, the traditional parental roles affect styles of interaction as well. Several observational studies have now shown that fathers are associated with playful—often vigorously stimulating—social interaction, whereas mothers are associated with caretaking. These social styles obviously reflect traditionally sex-stereotyped roles, and it has been suggested that they play an important role in the development of gender role and gender identity. We do not yet know whether the maternal and paternal styles are purely products of social influences or whether there are biological determinants as well.

We do know, however, that both mothers and fathers are capable of behaving sensitively and responsively in interaction with their children. With the exception of lactation, there is no evidence that women are biologically predisposed to be better parents than men are. Social conventions, not biological imperatives, underlie the traditional division of parental responsibilities.

REFERENCES

Aberle, D. F., and Naegele, F. D. 1952. Middle-class fathers' occupational role and attitude toward children. *American Journal of Orthopsychiatry* 22: 366-78.

Ainsworth, M. D. S. 1973. The development of infant-mother attachment. In B. M. Caldwell, and H. N. Ricciuti (eds.), *Review of Child Development Research,* vol, 3. Chicago: University of Chicago Press.

Ainsworth, M. D. S.; Bell, S. M.; and Stayton, D. J. 1974. Infant-mother attachment and social development: Socialisation as a product of reciprocal responsiveness to signals. In M. P. M. Richards (ed.), *The Integration of a Child into a Social World.* Cambridge: Cambridge University Press.

Ainsworth, M. D. S.; Blehar, M. C.; Waters, E.; and Wall, S. 1978. *Patterns of Attachment.* Hillsdale, N.J.: Lawrence Erlbaum Associates.

Ainsworth, M. D. S., and Wittig, B. A. 1969. Attachment and exploratory behavior of one-year-olds in a strange situation. In B. M. Foss (ed.), *Determinants of Infant Behavior,* vol. 4. London: Methuen.

Anderson, B. J., and Standley, K. 1976. A methodology for observation of the childbirth environment. Paper presented to the *American Psychological Association,* Washington, D.C., September.

Arnold, R.; Bulatas, R.; Buripakdi, C.; Ching, B. J.; Fawcett, J. T.; Iritani, T.; Lee, S. J., and Wu, T. S. 1975. *The Value of Children: Introduction and Comparative Analysis.* Honolulu HI: East West Population Institute.

Atkinson, J., and Endsley, R. C. 1976. Influence of sex of child and parent on parental reactions to hypothetical parent-child situations. *Genetic Psychology Monographs* 94: 131–47.

Ban, P., and Lewis, M. 1974. Mothers and fathers, girls and boys: Attachment behavior in the one-year-old. *Merrill-Palmer Quarterly* 20: 195-204.

Bearison, D. J. 1979. Sex-linked patterns of socialization. *Sex Roles,* 5: 11-18.

Belsky, J. 1979. Mother-father-infant interaction: A naturalistic observational study. *Developmental Psychology* 15: 601-07.

———. 1980. A family analysis of parental influence on infant exploratory competence. In F. A. Pedersen (ed.), *The Father-Infant Relationship: Observational Studies in a Family Context.* New York: Praeger.

———. 1981. Early human experience: A family perspective. *Developmental Psychology,* 17: 3-28.

Biller, H. B. 1971. *Father, Child, and Sex Role.* Lexington, MA: Heath.

———. 1974. *Paternal Deprivation: Family, School, Sexuality and Society.* Lexington, MA: Heath.

Biller, H. B., and Borstelmann, L. J. 1967. Masculine development: An integrative review. *Merrill-Palmer Quarterly.* 13: 253–94.

Bowlby, J. 1951. *Maternal Care and Mental Health.* Geneva: WHO.

———. 1969. *Attachment and Loss,* vol. 1, *Attachment.* New York: Basic Books.

Brazelton, T. B.; Yogman, M. W.; Als, H.; and Tronick, E. 1979. The infant as a focus for family reciprocity. In M. Lewis and L. A. Rosenblum (eds.), *The Child and Its Family.* New York: Plenum.

Bronfenbrenner, U. 1961. Some familial antecedents of responsibility and leadership in adolescents. In L. Petrullo and B. M. Bass (eds.), *Leadership and Interpersonal Behavior.* New York: Holt, Rinehart & Winston.

Clarke-Stewart, K. A. 1973. Interactions between mothers and their young children: Characteristics and consequences. *Monographs of the Society for Research in Child Development* 38: (serial number 153).

———. 1978. And daddy makes three: The father's impact on mother and young child. *Child Development* 49: 466–78.

Cohen, L. J., and Campos, J. J. 1974. Father, mother, and stranger as elicitors of attachment behaviors in infancy. *Developmental Psychology* 10: 146–54.

Devereux, E. C.; Bronfenbrenner, U.; and Rodgers, R. R. 1969. Child rearing in England and the United States: A cross-cultural comparison. *Journal of Marriage and the Family* 32: 257-70.

Devereux, E. C.; Bronfenbrenner, U.; and Suci, G. 1962. Patterns of parent behavior in the United States of America and the Federal Republic of Germany: A cross-cultural comparison. *International Social Science Journal* 14: 488-506.

Devereux, E. C.; Shouval, R.; Bronfenbrenner, U.; Rodgers, R. R.; Kav-Venaki, S.; and Kiely, E.

1974. Socialization practices of parents, teachers, and peers in Israel: The kibbutz versus the city. *Child Development* 45: 269-82.

Fagot, B. I. 1974. Sex differences in toddler's behavior and parental reaction. *Developmental Psychology* 10: 554-58.

Fagot, B. I. 1978. The influence of sex of child on parental reactions to toddler children. *Child Development* 49: 459-65.

Feldman, S. S., and Ingham, M. E. 1975. Attachment behavior: A validation study in two age groups. *Child Development* 46: 319-30.

Feldman, S. S., and Nash, S. C. 1977. The effect of family formation on sex stereotypic behavior: A study of responsiveness to babies. In W. Miller and L. Newman (eds.), *The First Child and Family Formation.* Chapel Hill, NC: Carolina Population Institute.

———. 1978. Interest in babies during young adulthood. *Child Development,* 49: 617-22.

Field, T. 1978. Interaction behaviors of primary versus secondary caretaker fathers. *Developmental Psychology* 14: 183-84.

Frodi, A. M., and Lamb, M. E. 1978. Sex differences in responsiveness to infants: A developmental study of psychophysiological and behavioral responses. *Child Development* 49: 1182-88.

Frodi, A. M.; Lamb, M. E.; Leavitt, L. A.; and Donovan, W. L. 1978. Fathers' and mothers' responses to infant smiles and cries. *Infant Behavior and Development* 1: 187-98.

Frodi, A. M.; Lamb, M. E.; Leavitt, L. A.; Donovan, W. L.; Neff, C.; and Sherry, D. 1978. Fathers' and mothers' responses to the faces and cries of normal and premature infants. *Developmental Psychology* 14: 490-98.

Gewirtz, H. B., and Gewirtz, J. L. 1968. Visiting and caretaking patterns for Kibbutz infants: Age and sex trends. *American Journal of Orthopsychiatry* 38: 427-43.

Glick, P. C., and Norton, A. J. 1979. *Marrying, Divorcing, and Living Together in the US Today.* Washington, D.C.: Population Reference Bureau.

Golinkoff, R. M., and Ames, G. J. 1979. A comparison of fathers' and mothers' speech with their young children. *Child Development* 50: 28-32.

Goodenough, E. W. 1957. Interest in persons as an aspect of sex difference in the early years. *Genetic Psychology Monographs* 55: 287-323.

Grossmann, K. E.; Grossmann, K.; Huber, F.; and Wartner, U. 1981. German children's behavior towards their mothers at 12 months and their fathers at 18 months in Ainsworth's strange situation. *International Journal of Behavioral Development* 4: 157-81.

Greenberg, M., and Morris, N. 1974. Engrossment: The newborn's impact upon the father. *American Journal of Orthopsychiatry* 44: 520-31.

Heilbrun, A. B. 1965. An empirical test of the modeling theory of sex-role learning. *Child Development* 36: 789-99.

Henneborg, W. J., and Cogan, R. 1975. The effect of husband participation in reported pain and the probability of medication during labor and birth. *Journal of Psychosomatic Research* 19: 215-22.

Hetherington, E. M. 1965. A developmental study of the effects of sex of the dominant parent on sex-role preference, identification, and imitation in chilldren. *Journal of Personality and Social Psychology* 2: 188-94.

———. 1967. The effects of familial variables on sex-typing on parent-child similarity and on imitation in children. In J. P. Hill (ed.), *Minnesota Symposia on Child Psychology I.* Minneapolis, MN: University of Minnesota Press.

Hetherington, E. M., and Brackbill, Y. 1963. Etiology and covariation of obstinacy, orderliness, and parsimony in young children. *Child Development* 34: 919-43.

Hetherington, E. M., and Deur, J. L. 1971. The effect of father absence on child development *Young Children* 26: 233-48.

Hoffman, L. W. 1977. Changes in family roles, socialization and sex differences. *American Psychologist* 32: 644-58.

Johnson, M. M. 1963. Sex role learning in the nuclear family. *Child Development* 34: 315-33.

Kagan, J.; Hosken, B.; and Watson, S. 1961. Child's symbolic conceptualization of parents. *Child Development* 32: 625-36.
Kemper, T. D., and Reichler, M. L. 1976. Father's work integration and types and frequencies of rewards and punishments administered by fathers and mothers to adolescent sons and daughters. *Journal of Genetic Psychology* 129: 207-19.
Klaus, M. H.; Trause, M. A.; and Kennell, J. H. 1975. Human maternal behavior following delivery: Is it species specific? Unpublished manuscript, Case Western Reserve University.
Kohn, M. L. 1959. Social class and the exercise of parental authority. *American Sociological Review* 24: 352-66.
———. 1969. *Class and Conformity: A Study in Values.* Homewood, IL: Dorsey.
———. 1979. The effects of social class on parental values and practises. In D. Reiss and H. A. Hoffman (eds.), *The American Family: Dying or Developing?* New York: Plenum.
Kohn, M. L., and Carroll, E. E. 1960. Social class and the allocation of parental responsibilities. *Sociometry* 23: 372-92.
Kotelchuck, M. 1972. *The nature of the child's tie to his father.* Unpublished doctoral dissertation, Harvard University, 1972.
———. 1975. Father caretaking characteristics and their influence on infant-father interaction. Paper presented to the *American Psychological Association,* Chicago, September.
———. 1976. The infant's relationship to the father: Experimental evidence. In M. E. Lamb (ed.), *The Role of the Father in Child Development.* New York: Wiley.
Kotelchuck, M.; Zelazo, P. R.; Kagan, J.; and Spelke, E. 1975. Infant reactions to parental separations when left with familiar and unfamiliar adults. *Journal of Genetic Psychology* 126: 255-62.
Lamb, M. E. 1975a. Fathers: Forgotten contributors to child development. *Human Development* 18: 245-66.
———. 1975b. Physiological mechanisms in the control of maternal behavior in rats: A review. *Psychological Bulletin* 82: 104-19.
———. 1976a. Effects of stress and cohort on mother- and father-infant interaction. *Developmental Psychology* 12: 435-43.
———. 1976b. Interaction between eight-month-old children and their fathers and mothers. In M. E. Lamb (ed.), *The Role of the Father in Child Development.* New York: Wiley.
———. 1976c. Interactions between two-year-olds and their mothers and fathers. *Psychological Reports* 38: 447-50.
———. 1976d. Parent-infant interaction in eight-month-olds. *Child Psychiatry and Human Development* 7: 56-63.
———. 1976e. The role of the father: An overview. In M. E. Lamb, (ed.), *The Role of the Father in Child Development.* New York: Wiley.
———. 1976f. Twelve-month-olds and their parents: Interaction in a laboratory playroom. *Developmental Psychology* 12: 237-44.
———. 1977a. The development of mother-infant and father-infant attachments in the second year of life. *Developmental Psychology* 13: 637-48.
———. 1977b. The development of parental preferences in the first two years of life. *Sex Roles:* 3:495-97.
———. 1977c. Father-infant and mother-infant interaction in the first year of life. *Child Development* 48: 167-81.
———. 1978a. Infant social cognition and "second-order" effects. *Infant Behavior and Development* 1: 1-10.
———. 1978b. Qualitative aspects of mother- and father-infant attachments. *Infant Behavior and Development* 1: 265-75.
———. 1978c. Social interaction in infancy and the development of personality. In M. E. Lamb (ed.), *Social and Personality Development.* New York: Holt, Rinehart & Winston.

———. 1979. Separation and reunion behaviors as criteria of attachment to mothers and fathers. *Early Human Development* 3: 329-39.

———. 1981a. The development of social expectations in the first year of life. In M. E. Lamb and L. R. Sherrod (eds.), *Infant Social Cognition: Empirical and Theoretical Considerations.* Hillsdale NJ: Lawrence Erlbaum Associates.

———. 1981b. Developing trust and perceived effectance in infancy. In L. P. Lipsitt (ed.), *Advances in Infancy Research,* vol. 1. Norwood, NJ: Ablex.

———. 1981c. Fathers and child development: An integrative overview. In M. E. Lamb (ed.), *The Role of the Father in Child Development.* New York: Wiley.

Lamb, M. E. (ed.) 1981d. *The Role of the Father in Child Development.* 2nd ed. New York: Wiley.

Lamb, M. E., and Easterbrooks, M. A. 1981. Individual differences in parental sensitivity: Origins, components and consequences. In M. E. Lamb, and L. R. Sherrod (eds.), *Infant Social Cognition: Empirical and Theoretical Considerations.* Hillsdale, NJ: Lawrence Erlbaum Associates.

Lamb, M. E.; Frodi, A. M.; Frodi, M., and Hwang, C. P. 1982. Characteristics of maternal and paternal behavior in traditional and nontraditional Swedish families. *International Journal of Behavioral Development* 5: 131-141.

Lamb, M. E.; Frodi, A. M.; Hwang, C. P; and Frodi, M. 1982. Varying degrees of paternal involvement in infant care: Correlates and effects. In M. E. Lamb (ed.), *Nontraditional Families: Parenting and Child Development.* Hillsdale NJ: Lawrence Erlbaum Associates.

Lamb, M. E.; Frodi, A. M.; Hwang, C. P.; Frodi, M.; and Steinberg, J. 1982. Mother- and father-infant interaction involving play and holding in traditional and nontraditional Swedish families. *Developmental Psychology* 18: 215-221.

Lamb, M. E., and Goldberg, W. A. 1982. The father-child relationship: A synthesis of biological, evolutionary and social perspectives. In L. W. Hoffman, R. A. Gandelman and H. R. Schiffman (eds.), *Parenting: Its Causes and Consequences.* Hillsdale NJ: Lawrence Erlbaum Associates.

Lamb, M. E., and Lamb, J. E. 1976. The nature and importance of the father-infant relationship. *Family Coordinator* 25: 379-85.

Lamb, M. E., and Sherrod, L. R. (eds.) 1981. *Infant social cognition: Empirical and theoretical considerations.* Hillsdale NJ: Lawrence Erlbaum Associates.

Lamb, M. E.; Owen, M. R.; and Chase-Lansdale, L. 1979. The father-daughter relationship: Past, present and future. In C. B. Kopp and M. Kirkpatrick (eds.), *Becoming Female: Perspectives on Development.* New York: Plenum.

Lamb, M. E.; Frodi, A. M.; Hwang, C P., Frodi, M.; and Steinberg, J. 1982. Father-infant and mother-infant interaction in traditional and nontraditional Swedish families. In R. N. Emde and R. J. Harmon (eds.), *Attachment and Affiliative Systems.* New York: Plenum.

Langlois, J. H., and Downs, A. C. 1980. Mothers, fathers, and peers as socialization agents of sex-typed play behaviors in young children. *Child Development* 51: 1237-1247.

Lansky, L. M. 1967. The family structure also affects the model: Sex-role attitudes in parents of preschool children. *Merrill-Palmer Quarterly* 13: 139-50.

Lester, B. M.; Kotelchuck, M.; Spelke, E.; Sellers, M. J.; and Klein, R. E. 1974. Separation protest in Guatemalan infants: Cross-cultural and cognitive findings. *Developmental Psychology* 10: 79-85.

Lewis, M., and Ban, P. 1971. Stability of attachment behavior: A transformational analysis. Paper presented to the *Society for Research in Child Development.* Minneapolis, April.

Lewis, M.; Feiring, C.; and Weinraub, M., 1981. The father as a member of the child's social network. In M. E. Lamb (ed.), *The Role of the Father in Child Development.* New York: Wiley.

Lewis, M., and Weinraub, M. 1974. Sex of parent x sex of child: Socioemotional development. In R. Richart, R. Friedman, and R. Vande Wiele (eds.), *Sex Differences in Behavior.* New York: Wiley.

Lewis, M., and Weinraub, M. 1976. The father's role in the infant's social network. In M. E. Lamb (ed.), *The Role of the Father in Child Development.* New York: Wiley.

Lewis, M.; Weinraub, M.; and Ban, P. 1972. Mothers and fathers, girls and boys: Attachment behavior in the first two years of life. *Educational Testing Service Research Bulletin.* Princeton NJ.

Lind, R. 1974. Observations after delivery of communications between mother-infant-father. Paper presented to the *International Congress of Pediatrics,* Buenos Aires, October.

Lynn, D. B., and Maaske, M. 1970. Imitation versus similarity: Child to parent. Paper presented to the *Western Psychological Association,* Los Angeles, April.

Lynn, D. B., and Cross, A. R. 1974. Parent preference of preschool children. *Journal of Marriage and the Family* 36: 555-59.

Lytton, H. 1979. Disciplinary encounters between young boys and their mothers and fathers: Is there a contingency system? *Developmental Psychology* 15: 256-68.

Main, M. and Weston, D. R. 1981. Security of attachment to mother and father: Related to conflict behavior and the readiness to establish new relationships. *Child Development* 52: 932-40.

Marton, P. L., and Minde, K. 1980. Paternal and maternal behavior with premature infants. Paper presented to the *American Orthopsychiatric Association,* Toronto, April.

Minde, K.; Marton, P.; Manning, D.; and Hines, B. 1979. Some determinants of mother-infant interaction in the premature nursery. *Journal of the American Academy of Child Psychiatry.*

Money, J., and Ehrhardt, A. A. 1972. *Man and Woman, Boy and Girl.* Baltimore: Johns Hopkins University Press.

Nadelman, L. 1976. Perceptions of parents by London five-year-olds. Paper presented to the *American Psychological Association,* Washington D. C., September.

Nash, S. C. and Feldman, S. S. 1981. Sex role and sex related attributions: Constancy and change across the family life cycle. In M. E. Lamb and A. L. Brown (eds.), *Advances in Developmental Psychology,* Vol. 1. Hillsdale NJ: Lawrence Erlbaum Associates.

Noller, P. 1980. Cross gender effects in two-child families. *Developmental Psychology* 16: 159-60.

Owen, M. T.; Chase-Lansdale, L., and Lamb, M. E. 1983. The relationship between mother-infant and father-infant attachments. Manuscript in preparation.

Parke, R. D., and Tinsley, B. R. 1981. The father's role in infancy: Determinants of involvement in feeding and playing. In M. E. Lamb (ed.), *The Role of the Father in Child Development.* New York: Wiley.

Parke, R. D., and O'Leary, S. E. 1976. Father-mother-infant interaction in the newborn period: Some findings, some observations and some unresolved issues. In K. Riegel and J. Meacham (eds.), *The Developing Individual in a Changing World,* vol. 2, *Social and Environmental Issues.* The Hague: Mouton.

Parke, R. D., and Sawin, D. B. 1975. Infant characteristics and behavior as elicitors of maternal and paternal responsibility in the newborn period. Paper presented to the *Society for Research in Child Development,* Denver, April.

Parke, R. D., and Sawin, D. B. 1980. The family in early infancy: Social interactional and attitudinal analyses. In F. A. Pedersen (ed.), *The Father-Infant Relationship: Observational Studies in a Family Context.* New York: Praeger.

Parke, R. D.; Power, T. G.; and Gottman, J. 1979. Conceptualizing and quantifying influence patterns in the family triad. In M. E. Lamb; S. J. Suomi; and G. R. Stephenson (eds.), *Social Interaction Analysis: Methodological Issues.* Madison: University of Wisconsin Press.

Pedersen F. A. 1975. Mother, father and infant as an interactive system. Paper presented to the *American Psychological Association,* Chicago, September.

Pedersen, F. A.; Anderson, B.; and Cain, R. 1977. An approach to understanding linkages between the parent-infant and spouse relationships. Paper presented to the *Society for Research in Child Development,* New Orleans, March.

Pedersen, F. A.; Cain; R. L.; Zaslow, M. J.; and Anderson, B. J. 1982. Variations in infant experience associated with alternative family roles. In L. M. Laosa and I. E. Sigel (eds.). *Families as Learning Environments for Children.* New York: Plenum.

Pedersen, F. A., and Robson, K. 1969. Father participation in infancy. *American Journal of Orthopsychiatry.* 39: 466-72.

Pedersen, F. A.; Rubenstein, J. L.; and Yarrow, L. J. 1979. Infant development in father-absent families. *Journal of Genetic Psychology* 135: 51-61.

Pleck, J. H., and Rustad, M. 1981. Husbands' and wives' time in family work and paid work in the 1975-76 studies of time use. Unpublished manuscript, Wellesley College.

Power, T. G. and Parke, R. D. 1979. Toward a taxonomy of father-infant and mother-infant play patterns. Paper presented to the *Society for Research in Child Development,* San Francisco, March.

Price, G. 1977. Factors influencing reciprocity in early mother-infant interaction. Paper presented to the *Society for Research in Child Development,* New Orleans, March.

Radke, M. J. 1946. The relation of parental authority to children's behavior and attitudes. *University of Minnesota Institute of Child Welfare Monograph* #22.

Redican, W. K. 1975. A longitudinal study of behavioral interactions between adult male and infant rhesus monkeys *(Macaca mulatta).* Unpublished Ph.D. dissertation, University of California, Davis.

Redican, W. K., and Mitchell, G. 1973. A longitudinal study of parental behavior in adult male rhesus monkeys I. Observations on the first dyad. *Developmental Psychology* 8: 135-36.

Redican, W. K., and Mitchell, G. 1974. Play between adult male and infant rhesus monkeys. *American Zoologist* 14: 295-312.

Rendina, I., and Dickerscheid, J. D. 1976. Father involvement with first-born infants. *Family Coordinator* 25: 373-79.

Reuter, M. W. and Biller, H. B. 1973. Perceived paternal nurturance-availability and personality adjustment among college males. *Journal of Consulting and Clinical Psychology* 40: 339-42.

Richards, M. P. M.; Dunn, J. F.; and Antonis, B. 1975. Caretaking in the first year of life: The role of fathers' and mothers' social isolation. Unpublished manuscript, Cambridge University.

Ross, G.; Kagan, J.; Zelazo, P.; and Kotelchuck, M. 1975. Separation protest in infants in home and laboratory. *Developmental Psychology* 11: 256-57.

Rothbart, M. K., and Maccoby, E. E. 1966. Parents' differential reactions to sons and daughters. *Journal of Personality and Social Psychology* 4: 237-43.

Russell, G. 1978. The father role and its relation to masculinity, femininity, and androgyny. *Child Development* 49: 1174-81.

———. 1982. Shared caregiving families: An Australian study. In M. E. Lamb, (ed.), *Non-traditional Families.* Hillsdale, NJ: Lawrence Erlbaum Associates.

Rutter, M. 1979. Maternal deprivation, 1972-1978: New findings, new concepts, new approaches. *Child Development* 50: 283-305.

Schaffer, H. R., and Emerson, P. E. 1964. The development of social attachments in infancy. *Monographs of the Society for Research in Child Development* 29: whole number 94.

Schaffer, H. R. 1963. Some issues for research in the study of attachment behaviour. In B. M. Foss (ed.), *Determinants of Infant Behavior* vol. 2. London: Methuen.

Sears, R. R.; Maccoby, E. E.; and Levin, H. 1957. *Patterns of Child Rearing.* Evanston, Ill: Row Peterson.

Sears, R. R.; Rau, L.; and Alpert, R. 1965. *Identification and Child Rearing.* Stanford: Stanford University Press.

Sheehy, G. 1979. Introducing the postponing generation. *Esquire* 92(4): 25-33.

Spelke, E.; Zelazo, P.; Kagan, J.; and Kotelchuck, M. 1973. Father interaction and separation protest. *Developmental Psychology* 9: 83-90.

Stevenson, M. B., and Lamb, M. E. 1981. The effects of social experience and social style on cognitive competence and performance. In M. E. Lamb and L. R. Sherrod (eds.), *Infant Social Cognition.* Hillsdale, NJ: Lawrence Erlbaum Associates.

Tasch, R. J. 1952. The role of the father in the family. *Journal of Experimental Education* 20: 319-61.

Wachs, T.; Uzgiris, I.; and Hunt, J. 1971. Cognitive development in infants of different age levels and from different environmental backgrounds. *Merrill-Palmer Quarterly* 17: 283-317.

West, M. M., and Konner, M. J. 1976. The role of the father: an anthropological perspective. In M. E. Lamb (ed.), *The Role of the Father in Child Development.* New York: Wiley.

Willemsen, E.; Flaherty, D.; Heaton, C.; and Ritchey, G. 1974. Attachment behavior of one-year-olds as a function of mother vs father, sex of child, session, and toys. *Genetic Psychology Monographs,* 90: 305-24.

Yogman, M. J.; Dixon, S.; Tronick, E.; Als, H.; and Brazelton, T. B. 1977. The goals and structure of face-to-face interaction between infants and their fathers. Paper presented to the *Society for Research in Child Development,* New Orleans, March.

Zelazo, P. R.; Kotelchuck, M.; Barber, L.; and David, J. 1977. Fathers and sons: An experimental facilitation of attachment behaviors. Paper presented to the *Society for Research in Child Development,* New Orleans, March.

Index

Abbott, D. H., 16
Aberle, D. F., 420
Adoptions/orphans, 103, 105, 107-108, 120-121, 124, 143-144, 203, 247, 312-314, 381, 384, 385-387
Affiliation/affiliative interactions, 45-46, 89, 98, 103, 106-107, 134, 160-165, 170, 215-217, 219-235, 240, 392
Aggression/competition, 61, 66-67, 69, 71-76, 80-83, 89, 98, 115-117, 119, 124, 177, 216-217, 223-224, 240-241, 267, 328-330, 348
Agonistic buffering, 52-53, 82, 109, 115, 129, 138, 147, 150, 160, 165, 177, 186-187, 189, 204-210, 216-217, 223-224, 240-241, 348, 379, 381-382, 389, 391-393
Ainsworth, M. D. S., 408, 409, 421
Alcock, J., 267
Aldrich-Blake, F. P. G., 329, 395
Alexander, B. K., 56, 107, 128, 129, 142, 386
Alexander, R. D., 114, 255, 262, 273, 283, 293, 294, 310, 319, 321, 323, 325, 326, 327, 328, 329, 330
Allen, G. M., 297
Alley, T., 165, 187
Allomothering, 88-89, 141, 360, 364
Altmann, J., 94, 107, 131, 146, 147, 193, 214, 217, 228, 231, 241, 315, 317, 329, 392
Altmann, S. A., 59, 95, 206, 230, 262, 284, 311, 312, 318, 325, 327, 329
Altriciality, 287, 290, 294
Altruism/reciprocal altruism, 115, 181, 240-241, 390
Alvarez del Toro, M., 298
Amphibians, 273
Anderson, B. J., 422
Anderson, C., 328, 330
Anisogamy, 283-284, 299-300
Arnold, R., 418
Arnold, S. J., 261, 263, 418
Atkinson, J., 420
Attachment, 233-241, 412-415
Auerbach, K. G., 378

Bachmann, C., 269
Baldwin, J. D., 363, 365
Bales, K. B., 213
Ban, P., 413
Barash, D., 143, 276, 281, 325

Bateman, A. J., 261, 262, 264, 277, 284
Baxter, J., 129
Bearder, S. K., 359
Bearison, D. J., 420
Beecher, M. D., 282, 283
Bell, R. H. V., 289
Belsky, J., 416, 421, 422
Benshoof, V., 325
Bent, A. C., 283
Berenstain, J., 121, 363, 379, 382
Berkson, G., 394
Berman, C. M., 116, 387
Bernstein, I. S., 89, 108, 114, 205, 284, 285
Bertram, B. C. R., 51, 283, 288, 363, 382
Bertrand, M., 60, 77, 81, 89, 94, 107
Biller, H. B., 419
Biosocial variables, 214, 218, 232-241, 408-410, 417, 423
Biparentalism, 2, 280, 282, 290, 292, 294-300, 365-369
Birds, 268, 269, 271, 276, 279, 282-284, 294-295, 297, 300
Bishop, Y. M. M., 98
Blaffer-Hrdy, S. *See* Hrdy, S. B.
Blankenship, L., 147
Blumer, L. S., 260, 273, 276
Body size, 16-17, 262, 288-295, 336-340, 366
Bogess, J., 213
Boggs, C. L., 261
Booth, C., 378
Boucher, D. H., 299
Bowden, D., 363
Bowlby, J., 408, 414, 422
Box, H. O., 2, 5, 367
Brace, C. L., 321
Bradbury, J. W., 297, 298
Bramblett, C. A., 350, 363, 366
Brandt, E. M., 56, 77, 89, 106, 382
Brazelton, T. B., 417
Breder, C. M., 260, 273, 275, 276, 287, 359
Breuggeman, J. A., 386
Bridgwater, D. D., 336
Brockelman, W. Y., 395
Bronfenbrenner, U., 417, 420
Bronson, F. H., 16
Brood patch, 280, 282
Brown, J. L., 20, 267, 268, 291
Brunning, D. F., 268, 269, 271, 276, 278
Budd, P. L., 273
Burley, N., 265, 325
Burton, F. D., 89, 380
Buskirk, W. H., 208, 286
Busse, C., 122, 146, 167, 174, 177, 179, 181, 182, 187, 189, 194, 204, 205, 207, 208, 209, 286, 311, 314, 315, 326, 327, 330, 390, 391, 393
Bygott, J. D., 246, 395

Caine, N., 460
Campbell, B. G., 260, 283, 321
Carey, M., 268
Carpenter, C. R., 275, 394
Carpenter, F. L., 295
Castro, R., 1, 368
Ceboids
- *Aloutta,* 352-353, 363
- *Aotus,* 340-342, 352, 363
- *Ateles,* 341-342, 353-354, 365
- *Callicebus,* 340-342, 354, 356, 366
- *Callithrix,* 2, 16-17, 21, 294, 336-344, 354-356, 365-369
- *Callimico,* 338, 340-341, 343, 354, 356, 366
- *Cebus,* 341-342, 353, 363
- *Leontopithecus,* 6, 338-340, 355-357, 367-369
- *Pithecia,* 340-342, 354, 365
- *Saguinus,* 1-2, 5-6, 10, 16-18, 338-340, 354-355, 357, 367
- *Saimiri,* 337-338, 342, 353, 363-364

Cebul, M. S., 6, 7, 8, 10
Census, 188, 217
Cercopithecoids
- *Cercocebus,* 378
- *Cercopithecus,* 56, 78
- Colobines, 207, 378
- Geledas, 379
- Macaques
  - *arctoides,* 56-86, 89-109, 382-385
  - *fuscata,* 56, 108-109, 127-144, 146, 385-386
  - *mulatta,* 26, 51, 56, 77-78, 89, 113-124, 347, 386-389
  - *nemestrina,* 378
  - *radiata,* 56, 347, 379
  - *sinica,* 56
  - *sylvanus,* 20-53, 89, 146-147, 151, 379-382
- *Papio* (baboons)
  - *anubis,* 56, 147-184, 193, 209-210, 329, 389-393
  - *cynocephalus,* 193, 208, 213-241, 391-392
  - *hamadryas,* 268-269, 312, 324, 379
  - *ursinus,* 122, 190-210, 312, 390

Chalmers, N., 379
Chapman, M., 207
Charles-Dominique, P., 290, 349, 358, 359, 360
Cheney, D. L., 286
Chevalier-Skolnikoff, S., 94
Chivers, D. J., 275, 283, 286, 292, 322, 394, 395
Christen, A., 5, 369
Clark, T. W., 130
Clarke-Stewart, K. A., 411, 416, 417, 420
Clutton-Brock, T. H., 294, 326, 338, 341
Cohen, L. J., 413
Coimbra-Filho, A. F., 369
Collias, N., 363
Collins, D., 207
Complementarity, 312-314, 318, 320
Connection, 216, 219-222, 240
Consort. *See* Reproduction/mating
Contact, 61, 69-71, 67-68, 72-74, 79, 106, 168-170, 171, 193, 215, 219-223, 240, 250
Cope's Law, 337
Copulation. *See* Reproduction/mating
Countercarrying, 189, 314-315, 318
Cox, C. R., 267
Crandall, L. S., 2, 5
Crook, J. H., 129, 262
Cubbicciotti, D. D., 286
Cuckoldry/anticuckoldry, 269, 272, 276, 280-281, 285-286, 300

Daly, M., 261, 280
Darwin, C., 262, 264
Davies, N. B., 291
Davis, W. H., 297
Dawkins, R., 274, 299
Dawson, G. A., 1, 16, 367
Deag, J. M., 21, 22, 26, 27, 43, 45, 46, 56, 89, 109, 115, 129, 146, 147, 151, 165, 189, 380, 381, 382
Defense, 108-109, 114, 169, 171, 263, 284, 315-316, 317
Defler, T. R., 349, 363
Demography, 3, 22, 57-58, 91-92, 93, 95, 106, 121, 130-133, 148-150, 155, 188, 191-192, 214
Development (socialization), 10-15, 77-78, 82, 88, 106-109, 113, 134, 204, 219, 241, 255, 266, 368, 380, 383, 386, 420-423
Devereux, E. C., 417
DeVore, I., 78, 263, 284, 389
Dickeman, M., 325
Dittus, W. P. J., 56
Dominance, 68, 71-72, 76, 78, 81-82, 83, 200-203, 209, 218, 233, 236, 240-241, 284
Doyle, G. A., 358, 359
Drickamer, L. C., 114, 115, 120, 123
Dunbar, R. I. M., 379

Eaton, G. G., 128
Ecological considerations, 259ff., 336ff.
Ehrlich, A., 359
Eisenberg, J. F., 262, 288, 337, 340, 362, 363, 365, 399
Ellefson, J. O., 275, 394
Emlen, S. T., 114, 262, 267, 268, 272, 284, 287, 295
Energy costs, 15-17, 290, 295, 312-313, 337, 342
Enomoto, T., 129
Epple, G., 1, 2, 5, 6, 16, 17, 338, 367, 368, 369
Erickson, C. J., 276, 281
Estrada, A., 57, 60, 71, 76, 77, 78, 79, 83, 84, 89, 98, 106, 384
Estrus/estrous cycle. *See* Reproduction/mating
Extragametic effort, 263, 272-277, 280-282, 286-288, 290-295, 299-301

Fagen, R. M., 98
Fagot, B. I., 420
Fedigan, L. M., 128, 129, 139
Feeding/nutrition, 216-217, 228-231, 241, 288-290, 301, 316-317, 342, 409
Feldman, S. S., 409, 410, 413
Female(s)
- affiliative relations with, 71, 83, 103-105, 107-108, 129, 173, 250, 358, 361, 366, 410
- age, 132-133, 410
- choice, 50-52, 139, 265, 267
- exploitation of, 171-172, 184
- protection of infants by, 81-82, 205
- rank, 16, 60, 68, 76, 78, 81, 116, 120-121, 132-133, 141, 200, 218, 240, 247, 317, 330

Field, T., 418
Fish, 276
Fisher, R. A., 261, 265, 266, 326

Fitness/inclusive fitness, 115, 122, 128, 174, 255-256, 263, 271-272, 278-279, 284, 311, 330, 348
Fitzgerald, A., 2, 367
Fogden, M., 358
Foltz, D. W., 293
Food sharing, 6-10, 217, 229-230, 231, 256
Ford, S. M., 17
Fossey, D., 292
Fox, G., 395
Fragaszy, D. M., 366
Franz, J., 2
French, J. A., 17
Frodi, A. M., 410

Gadgil, M., 261, 266
Galdikas, B. M. F., 460
Gaulin, S. J. C., 283, 289, 290, 291, 297
Geist, V., 288, 289, 325
Gengozian, N., 340
Gewirtz, H. B., 419
Ghiselin, M. T., 273
Gifford, D. P., 378
Gillman, J., 206, 218
Gilmore, H. B., 109, 146, 147, 162, 165, 166, 170, 173, 177, 182, 186, 189
Glander, K. E., 363
Glick, P. C., 412
Golinkoff, R. M., 411
Goodall, J. van Lawick, 245, 246, 328, 395, 397, 398
Goodenough, E. W., 420
Goodman, D., 261
Gould, S. J., 124, 330
Gouzoules, H., 57, 78, 81, 82, 106, 109, 130, 131, 134, 141, 382
Grafen, A., 277, 281
Graul, W. D., 260, 273, 276
Greenberg, M., 409
Grewall, B. S., 386
Griffin, D. R., 297
Grooming, 15, 23, 60, 64-65, 78, 116-117, 119, 120, 122, 135-136, 151, 156, 173, 214, 216, 232, 235, 240, 246, 253, 317, 351, 359-360, 361
Grossmann, K. E., 421
Gubernick, D. J., 377

Hall, R. R. L., 65, 329, 378
Halliday, T. R., 267
Hamilton, W. D., 79, 272
Hamilton, W. J., III, 189, 204, 206, 208, 286, 312, 313, 316, 317, 329, 330
Hamilton's Rule, 271-272
Hampton, S. H., 337, 340, 368
Hanif, M., 365
Hansen, E. W., 77
Harassment, 81, 172, 216, 231-234, 241, 315-316
Harcourt, A. H., 396, 398
Harding, R. S. O., 148
Harlow, H., 168, 347, 388
Harrington, J. E., 361
Harvey, P. H., 114
Hasegawa, T., 107, 143, 144, 385
Hausfater, G., 190, 214, 217, 238, 263, 286, 315, 316, 327
Hearn, J. P., 2, 16
Heilbrun, A. B., 419, 420
Heltne, P., 341, 342, 366, 367
Hemmingsen, A. M., 288
Hendrickx, A. G., 206, 218, 326
Hendy-Neely, H., 56, 60, 77, 78, 89, 106, 108, 383
Henneborg, W. J., 422
Hershkovitz, P., 17, 336, 338
Hetherington, E. M., 419, 420
Hill, W. C. O., 2, 5, 295
Hinde, R. A., 56
Hiraiwa, M., 385
Hoage, R. J., 2, 5, 6, 17, 369
Hoffman, L. W., 412, 418
Hooker, T., 276
Hormones, 124, 128, 280, 364-365, 409
Horn, H. S., 261
Horr, D., 399
Hrdy, S. B., 56, 82, 84, 88, 128, 129, 146, 165, 173, 174, 180, 183, 186, 207, 213, 245, 260, 277, 281, 284, 286, 314, 315, 316, 347, 348, 363, 377
Humans, 261, 277, 283, 407-423
  care taking
    attachments, 412-415
    influences on development, 420-423
    paternal competence, 408-410
    sex differences, 418-420
    types, 227, 283, 415-418
  social organization, 319-322, 419
Hunt, S. M., 364
Huxley, J. S., 266
Hylobatids, 21, 275-276, 292, 322, 342, 394-395

Incest taboo, 2
Infanticide, 17, 129, 167, 178-179, 181, 186, 205-210, 246, 264, 266, 314-315, 348-349, 363
Infants
  aggression towards, 66-69, 71, 75-76, 80-85, 134
  cooperation by, 70-71, 79, 83, 89, 105, 108, 119, 122, 142, 151, 162, 170, 172, 178, 223, 240-241
  distress of, 81, 82, 84, 108, 136, 168, 204, 209, 254, 347, 361, 364-365, 388, 412, 413
  exploitation of, 80, 129, 146-147, 150, 154, 160, 165-170, 181, 183, 204, 348, 381-382, 389-393
  protection of, 20, 26, 71, 108-109, 121-122, 124, 127, 136, 177, 178-179, 181, 186, 205-210, 216, 315-317, 358-359, 360-361, 390-391
  specificity/preferences for, by males, 36-41, 48, 73-74, 76, 83, 85, 96, 103, 105, 107, 135-139, 156-164, 194-196, 205
  sociometric distribution of interactions with, by males, 27-41, 62-76, 101-106, 116-119, 131-132, 152-160, 194-200
Ingram, J. C., 2, 5, 367
Invertebrates, 272-273
Itani, J., 27, 56, 77, 108, 109, 127, 128, 129, 140, 141, 142, 144, 146, 147, 165, 167, 385
Izawa, K., 1, 16, 342, 368

Jarman, P. J., 289, 193
Jarman-Bell Principle, 288-290
Jenni, D. A., 260, 276
Johns, J. E., 280
Johnson, M. M., 420
Jolly, A., 288, 360, 361, 362
Jones, C., 328
Jones, N. G. B., 89, 382

Kagan, J., 417
Kaplan, J. R., 52, 115
Katz, Y., 2, 16
Kaufman, I. C., 378, 379
Kaufmann, J. H., 386
Kemper, T. D., 420
Kidnapping, 181, 189, 391
Kin/kinship, 3-15, 52-53, 66, 68-69, 79, 82, 122, 124, 187, 271, 276, 278-279, 316, 384
  selection, 52-53, 79, 124, 143, 181, 183, 271
King, N. E., 245, 246, 256
Kinzey, W. G., 366
Klaus, M. H., 409
Kleiber, M. A., 288
Kleiman, D., 1, 2, 75, 279, 283, 286, 293, 336, 338, 340, 342, 343, 346, 347, 348, 363, 369, 370, 371, 394
Klein, L., 365
Klopfer, D. H., 348, 358, 359, 360, 361
Kohn, M. L., 417, 420
Kotelchuck, M., 411, 413, 416, 419
Koyama, N., 130
Kummer, H., 77, 109, 323, 324, 378, 379
Kurland, J. A., 52, 129, 261, 263, 264, 272, 273, 281, 283, 284, 285, 286, 386

LaBov, J., 206
Lack, D., 262, 275, 279, 294, 321
Lactation/nursing 2, 3-5, 206, 279-281, 363, 423
Lahiri, R. K., 21
Lamb, M. E., 347, 377, 401, 408, 409, 410, 412, 413, 414, 415, 416, 417, 418, 419, 421
Lancaster, J. B., 56, 78, 256, 315
Langlois, J. H., 420
Lansky, L. M., 420
Lawick-Goodall, J. van. *See* Goodall, J. van Lawick
Lee, R. B., 321
Leen, N., 297
Leibrecht, B. C., 362
Lester, B. M., 413
Leutenegger, W., 293, 294, 321, 336, 337, 342, 343, 366
Lewis, M., 411, 413, 421
Lind, R., 411
Lindburg, D. G., 89, 116, 286, 386
Lovejoy, C. O., 310, 319, 321
Low, B. S., 261, 264, 265, 287
Loy, J., 280
Lucas, N. S., 2
Lydekker, R., 379
Lyman, C. P., 297
Lynn, D. B., 347, 417, 420
Lytton, H., 422

McCann, C., 394
McClure, H. E., 394
McCracken, G. F., 298
McGinnis, P., 256
McGrew, W., 256
McKenna, J. J., 347
MacKinnon, J., 400
MacLennan, A. H., 206
MacRoberts, M. H., 21, 379
Mains, M., 421
Magnuson, J. J., 329
Makwana, S. C., 89
Males
  age, 52, 65-68, 71-76, 127, 238-240
  aggression by, 45-46, 66-69, 71-72, 80-85, 117, 134, 216-218, 223-224, 240-241, 245
  castrated, 122
  defense of infants by, 83, 89, 99, 108, 136, 144, 173, 231-232, 241, 363
  dominance/rank, 43-45, 49-50, 65, 68, 71, 78, 100-101, 114, 119-120, 127, 148, 167, 177, 190, 199-201, 205-207, 218, 233, 236, 240-241, 247, 311, 317
  exploitation of infants by, 83, 129, 150
  life history, 98-99, 127, 210, 245-246
  migration, 16, 114, 130, 191-192, 194, 199, 204, 210, 214, 241
Mammals, nonprimate, 130, 143, 273, 282-284, 288-289, 293, 297-298
Manning, A., 165
Marik, M., 5
Marler, P., 168
Marney-Petix, V. C., 351
Marshall, A. J., 281
Martin, R. D., 358, 360
Marton, P. L., 415
Mason, W. A., 342, 366
Maternal employment, 412, 417
Maternal restrictiveness, 5, 77, 79, 151, 225-226, 245-246, 248, 361, 383
Maynard Smith, J., 20, 268, 273, 277, 280, 285, 287, 293, 294, 312
Mayr, E., 266
Meikle, D. B., 114, 122, 387
Meritt, D. A., 363, 366
Merz, E., 21
Milne, L. J., 275
Mind, K., 422
Mitchell, G. D., 1, 20, 23, 26, 56, 77, 78, 88, 213, 245, 347, 348, 377
Mobbing, 169, 209-210
Money, J., 419
Monogamy, 1-2, 16, 18, 20, 283, 321-323, 329, 336-343, 350, 358-359, 362, 366, 370
Moodie, G. E. E., 276
Mori, U., 379
Morris, D., 269
Mortality, 28, 173, 208, 292, 296, 299, 301
Moynihan, M., 341, 362, 367
Mullins, D. E., 261
Munro, H. N., 288
Murdock, G. P., 319, 321
Myers, P., 329

Nadelman, L., 417
Napier, J. R., 2, 114, 338, 340, 341, 359
Nash, L. T., 107, 109
Nash, S. C., 409
Natal coat, 78, 82, 84, 150, 154, 160, 165-166, 175-176, 187, 189, 209, 210
Neely, H. H. *See* Hendy-Neely, H.
Neville, M. K., 363
Neyman, P. F., 1, 16
Nice, M., 283
Niemitz, C., 358
Nisbet, I. C. T., 264
Nishida, T., 245
Noller, P., 420

O'Donald, P., 265, 266
Odum, H. T., 289
Orians, G. H., 114, 261, 265, 268, 284, 287, 322
Orr, R. T., 297
Owen, M. T., 421

Packer, C., 146, 147, 165, 167, 173, 174, 177, 179, 180, 182, 186, 190, 193, 204, 205, 208, 209, 210, 263, 286, 311, 314, 315, 327, 378, 390, 391, 393
Parental investment, 16-17, 114, 122-124, 181, 261-265, 277-284, 421-424
Parity, 17, 116, 121, 409
Parke, R. D., 408, 409, 418, 419, 422
Parker, G. A., 300
Partridge, L., 265
Paternal(ism)
  benefits of, 229-232, 240-241
  competence, 408-410

functions of, 10-15, 50, 79, 108-109, 114-115, 119-124, 137, 144, 165-170, 178
influences on, 421-423
interactive styles, 415-418
interest, 131-134, 409
investment, 114, 122-124, 173-174, 411-412, 421-422
preferences, 36-41, 73-74, 103, 141, 414, 415, 418-420
responsiveness, 408-410, 416, 421
seniority, 409, 421
types of, 2-3, 6, 23-27, 29-31, 47, 60-61, 89, 98, 113-115, 118, 120, 127, 135, 140, 151-154, 171-173, 189, 215-216, 245-246, 311-312, 316, 320, 348-349, 378-400, 408-422
Paternity
certainty/assessment, 2, 139, 160, 166, 205-206, 217, 234-235, 238, 240, 255-256, 281, 337
kinship, 114, 139, 190-191, 194-199, 205-206, 217-218, 234, 238, 256, 273-284, 290-292, 295-301
Pedersen, F. A., 347, 348, 411, 412, 417, 420, 421, 422
Peffer-Smith, P. G., 94, 101, 103
Perrone, M., 273
Petter, J. J., 290, 358, 359, 360, 361
Phoenix, C. H., 124
Phyletic dwarfism, 16-18, 337-338, 342-344, 366
Play, 26, 60, 97-98, 117, 246, 253-254, 364, 413, 416-418
Pleck, J. H., 412
Poirier, F. E., 109
Pola, Y. V., 369
Pollock, J. I., 294, 295, 360
Polygamy, 42, 113-114, 268-270, 283-284, 309, 310, 321-324, 336, 342, 363
Pongids, 244-257, 292, 396-400
Pook, A. G., 340, 341, 342, 343, 366
Popp, J., 146, 147, 165, 166, 167, 174, 177, 181, 182, 186, 187, 189, 190, 206, 208, 209, 311, 314, 326, 391
Porter, K. R., 274, 287
Post, D., 214
Post-partum investment, 262, 269, 276-278, 282, 288, 290
Power, T. G., 416
Precociality, 287-288, 290-291, 300
Predation, 17, 84, 108, 114, 191, 284-285, 288, 292, 296, 298, 317-318, 328, 330
Pressley, P. H., 276
Prezygotic and postzygotic expenditures, 263-264, 269, 271, 277, 284
Price, G., 422
Primate species (nonhuman). *See specific species*
Prosimians
Cheirogaleids, 352, 360
Daubentonia, 290, 351, 359-360
Galago, 351, 359
Indris, 294, 348-350, 359-360
Lemur, 290, 352, 360-361
Loris, 351-352, 358-359
Tarsius, 351, 358
Protocultural difference, in pattern of male care, 128, 142
Proximity, 151, 154, 215, 219-221, 240-241, 248, 367
Pussey, A., 245, 396

Quiatt, D., 347

Rabb, G. B., 2
Radke, M. J., 417
Ralls, K., 266, 267, 283, 294
Ramirez, M. F., 342, 368
Ransom, T. W., 56, 78, 107, 146, 147, 162, 165, 173, 174, 177, 179, 182, 186, 187, 193, 204, 208, 286, 314, 389, 392, 393
Raphael, D., 245
Redican, W. K., 20, 21, 23, 26, 89, 109, 113, 124, 213, 245, 377, 378, 387, 398, 408
Redina, I., 416, 419
Rensch, B., 294
Relatedness, probability of, 52, 122, 187, 190-191, 205, 206, 271-272, 277-282, 286, 290-292
Relatedness effects, 290-291
direct/indirect, 296-301
Reproduction/mating
conception, 217-218, 235, 237-238, 240
consort, 83, 107, 114, 117, 134, 139, 160, 166, 167, 190, 193, 217, 238, 256
copulation, 117, 166, 191, 205, 218, 235, 237, 266-267, 273, 281
estrus/estrous cycle, 2, 16, 114, 177, 206, 256

mating effort, 113, 261, 286
exploitative competition, 263, 266, 269
interference competition, 262-263, 266
ovulation, concealed, 281, 325-326
reproductive effort, 17, 261, 262-271
RS, 114, 120, 125, 265
Reptiles, 130
Reuter, M. W., 419
Reynolds, P. C., 319, 325, 397
Rhine, R. J., 89, 284, 383
Richard, A., 360, 365
Richards, A. P. M., 416
Richdale, L. E., 298
Ricklefs, R. E., 296
Ridley, M., 273, 312
Rijsken, H. D., 400
Robinson, J., 366
Rodman, P. S., 400
Rosenblum, L. A., 17, 94, 108, 347, 364
Rosenson, L. M., 359
Ross, G., 413
Roth, H. H., 2
Rothbart, M. K., 420
Rothe, H., 5, 340
Rowell, T. E., 78, 108, 109, 378, 387
Rudran, R., 363
Russell, C., 109
Russell, G., 412
Rutter, M., 422

Saayman, G. S., 378
Sanderson, I. T., 5
Sauer, E. G., 359
Schaffer, H. R., 411, 412
Schoener, T. W., 265
Sears, R. R., 419
Selander, R. K., 267, 268
Selfishness, 80, 114, 115, 119
Sex(ual)
differences, 10-13, 63-65, 66-68, 76-77, 80, 122, 367, 369, 384, 410, 414, 418-420
dimorphism, 114, 273, 283, 286, 294-295, 301
inhibition, 2, 16, 18
selection, 261, 264, 267, 283, 294-295, 299, 301
skin swelling, 206, 217-218, 256, 310, 321, 325-328
Seyfarth, R. M., 107
Sheehy, G., 412
Siblings, 5, 7, 10, 13, 16, 17, 52, 66, 68-69, 73, 77, 79, 85, 122, 124, 140, 143-144, 203, 209, 368, 384
Siegel, S., 60, 101
Simonds, P., 379
Sleeper, P., 366
Smith, D. G., 51, 114, 120
Smith, E. O., 94, 381, 383
Smith, R. L., 300
Smuts, B., 190, 193, 204, 209, 210, 392, 393
Snyder, P. A., 369
Social strategies, 79, 180-184
Sociobiology, 79, 127-129, 214, 218, 232-241, 255-256, 259ff., 338
Socioecology, 259ff., 336ff., 341, 370
Southwick, C. H., 89, 107, 380
Spelke, E., 413
Spencer-Booth, Y., 56, 89, 245, 347, 379, 387
Srikosamatara, S., 322
Stanley, S. M., 330
Stearns, S. C., 261, 296
Stein, D. M., 186, 193, 204, 208, 209, 210, 213, 214, 240, 377, 391, 393
Stellar, E., 367
Stevenson, M. B., 420
Stevenson, M. F., 367
Stoltz, L. P., 109, 320
Strassmann, B. I., 310, 319, 326, 327, 328
Struhsaker, T. T., 214
Strum, S. C., 148, 151, 167, 168, 171, 172, 180, 181, 190, 193, 204, 208, 209, 315
Subgroups, 131, 214, 217, 239-240, 328-331, 342
Sugiyama, Y., 379
Suomi, S. J., 113, 124, 378, 388
Sussman, R. W., 361
Suzuki, A., 246, 328, 396
Syngamy, 260, 262
Symons, D., 325

Tail carriage, 142, 194, 204-205, 216
Tasch, R. J., 420
Tattersall, I., 358
Taub, D. M., 20, 21, 22, 26, 32, 41, 43, 49, 50, 51, 52, 53, 56, 89, 146, 147, 165, 189, 205, 377, 378, 381, 382
Taylor, C. R., 107, 121, 312, 313
Taylor, H., 89, 113, 116, 387
Teleki, G., 396
Tembrock, G., 276

Tenaza, R., 394, 395
Tenure, 114, 119, 120, 124, 148, 154, 160, 179, 192, 207
Terborg, F., 1, 16, 17
Thornhill, R., 260, 261, 264, 267
Thorpe, W. H., 276
Tilford, B. A., 399
Triads/triadic interactions, 26-27, 52, 186-210, 380-382, 390
Trivers, R. L., 20, 51, 113, 114, 181, 255, 260, 261, 262, 264, 265, 266, 267, 268, 272, 273, 276, 283, 287, 291, 294, 295, 299, 310, 312, 348
Tutin, C. E. G., 256, 263, 296, 397, 400
Twins, 4, 7, 10, 336-343, 366

Ullrich, W., 400
Uniparental care, 287, 290-291, 301

Vaitl, E. A., 364
Vandenbergh, J. G., 115, 364
van den Berghe, P. L., 316, 325
Van Hooff, J. A. R. A. M., 23
van Lawick-Goodall, J. *See* Goodall, J. van Lawick
Vaughan, T. A., 280
Veit, P. G., 398
Vessey, S. H., 114
Vogt, J. L., 2, 5, 347, 364, 365, 368
vom Saal, F. S., 206

Wachs, T., 420
Wade, M. J., 261, 262, 265
Walker, E. P., 273
Wallace, G. D., 280, 281
Walters, J., 214
Warren, D. C., 277, 278, 281
Wells, K. D., 262, 275, 287
Welty, J. C., 260, 280, 281, 282
Wendt, H., 2, 368
Werren, J. H., 271, 281
West, M. M., 419
Western, J., 183, 214
Whiten, A., 381
Wildt, D. E., 206
Wiley, R. H., 260, 294
Willemsen, E., 413
Williams, G. C., 122, 261, 265, 266, 267, 273, 284, 285, 286, 287, 291, 293, 294, 296, 297
Williams, L., 365
Wilson, A. P., 88, 122
Wilson, E. O., 128, 129, 143, 267, 273, 282, 338, 341, 377
Wilson, M. E., 399, 400
Wing, L. W., 281
Wittenberger, J. F., 20, 114, 266, 268, 283, 287, 321, 322
Wolters, H. J., 2, 5, 6, 17
Wooton, R. J., 268, 276
Wrangham, R. W., 327, 329, 397

Xenophobia, 123

Yogman, M. J., 415

Zajonc, R. B., 52
Zelazo, P. R., 410, 413
Zenone, P. G., 276, 281
Zereloff, S. I., 325
Zukowsky, L., 2, 5

About the Editor

DAVID MILTON TAUB is the associate director of the Yemassee Primate Center in South Carolina and an adjunct professor in the Department of Psychiatry and Behavioral Sciences at the Medical University of South Carolina in Charleston. He received the B.A. from the University of Texas at Austin, the M.A. from the University of California, Los Angeles, and the Ph.D. in biological anthropology from the University of California at Davis. Dr. Taub's field researches have included several studies in Morocco on the social behavior of the only African macaque, the Barbary macaque. These studies initiated his research interests in male-infant relationships. The author of several papers on this subject, he was recently invited to participate in a symposium on paternal behavior sponsored by the Animal Behavior Society, which included such noted scientists as Robert Trivers and Donald Dewsbury.

Dr. Taub has been involved in a variety of primate behavioral researches during the past ten years, including several years on the faculty of the Bowman Gray School of Medicine. There he pioneered novel biomedical studies of the role of psychosocial factors in the etiology of cardiovascular disease using nonhuman primate models. Dr. Taub has served on review committees for the National Institutes of Mental Health and for the Food and Drug Administration, as well as having served as an ad hoc reviewer for the National Science Foundation and the National Institutes of Health. The author or coauthor of a wide range of scientific papers on primate social behavior, he is a consulting editor for the *American Journal of Primatology,* and has also served as an invited reviewer for a number of other scientific journals. Dr. Taub was a charter member of the American Society of Primatologists, and has served as its executive secretary, editor of the *ASP Bulletin,* and most recently as the Program Chair for their scientific meetings. Currently, Dr. Taub is the principal investigator of the world's largest free-ranging, island breeding colony of rhesus monkeys, where he continues his researches into the sociobiology of that species. With his colleagues at MUSC, he is also developing biomedical researches into the behavioral effects of drug use with nonhuman primates. Active in community affairs, Dr. Taub is the only nonveterinarian to have served on the South Carolina State Board of Veterinary Medical Examiners, having been appointed by the governor for a six-year term in 1982.